# Data Analysis of Asymmetric Structures

# STATISTICS: Textbooks and Monographs

D. B. Owen
*Founding Editor, 1972–1991*

## *Associate Editors*

**Statistical Computing/**
**Nonparametric Statistics**
Professor William R. Schucany
*Southern Methodist University*

**Probability**
Professor Marcel F. Neuts
*University of Arizona*

**Multivariate Analysis**
Professor Anant M. Kshirsagar
*University of Michigan*

**Quality Control/Reliability**
Professor Edward G. Schilling
*Rochester Institute of Technology*

## *Editorial Board*

**Applied Probability**
Dr. Paul R. Garvey
*The MITRE Corporation*

**Economic Statistics**
Professor David E. A. Giles
*University of Victoria*

**Experimental Designs**
Mr. Thomas B. Barker
*Rochester Institute of Technology*

**Multivariate Analysis**
Professor Subir Ghosh
*University of California–Riverside*

**Statistical Distributions**
Professor N. Balakrishnan
*McMaster University*

**Statistical Process Improvement**
Professor G. Geoffrey Vining
*Virginia Polytechnic Institute*

**Stochastic Processes**
Professor V. Lakshmikantham
*Florida Institute of Technology*

**Survey Sampling**
Professor Lynne Stokes
*Southern Methodist University*

**Time Series**
Sastry G. Pantula
*North Carolina State University*

1. The Generalized Jackknife Statistic, *H. L. Gray and W. R. Schucany*
2. Multivariate Analysis, *Anant M. Kshirsagar*
3. Statistics and Society, *Walter T. Federer*
4. Multivariate Analysis: A Selected and Abstracted Bibliography, 1957–1972, *Kocherlakota Subrahmaniam and Kathleen Subrahmaniam*
5. Design of Experiments: A Realistic Approach, *Virgil L. Anderson and Robert A. McLean*
6. Statistical and Mathematical Aspects of Pollution Problems, *John W. Pratt*
7. Introduction to Probability and Statistics (in two parts), Part I: Probability; Part II: Statistics, *Narayan C. Giri*
8. Statistical Theory of the Analysis of Experimental Designs, *J. Ogawa*
9. Statistical Techniques in Simulation (in two parts), *Jack P. C. Kleijnen*
10. Data Quality Control and Editing, *Joseph I. Naus*
11. Cost of Living Index Numbers: Practice, Precision, and Theory, *Kali S. Banerjee*
12. Weighing Designs: For Chemistry, Medicine, Economics, Operations Research, Statistics, *Kali S. Banerjee*
13. The Search for Oil: Some Statistical Methods and Techniques, *edited by D. B. Owen*
14. Sample Size Choice: Charts for Experiments with Linear Models, *Robert E. Odeh and Martin Fox*
15. Statistical Methods for Engineers and Scientists, *Robert M. Bethea, Benjamin S. Duran, and Thomas L. Boullion*
16. Statistical Quality Control Methods, *Irving W. Burr*

17. On the History of Statistics and Probability, *edited by D. B. Owen*
18. Econometrics, *Peter Schmidt*
19. Sufficient Statistics: Selected Contributions, *Vasant S. Huzurbazar (edited by Anant M. Kshirsagar)*
20. Handbook of Statistical Distributions, *Jagdish K. Patel, C. H. Kapadia, and D. B. Owen*
21. Case Studies in Sample Design, *A. C. Rosander*
22. Pocket Book of Statistical Tables, *compiled by R. E. Odeh, D. B. Owen, Z. W. Birnbaum, and L. Fisher*
23. The Information in Contingency Tables, *D. V. Gokhale and Solomon Kullback*
24. Statistical Analysis of Reliability and Life-Testing Models: Theory and Methods, *Lee J. Bain*
25. Elementary Statistical Quality Control, *Irving W. Burr*
26. An Introduction to Probability and Statistics Using BASIC, *Richard A. Groeneveld*
27. Basic Applied Statistics, *B. L. Raktoe and J. J. Hubert*
28. A Primer in Probability, *Kathleen Subrahmaniam*
29. Random Processes: A First Look, *R. Syski*
30. Regression Methods: A Tool for Data Analysis, *Rudolf J. Freund and Paul D. Minton*
31. Randomization Tests, *Eugene S. Edgington*
32. Tables for Normal Tolerance Limits, Sampling Plans and Screening, *Robert E. Odeh and D. B. Owen*
33. Statistical Computing, *William J. Kennedy, Jr., and James E. Gentle*
34. Regression Analysis and Its Application: A Data-Oriented Approach, *Richard F. Gunst and Robert L. Mason*
35. Scientific Strategies to Save Your Life, *I. D. J. Bross*
36. Statistics in the Pharmaceutical Industry, *edited by C. Ralph Buncher and Jia-Yeong Tsay*
37. Sampling from a Finite Population, *J. Hajek*
38. Statistical Modeling Techniques, *S. S. Shapiro and A. J. Gross*
39. Statistical Theory and Inference in Research, *T. A. Bancroft and C.-P. Han*
40. Handbook of the Normal Distribution, *Jagdish K. Patel and Campbell B. Read*
41. Recent Advances in Regression Methods, *Hrishikesh D. Vinod and Aman Ullah*
42. Acceptance Sampling in Quality Control, *Edward G. Schilling*
43. The Randomized Clinical Trial and Therapeutic Decisions, *edited by Niels Tygstrup, John M Lachin, and Erik Juhl*
44. Regression Analysis of Survival Data in Cancer Chemotherapy, *Walter H. Carter, Jr., Galen L. Wampler, and Donald M. Stablein*
45. A Course in Linear Models, *Anant M. Kshirsagar*
46. Clinical Trials: Issues and Approaches, *edited by Stanley H. Shapiro and Thomas H. Louis*
47. Statistical Analysis of DNA Sequence Data, *edited by B. S. Weir*
48. Nonlinear Regression Modeling: A Unified Practical Approach, *David A. Ratkowsky*
49. Attribute Sampling Plans, Tables of Tests and Confidence Limits for Proportions, *Robert E. Odeh and D. B. Owen*

50. Experimental Design, Statistical Models, and Genetic Statistics, *edited by Klaus Hinkelmann*
51. Statistical Methods for Cancer Studies, *edited by Richard G. Cornell*
52. Practical Statistical Sampling for Auditors, *Arthur J. Wilburn*
53. Statistical Methods for Cancer Studies, *edited by Edward J. Wegman and James G. Smith*
54. Self-Organizing Methods in Modeling: GMDH Type Algorithms, *edited by Stanley J. Farlow*
55. Applied Factorial and Fractional Designs, *Robert A. McLean and Virgil L. Anderson*
56. Design of Experiments: Ranking and Selection, *edited by Thomas J. Santner and Ajit C. Tamhane*
57. Statistical Methods for Engineers and Scientists: Second Edition, Revised and Expanded, *Robert M. Bethea, Benjamin S. Duran, and Thomas L. Boullion*
58. Ensemble Modeling: Inference from Small-Scale Properties to Large-Scale Systems, *Alan E. Gelfand and Crayton C. Walker*
59. Computer Modeling for Business and Industry, *Bruce L. Bowerman and Richard T. O'Connell*
60. Bayesian Analysis of Linear Models, *Lyle D. Broemeling*
61. Methodological Issues for Health Care Surveys, *Brenda Cox and Steven Cohen*
62. Applied Regression Analysis and Experimental Design, *Richard J. Brook and Gregory C. Arnold*
63. Statpal: A Statistical Package for Microcomputers—PC-DOS Version for the IBM PC and Compatibles, *Bruce J. Chalmer and David G. Whitmore*
64. Statpal: A Statistical Package for Microcomputers—Apple Version for the II, II+, and IIe, *David G. Whitmore and Bruce J. Chalmer*
65. Nonparametric Statistical Inference: Second Edition, Revised and Expanded, *Jean Dickinson Gibbons*
66. Design and Analysis of Experiments, *Roger G. Petersen*
67. Statistical Methods for Pharmaceutical Research Planning, *Sten W. Bergman and John C. Gittins*
68. Goodness-of-Fit Techniques, *edited by Ralph B. D'Agostino and Michael A. Stephens*
69. Statistical Methods in Discrimination Litigation, *edited by D. H. Kaye and Mikel Aickin*
70. Truncated and Censored Samples from Normal Populations, *Helmut Schneider*
71. Robust Inference, *M. L. Tiku, W. Y. Tan, and N. Balakrishnan*
72. Statistical Image Processing and Graphics, *edited by Edward J. Wegman and Douglas J. DePriest*
73. Assignment Methods in Combinatorial Data Analysis, *Lawrence J. Hubert*
74. Econometrics and Structural Change, *Lyle D. Broemeling and Hiroki Tsurumi*
75. Multivariate Interpretation of Clinical Laboratory Data, *Adelin Albert and Eugene K. Harris*
76. Statistical Tools for Simulation Practitioners, *Jack P. C. Kleijnen*
77. Randomization Tests: Second Edition, *Eugene S. Edgington*

78. A Folio of Distributions: A Collection of Theoretical Quantile-Quantile Plots, *Edward B. Fowlkes*
79. Applied Categorical Data Analysis, *Daniel H. Freeman, Jr.*
80. Seemingly Unrelated Regression Equations Models: Estimation and Inference, *Virendra K. Srivastava and David E. A. Giles*
81. Response Surfaces: Designs and Analyses, *Andre I. Khuri and John A. Cornell*
82. Nonlinear Parameter Estimation: An Integrated System in BASIC, *John C. Nash and Mary Walker-Smith*
83. Cancer Modeling, *edited by James R. Thompson and Barry W. Brown*
84. Mixture Models: Inference and Applications to Clustering, *Geoffrey J. McLachlan and Kaye E. Basford*
85. Randomized Response: Theory and Techniques, *Arijit Chaudhuri and Rahul Mukerjee*
86. Biopharmaceutical Statistics for Drug Development, *edited by Karl E. Peace*
87. Parts per Million Values for Estimating Quality Levels, *Robert E. Odeh and D. B. Owen*
88. Lognormal Distributions: Theory and Applications, *edited by Edwin L. Crow and Kunio Shimizu*
89. Properties of Estimators for the Gamma Distribution, *K. O. Bowman and L. R. Shenton*
90. Spline Smoothing and Nonparametric Regression, *Randall L. Eubank*
91. Linear Least Squares Computations, *R. W. Farebrother*
92. Exploring Statistics, *Damaraju Raghavarao*
93. Applied Time Series Analysis for Business and Economic Forecasting, *Sufi M. Nazem*
94. Bayesian Analysis of Time Series and Dynamic Models, *edited by James C. Spall*
95. The Inverse Gaussian Distribution: Theory, Methodology, and Applications, *Raj S. Chhikara and J. Leroy Folks*
96. Parameter Estimation in Reliability and Life Span Models, *A. Clifford Cohen and Betty Jones Whitten*
97. Pooled Cross-Sectional and Time Series Data Analysis, *Terry E. Dielman*
98. Random Processes: A First Look, Second Edition, Revised and Expanded, *R. Syski*
99. Generalized Poisson Distributions: Properties and Applications, *P. C. Consul*
100. Nonlinear $L_p$-Norm Estimation, *Rene Gonin and Arthur H. Money*
101. Model Discrimination for Nonlinear Regression Models, *Dale S. Borowiak*
102. Applied Regression Analysis in Econometrics, *Howard E. Doran*
103. Continued Fractions in Statistical Applications, *K. O. Bowman and L. R. Shenton*
104. Statistical Methodology in the Pharmaceutical Sciences, *Donald A. Berry*
105. Experimental Design in Biotechnology, *Perry D. Haaland*
106. Statistical Issues in Drug Research and Development, *edited by Karl E. Peace*
107. Handbook of Nonlinear Regression Models, *David A. Ratkowsky*

108. Robust Regression: Analysis and Applications, *edited by Kenneth D. Lawrence and Jeffrey L. Arthur*
109. Statistical Design and Analysis of Industrial Experiments, *edited by Subir Ghosh*
110. *U*-Statistics: Theory and Practice, *A. J. Lee*
111. A Primer in Probability: Second Edition, Revised and Expanded, *Kathleen Subrahmaniam*
112. Data Quality Control: Theory and Pragmatics, *edited by Gunar E. Liepins and V. R. R. Uppuluri*
113. Engineering Quality by Design: Interpreting the Taguchi Approach, *Thomas B. Barker*
114. Survivorship Analysis for Clinical Studies, *Eugene K. Harris and Adelin Albert*
115. Statistical Analysis of Reliability and Life-Testing Models: Second Edition, *Lee J. Bain and Max Engelhardt*
116. Stochastic Models of Carcinogenesis, *Wai-Yuan Tan*
117. Statistics and Society: Data Collection and Interpretation, Second Edition, Revised and Expanded, *Walter T. Federer*
118. Handbook of Sequential Analysis, *B. K. Ghosh and P. K. Sen*
119. Truncated and Censored Samples: Theory and Applications, *A. Clifford Cohen*
120. Survey Sampling Principles, *E. K. Foreman*
121. Applied Engineering Statistics, *Robert M. Bethea and R. Russell Rhinehart*
122. Sample Size Choice: Charts for Experiments with Linear Models: Second Edition, *Robert E. Odeh and Martin Fox*
123. Handbook of the Logistic Distribution, *edited by N. Balakrishnan*
124. Fundamentals of Biostatistical Inference, *Chap T. Le*
125. Correspondence Analysis Handbook, *J.-P. Benzécri*
126. Quadratic Forms in Random Variables: Theory and Applications, *A. M. Mathai and Serge B. Provost*
127. Confidence Intervals on Variance Components, *Richard K. Burdick and Franklin A. Graybill*
128. Biopharmaceutical Sequential Statistical Applications, *edited by Karl E. Peace*
129. Item Response Theory: Parameter Estimation Techniques, *Frank B. Baker*
130. Survey Sampling: Theory and Methods, *Arijit Chaudhuri and Horst Stenger*
131. Nonparametric Statistical Inference: Third Edition, Revised and Expanded, *Jean Dickinson Gibbons and Subhabrata Chakraborti*
132. Bivariate Discrete Distribution, *Subrahmaniam Kocherlakota and Kathleen Kocherlakota*
133. Design and Analysis of Bioavailability and Bioequivalence Studies, *Shein-Chung Chow and Jen-pei Liu*
134. Multiple Comparisons, Selection, and Applications in Biometry, *edited by Fred M. Hoppe*
135. Cross-Over Experiments: Design, Analysis, and Application, *David A. Ratkowsky, Marc A. Evans, and J. Richard Alldredge*
136. Introduction to Probability and Statistics: Second Edition, Revised and Expanded, *Narayan C. Giri*

137. Applied Analysis of Variance in Behavioral Science, *edited by Lynne K. Edwards*
138. Drug Safety Assessment in Clinical Trials, *edited by Gene S. Gilbert*
139. Design of Experiments: A No-Name Approach, *Thomas J. Lorenzen and Virgil L. Anderson*
140. Statistics in the Pharmaceutical Industry: Second Edition, Revised and Expanded, *edited by C. Ralph Buncher and Jia-Yeong Tsay*
141. Advanced Linear Models: Theory and Applications, *Song-Gui Wang and Shein-Chung Chow*
142. Multistage Selection and Ranking Procedures: Second-Order Asymptotics, *Nitis Mukhopadhyay and Tumulesh K. S. Solanky*
143. Statistical Design and Analysis in Pharmaceutical Science: Validation, Process Controls, and Stability, *Shein-Chung Chow and Jen-pei Liu*
144. Statistical Methods for Engineers and Scientists: Third Edition, Revised and Expanded, *Robert M. Bethea, Benjamin S. Duran, and Thomas L. Boullion*
145. Growth Curves, *Anant M. Kshirsagar and William Boyce Smith*
146. Statistical Bases of Reference Values in Laboratory Medicine, *Eugene K. Harris and James C. Boyd*
147. Randomization Tests: Third Edition, Revised and Expanded, *Eugene S. Edgington*
148. Practical Sampling Techniques: Second Edition, Revised and Expanded, *Ranjan K. Som*
149. Multivariate Statistical Analysis, *Narayan C. Giri*
150. Handbook of the Normal Distribution: Second Edition, Revised and Expanded, *Jagdish K. Patel and Campbell B. Read*
151. Bayesian Biostatistics, *edited by Donald A. Berry and Dalene K. Stangl*
152. Response Surfaces: Designs and Analyses, Second Edition, Revised and Expanded, *André I. Khuri and John A. Cornell*
153. Statistics of Quality, *edited by Subir Ghosh, William R. Schucany, and William B. Smith*
154. Linear and Nonlinear Models for the Analysis of Repeated Measurements, *Edward F. Vonesh and Vernon M. Chinchilli*
155. Handbook of Applied Economic Statistics, *Aman Ullah and David E. A. Giles*
156. Improving Efficiency by Shrinkage: The James-Stein and Ridge Regression Estimators, *Marvin H. J. Gruber*
157. Nonparametric Regression and Spline Smoothing: Second Edition, *Randall L. Eubank*
158. Asymptotics, Nonparametrics, and Time Series, *edited by Subir Ghosh*
159. Multivariate Analysis, Design of Experiments, and Survey Sampling, *edited by Subir Ghosh*
160. Statistical Process Monitoring and Control, *edited by Sung H. Park and G. Geoffrey Vining*
161. Statistics for the 21st Century: Methodologies for Applications of the Future, *edited by C. R. Rao and Gábor J. Székely*
162. Probability and Statistical Inference, *Nitis Mukhopadhyay*

163. Handbook of Stochastic Analysis and Applications, *edited by D. Kannan and V. Lakshmikantham*
164. Testing for Normality, *Henry C. Thode, Jr.*
165. Handbook of Applied Econometrics and Statistical Inference, *edited by Aman Ullah, Alan T. K. Wan, and Anoop Chaturvedi*
166. Visualizing Statistical Models and Concepts, *R. W. Farebrother and Michael Schyns*
167. Financial and Actuarial Statistics, *Dale Borowiak*
168. Nonparametric Statistical Inference, Fourth Edition, Revised and Expanded, *edited by Jean Dickinson Gibbons and Subhabrata Chakraborti*
169. Computer-Aided Econometrics, *edited by David EA. Giles*
170. The EM Algorithm and Related Statistical Models, *edited by Michiko Watanabe and Kazunori Yamaguchi*
171. Multivariate Statistical Analysis, Second Edition, Revised and Expanded, *Narayan C. Giri*
172. Computational Methods in Statistics and Econometrics, *Hisashi Tanizaki*
173. Applied Sequential Methodologies: Real-World Examples with Data Analysis, *edited by Nitis Mukhopadhyay, Sujay Datta, and Saibal Chattopadhyay*
174. Handbook of Beta Distribution and Its Applications, *edited by Richard Guarino and Saralees Nadarajah*
175. Survey Sampling: Theory and Methods, Second Edition, *Arijit Chardhuri and Horst Stenger*
176. Item Response Theory: Parameter Estimation Techniques, Second Edition, *edited by Frank B. Baker and Seock-Ho Kim*
177. Statistical Methods in Computer Security, *edited by William W. S. Chen*
178. Elementary Statistical Quality Control, Second Edition, *John T. Burr*
179. Data Analysis of Asymmetric Structures, *Takayuki Saito and Hiroshi Yadohisa*

***Additional Volumes in Preparation***

# Data Analysis of Asymmetric Structures

## Advanced Approaches in Computational Statistics

**Takayuki Saito**
*Tokyo Institute of Technology*
*Tokyo, Japan*

**Hiroshi Yadohisa**
*Kagoshima University*
*Kagoshima, Japan*

CRC Press
Taylor & Francis Group
Boca Raton London New York

CRC Press is an imprint of the
Taylor & Francis Group, an **informa** business

CRC Press
Taylor & Francis Group
6000 Broken Sound Parkway NW, Suite 300
Boca Raton, FL 33487-2742

First issued in paperback 2019

© 2005 by Taylor & Francis Group, LLC
CRC Press is an imprint of Taylor & Francis Group, an Informa business

ISBN-13: 978-0-8247-5398-6 (hbk)
ISBN-13: 978-0-367-39337-3 (pbk)

This book contains information obtained from authentic and highly regarded sources. Reasonable efforts have been made to publish reliable data and information, but the author and publisher cannot assume responsibility for the validity of all materials or the consequences of their use. The authors and publishers have attempted to trace the copyright holders of all material reproduced in this publication and apologize to copyright holders if permission to publish in this form has not been obtained. If any copyright material has not been acknowledged please write and let us know so we may rectify in any future reprint.

Except as permitted under U.S. Copyright Law, no part of this book may be reprinted, reproduced, transmitted, or utilized in any form by any electronic, mechanical, or other means, now known or hereafter invented, including photocopying, microfilming, and recording, or in any information storage or retrieval system, without written permission from the publishers.

For permission to photocopy or use material electronically from this work, please access www.copyright.com (http://www.copyright.com/) or contact the Copyright Clearance Center, Inc. (CCC), 222 Rosewood Drive, Danvers, MA 01923, 978-750-8400. CCC is a not-for-profit organization that provides licenses and registration for a variety of users. For organizations that have been granted a photocopy license by the CCC, a separate system of payment has been arranged.

**Trademark Notice:** Product or corporate names may be trademarks or registered trademarks, and are used only for identification and explanation without intent to infringe.

**Library of Congress Cataloging-in-Publication Data**
A catalog record for this book is available from the Library of Congress.

**Visit the Taylor & Francis Web site at**
**http://www.taylorandfrancis.com**

**and the CRC Press Web site at**
**http://www.crcpress.com**

# Preface

Asymmetry is the lack of symmetry. Symmetry originally meant balanced proportion in geometrical properties such as size, shape, position, and so on. Symmetry is usually considered an important feature of beauty, one of aesthetic values. However, some irregularity or imbalance of proportions are often considered a source of beauty in spatial arrangement, and can become a different value. Attention to symmetry or asymmetry in representing beauty would depend on artistic styles or, in a broad sense, on cultural characteristics. For example, it may be remarked that the garden of Versailles Palace in France is an example of symmetrical beauty, while the garden of Katsura Imperial Villa in Japan is representative of asymmetrical beauty.

Researchers involved in data analysis are often interested in the asymmetry observed in data, which appears in a variety of disciplines, such as psychology, sociology, marketing research, behavioral sciences, and ecological sciences. In these disciplines the asymmetry is based on a concept in matrix theory and refers to some imbalance of the relationships in data. The subject of asymmetry becomes a source of interest for research. In order to explore structure, pattern, or some meaningful information inherent in the asymmetric data, researchers have developed many models, methods, and theories for data analysis.

This book is a comprehensive exposition of those studies, focusing on the asymmetry. It is intended not only for research workers and professionals but also undergraduate and graduate students who are interested in asymmetry.

We would like to express our sincere appreciation to Kiwamu Aoyama (Kagoshima University) for his mathematical review of some sections in Chapters 4 and 5. We also thank Katsuyuki Suenaga (Kagoshima Immaculate Heart College) for preparing complex original figures, and Yoshiki Mihara (Kagoshima University) who helped us greatly editing our LaTeX manuscripts.

Acknowledgment is also due to Marcel Dekker, who provided us with the opportunity to publish this book. Thanks are extended to all the staff involved in the publication process, in particular Acquisitions Editor Taisuke Soda, for their great support.

Takayuki Saito
Hiroshi Yadohisa

# Contents

*Preface* *iii*

**1. Introduction** **1**

1.1. What is Asymmetry? 1
1.2. Asymmetric Data 1
1.3. Analysis of Asymmetric Structures 3
1.4. Overview of the Book 3
1.5. Prospective Fields and Readers 5
1.6. Suggestions for Reading 6
1.7. Notation 6

**2. Paired Comparisons with Asymmetry** **8**

2.1. Overview and Preliminaries 8
2.2. Detection of Ordinal Structure 9
2.3. Analysis of Variance 15
2.4. Psychological Scaling 21
2.5. Operational Scaling 41
2.6. Summary 49

**3. Graphical Representation of Asymmetric Data** **51**

3.1. Overview and Preliminaries 51
3.2. Gower's Procedure 52
3.3. Escoufier and Grorud's Procedure 69
3.4. Vector Model 80
3.5. Vector Field Model 89

**4. Multidimensional Scaling of Asymmetric Data** **103**

4.1. Overview and Preliminaries 103
4.2. Generalization of Scalar Product Models 109
4.3. Similarity and Bias Model 124
4.4. Generalization of Distance Models 128
4.5. Feature-Matching Model and TSCALE 159

**5. Cluster Analysis of Asymmetric Data** **164**

5.1. Overview and Preliminaries 164
5.2. Hubert Algorithms and their Extensions (One-Mode Approach) 165
5.3. Classic (One-Mode Approach) 173
5.4. Brossier Algorithm (One-Mode Approach) 180
5.5. De Soete et al. Algorithm (Two-Mode Approach) 185
5.6. Bond Energy Algorithm (Two-Mode Approach) 188
5.7. Centroid Effect Algorithm (Two-Mode Approach) 190
5.8. Gennclus (Alternating Least Squares Approach) 194

**6. Network Analysis of Asymmetric Data** **199**

6.1. Overview and Preliminaries 199
6.2. Detection of Cohesive Groups 201
6.3. Network Scaling 206
6.4. Statistical Models for Social Network 208

**7. Multivariate Analysis of Asymmetry Between Data Sets** **215**

7.1. Overview and Preliminaries 215
7.2. Redundancy Analysis 220
7.3. Multivariate Regression on Composite Variates 225

7.4. Comparision of Related Procedures 229
7.5. Numerical Example 234

**Bibliography** **243**

**Index** **252**

# 1

# Introduction

## 1.1. WHAT IS ASYMMETRY?

Suppose a pair is formed of two persons, A and B. In a case where A likes B, B may not like A. In another case where A likes B very much, B may like A fairly well. In either case, the relation from A to B is not the same as the relation from B to A, and we will call the relationship asymmetric. Suppose a pair is formed of two brands of cars, and consider users of car brand A (say, A-users) and users of car brand B (say, B-users) at a certain time. After some time, some of the A-users will switch to brand B, and also some B-users will switch to brand A. The number of people switching from A to B is usually not the same as the number of people switching from B to A. In such a case, we call the relationship between brands A and B asymmetric. Consider trade flow between two countries, A and B. The amount of trade that A imports from B may be larger or smaller than that which B imports from A. In our terminology, we call such a relationship asymmetric. When the degree of asymmetry is large, the situation could become a political issue of trade imbalance.

## 1.2. ASYMMETRIC DATA

In a variety of disciplines such as psychology, social psychology, sociology, marketing research, behavioral sciences, ecological sciences, and so on, there are situations in which an observation represents the measure of relationship between a pair of objects and the relationship is asymmetric. The measure is known by many terms, including proximity, closeness, similarity, dissimilarity, affinity, association, confusion, mutual interaction, exchange, psychological distance, social mobility, migration, brand switching, substitutability, social network, dependence, and so on. Those observations for a set of objects are called asymmetric data across the disciplines. In this book, we may often use the proximity as a representative term, for convenience of description.

Since the asymmetry embodies some important information contained in the data, it is useful to extract such information by analyzing the asymmetric data. Then investigators need to explore any structure in the data, which may be a spatial structure, a cluster structure, a tree structure, or a network structure, depending on the motivation of data analysis. Figure 1.1 illustrates some of the areas and fields in which asymmetric data are considered.

Let us consider the form of data. Given observations of proximity for a set of $n$ objects, the data take the form of a square matrix with $n$ rows and $n$ columns, each element $o_{jk}$ of which indicates the observation of proximity for pair $(j, k)$. In contrast, one might be familiar with multivariate data, which are given in the form of a rectangular matrix with $m$ rows and $p$ columns, each element $x_{ij}$ of which indicates the $i$th observation of the $j$th variable. In this respect, the proximity data matrix is different from the customary multivariate data matrix.

Sometimes we call the proximity data $o_{jk}$ one-mode two-way data, in comparison with data of the following other form. When $o_{jk}$ is observed $m$

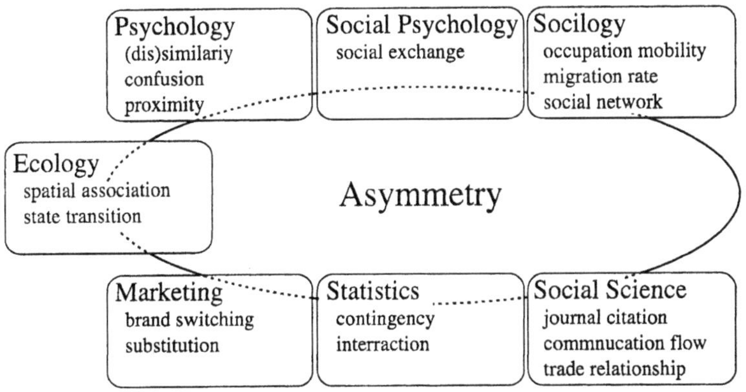


**FIGURE 1.1** Asymmetry appears in various fields

times, for example, on $m$ different conditions, we denote the observation by $o_{ijk}$ $(i = 1, \ldots, m)$; the data of this form are called two-mode three-way data. Generally we will not treat methods or procedures based on data of this sort in this book.

## 1.3. ANALYSIS OF ASYMMETRIC STRUCTURES

As mentioned above, asymmetric data are collected in many ways, by experiments, observations, surveys, or through statistical or archival records, in many fields. A variety of models, methods, and procedures have been investigated and proposed across those fields, for analysis of asymmetric structures involved in the data.

Now let us turn to methodological aspects concerned with analysis of asymmetry. Muitidimensional scaling, cluster analysis, graphical representation techniques, and redundancy analysis have developed in the area of multivariate analysis, some also with a focus on asymmetry. Methods of paired comparison and network analysis have developed in their own rights, but also with a focus on asymmetry. Figure 1.2 illustrates the relationships of those methodological aspects.

## 1.4. OVERVIEW OF THE BOOK

This book covers a variety of theories, methods, and models for analysis of asymmetry, as well as their applications. A description is provided for both theoretical and practical interests. Throughout the chapters, almost all sections are provided with examples. To enable comparison across those examples, results have been arranged using common sets of data as far as possible.

To begin, we treat in Chapter 2, methods and models for analyzing asymmetry related to paired comparisons. The method of paired comparisons has been studied over a long period, mainly in the fields of statistics or psychometrics, with applications. However, interest has traditionally been focused on statistical analysis or unidimensional psychological scaling, and the asymmetry inherent in data was not of principal concern. With a focus on asymmetry, we present detection of ordinal structure in dominance data, analysis of variance, and a method of psychological scaling for measurement of asymmetry in psychological judgment. In addition, we discuss operational scaling.

Chapters 3 and 4 are mutually related. In a different viewpoint from ours, subjects covered by these two chapters may be described in the category

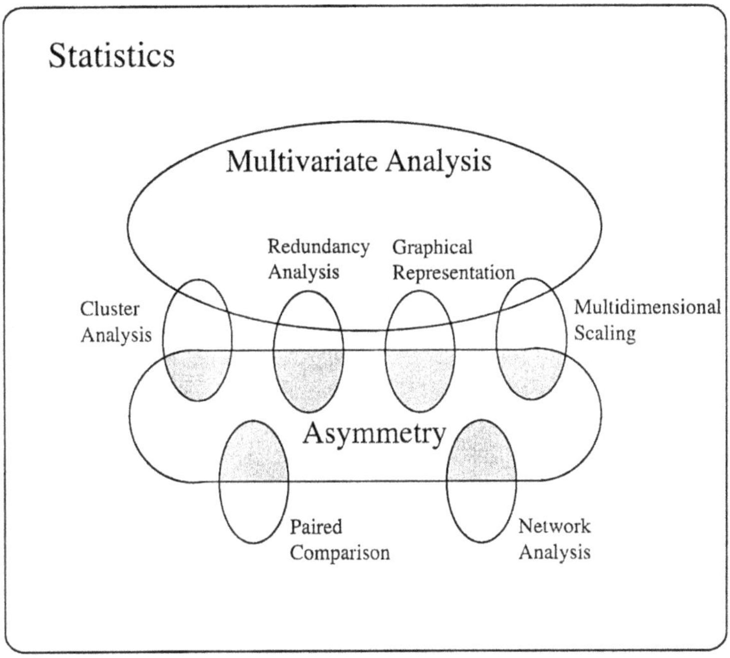


**FIGURE 1.2** Analysis of asymmetry in statistics

of asymmetric multidimensional scaling. In Chapter 3, we present models and procedures for graphical representation of asymmetric data. These are based on decomposition of a data matrix in linear algebra, and the models are formulated for data reduction rather than accounting for phenomena. They are essentially applicable to any kind of proximity measure, because they are used without regard to the meaning of the measure, even the direction of measurement, similarity, or dissimilarity.

In Chapter 4 we deal with models, methods, and procedures categorized in multidimensional scaling (MDS) for asymmetric data. After we give an overview of the subjects of asymmetry in MDS and a brief history, we present generalization of scalar product models, and then similarity and bias models in comparison with analysis of contingency tables. Next we describe the generalization of distance models, and, feature matching models. In a similar way to Borg and Groenen (1997), who mentioned four major roles for MDS, we may summarize the roles of asymmetric MDS in four categories: exploring data reduction, testing structural hypotheses for asymmetric data, exploring psychological structures or processes, and modeling judgment of proximity (similarity, dissimilarity).

Chapter 5 is devoted to cluster analysis of asymmetric data. Although some clustering methods have been proposed principally for multivariate data, some have been investigated on the basis of proximity data, with a close relationship with MDS. As a natural development in this regard, cluster analysis of asymmetric data has developed. Here we present agglomerative clustering methods designed for asymmetric data that developed from Hubert (1973), methods for direct clustering such as the bond energy and centroid effect algorithms, and methods for fitting tree structures such as GENNCLUS.

In Chapter 6, we present network analysis of asymmetric data. The network analysis is generally applicable for a variety of types of research that deal with phenomena in terms of network. As examples we take up some models and methods that are involved in sociological and psychological research. First we give a brief review on the concept of social network and the terms of graph theory, then we describe detection of cohesive groups, network scaling, and finally statistical models for social network.

Finally, in Chapter 7, we provide analysis of asymmetry of the relationship between two sets of multivariate data. The asymmetry is discussed in terms of the dependence of one set of variables on the other, and it is revealed by comparing the dependence in two directional analyses. The description is centered on redundancy analysis, in contrast to canonical regression analysis. In comparison, we discuss related procedures such as principal component analysis of instrumental variables and reduced rank regression analysis.

## 1.5. PROSPECTIVE FIELDS AND READERS

The disciplines addressed by this book are statistical data analysis, psychometrics, social psychology, sociology, marketing research, behavioral sciences, information sciences, and ecological sciences. The readers are intended to be not only students, graduate students, and research workers involved in those fields, but also in other fields who would find the methodology of the book relevant for the analysis of their data.

In comparison with the common methodology of statistical data analysis, the methodology of analysis of asymmetry is not well known even to those people involved in data analysis. In this regard, the book will suggest new points to people who need to analyze asymmetric data, but who do not know the methodology to meet the need. We believe that the comprehensive presentation of topics is useful and attractive even to people who have been concerned with a specified approach to the analysis of asymmetry in a particular area. Such people are often not aware of the approaches used in different areas, so the book will enable them to find new useful approaches.

## 1.6. SUGGESTIONS FOR READING

Elementary knowledge about linear algebra and statistical data analysis is assumed. Chapters 2, 3, 5, and 6 may be followed independently for reading, after Chapter 1. Chapter 4 is recommemmded for reading after Chapter 3, because the subjects of Chapter 3 are closely related to part of Chapter 4 and so references are often given from Chapter 4 to Chapter 3. Some topics of Chapter 2 are mentioned in Chapter 4. Chapter 5 has some relations to Chapter 3 and Chapter 4. Chapter 6 treats multivariate analysis of asymmetry on a standpoint quite different from those of the other chapters, and deals with data being different from proximity data. For this reason, it may be followed independently of the other chapters. Figure 1.3 shows these suggestions for reading.

## 1.7. NOTATION

We use upper case bold letters for matrices like $\boldsymbol{A}$, and lower case bold letters for vectors like $\boldsymbol{x}$. Transposition is usually designated by prime, for example, $\boldsymbol{A}'$ or $\boldsymbol{x}'$. Vector representations indicate column vectors, unless specified otherwise. We denote a matrix $\boldsymbol{A}$ with elements $a_{ij}$, which consists of $n$ columns by

$$\boldsymbol{A} = (a_{ij}) = (\mathbf{a}_1, \mathbf{a}_2, \ldots, \mathbf{a}_n) \tag{1.1}$$

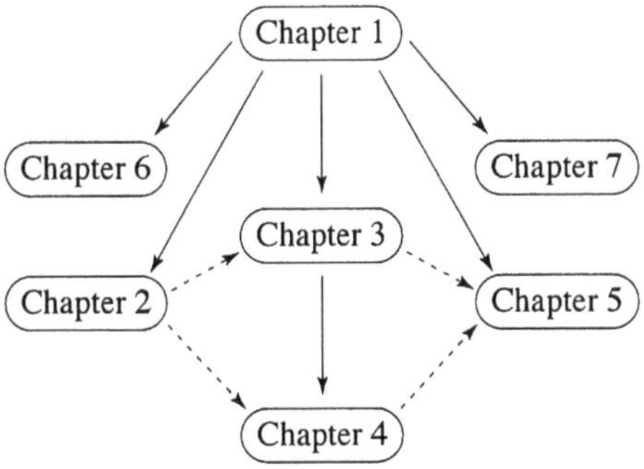


**FIGURE 1.3** Suggestions for reading

and a vector with $p$ elements by

$$\boldsymbol{x} = (x_i) = (x_1, x_2, \ldots, x_p)' \tag{1.2}$$

We denote the inverse of $\boldsymbol{A}$ by $\boldsymbol{A}^{-1}$, and a diagomial mnatrix $\boldsymbol{B}$ with $p$ elements on the diagonal by diag$(b_1, b_2, \ldots, b_p)$. The trace of $\boldsymbol{A}$ is represented as

$$\mathrm{tr}(\boldsymbol{A}) = \sum_{i=1}^{n} a_{ii} \tag{1.3}$$

We write the norm of an $m \times n$ matrix $\boldsymbol{A}$ and the norm of a $p$-dimensional vector $\boldsymbol{x}$ as

$$\|\boldsymbol{A}\| = \left(\sum_{i=1}^{m}\sum_{j=1}^{n} a_{ij}^2\right)^{1/2} \quad \text{and} \quad \|\boldsymbol{x}\| = \left(\sum_{i=1}^{p} x_i^2\right)^{1/2} \tag{1.4}$$

respectively. This notation will be used in a different definition in Secs. 3.3 and 4.2. The dot notation on a data matrix usually denotes row means, column means, and the total mean, for example,

$$a_{i\cdot} = \frac{1}{n}\sum_{j=1}^{n} a_{ij}, \qquad a_{\cdot j} = \frac{1}{m}\sum_{i=1}^{m} a_{ij} \qquad \text{and}$$

$$a_{\cdot\cdot} = \frac{1}{mn}\sum_{i=1}^{m}\sum_{j=1}^{n} a_{ij} \tag{1.5}$$

We write product as $\boldsymbol{AB}$ for matrices, and scalar product as $\boldsymbol{x}'\boldsymbol{y}$ for vectors. We denote the Cartesian product of two sets $P$ and $Q$ by $P \times Q$.

Expectations of a random variable $x$, a vector of random variables $\boldsymbol{x}$, and a matrix of random variables $\boldsymbol{X}$ are denoted by E$(x)$, E$(\boldsymbol{x})$, and E$(\boldsymbol{X})$, respectively. Let $\delta_{jk}$ denote Kroneckers' delta, which represents $\delta_{jk} = 1$ if $j = k$ and $\delta_{jk} = 0$ if $j \neq k$.

# 2

# Paired Comparisons with Asymmetry

## 2.1. OVERVIEW AND PRELIMINARIES

The method of paired comparisons has been widely investigated in psychological and statistical research, and has been utilized in practical applications. Given paired comparisons for a set of objects (stimuli), the data often take the form of a square matrix of measurements for the pairs of objects. The entries may be stated in terms of binary values, numerical values, or ordered categories, depending on the type of data collection.

This chapter will be confined to subjects related to the analysis of asymmetry. In Sec. 2.2 is treated detection of ordinal structure in dominance matrices by a statistical consideration (Kendall, 1962). Then, we turn to probability matrices observed in paired comparisons (Scheffé, 1952). In Sec. 2.3, analysis of variance is described under a stochastic model for comparative judgment. In Sec. 2.4, we describe psychological scaling under a judgment model (Saito, 1994), incorporating asymmetry, as an extension of Thurstone's scaling method (Guilford, 1954). In Sec. 2.5, we deal with construction of ratio scales from paired comparisons (Saaty, 1980), in which a strong condition of asymmetry is imposed in subjective judgments. Lastly, we give a summary of this chapter.

## 2.2. DETECTION OF ORDINAL STRUCTURE

In this section we describe how to detect ordinal structure in an asymmetric matrix for two situations: one in which observations are given in the form of a dominance matrix $\boldsymbol{C}$ and the other in which an asymmetric matrix $\boldsymbol{O}$ is originally observed and a dominance matrix $\boldsymbol{C}$ is constructed by pairwise comparison of two observations. By ordinal structure we mean a rank order of objects, which is derived from those observations. The rank order might be provided for the entire set or a subset of the objects. If the rank order is derived for the entire set, it is said that one can construct a ranking of all the objects or an ordinal scale of them from data (Coombs, 1964). We might consider that the derived scale represents a unidimensional latent scale in human preference or choice behavior, but might not in other cases such as birds' pecking behavior. Hereafter, we let the rank order refer to the entire set and so the detection means examination of the possibility of deriving the ordinal scale.

### 2.2.1. Dominance Matrix Given by Paired Comparisons

Let us consider an individual's preference judgment for a set of objects $\{S_1, S_2, \ldots, S_m\}$. Excluding pairs of identical stimuli, we arrange all possible $m(m-1)/2$ pairs of objects. For random presentation of each pair, the individual is required to state his preference for one member of the pair, or to make comparative judgment in binary response, in such a way as "yes" or "no." No choice is allowed in the response. If he prefers $S_j$ to $S_k$, denote the response by $c_{jk} = 1$. If he does not, denote it by $c_{jk} = 0$. Suppose that his judgment does not depend on the order of objects in the pair. Then $c_{jk} = 1$ means $c_{kj} = 0$. The total judgment is provided in terms of a dominance matrix $\boldsymbol{C} = (c_{jk})$, where $c_{jj}$ is undefined. Table 2.1 shows illustrative data, an individual's preference for six brands of icecream.

When $S_i$ is preferred to $S_j$, we may write $S_i \rightarrow S_j$ or $S_j \leftarrow S_i$. If paired comparisons for three objects $S_i$, $S_j$, $S_k$ are stated as $S_i \rightarrow S_j \rightarrow S_k \rightarrow S_i$ (that is, $c_{ij} + c_{jk} + c_{ki} = 3$) or $S_i \leftarrow S_j \leftarrow S_k \leftarrow S_i$ (that is, $c_{ij} + c_{jk} + c_{ki} = 0$), let us call the triad $S_iS_jS_k$ circular.

Figure 2.1 illustrates a diagram of dominance relations of the data of Table 2.1. Four circular triads, *ABD*, *DAE*, *AFD*, and *BEF* are revealed in the figure. Those triads may arise from the fluctuation in judgment or due to the subject's attention to more than one factor associated with the objects. In the rank order on unidimensional judgment, if $P \rightarrow Q$ and $Q \rightarrow R$, then $P \rightarrow R$, accordingly no circular triad should arise. Hence, if the rank ordering

**TABLE 2.1** Preference for Six Brands of Icecream

| | A | B | C | D | E | F |
|---|---|---|---|---|---|---|
| A | — | 1 | 0 | 0 | 1 | 1 |
| B | 0 | — | 0 | 1 | 1 | 0 |
| C | 1 | 1 | — | 1 | 1 | 1 |
| D | 1 | 0 | 0 | — | 0 | 0 |
| E | 0 | 0 | 0 | 1 | — | 1 |
| F | 0 | 1 | 0 | 1 | 0 | — |

in preference is possible, it is necessary and sufficient that no circular triad exists. In the fields of psychology and sensory inspection, researchers often find it interesting to detect an ordinal scale based on the dominance matrix.

### 2.2.2. Dominance Matrix Derived from Asymmetric Data

Let us denote the relationship between $S_j$ and $S_k$ by $o_{jk}$, and define $\boldsymbol{O} = (o_{jk})$, which is asymmetric. If one is interested in the dominance relation among the

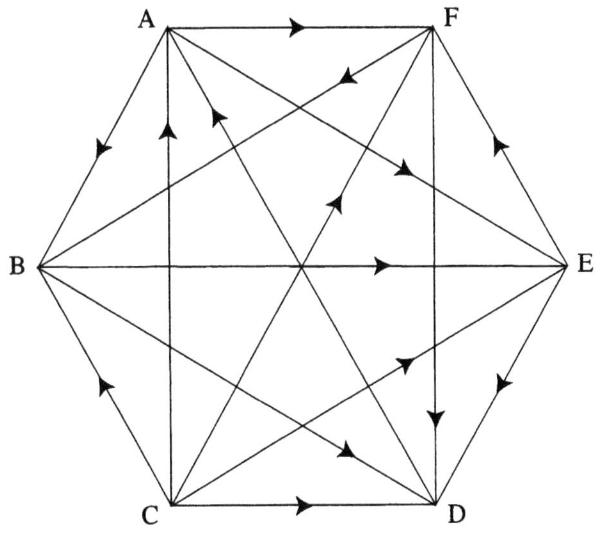


**FIGURE 2.1** Diagram of dominance

objects, one may construct a dominance matrix $\boldsymbol{C} = (c_{jk})$ by comparing $o_{jk}$ with $o_{kj}$. Define $c_{jk}$ $(j \neq k)$ in the following way:

$$\text{If} \quad o_{jk} > o_{kj}, \qquad \text{let} \quad c_{jk} = 1, \;\; c_{kj} = 0 \tag{2.1}$$

Let $c_{jj}$ be undefined. If $o_{jk} = o_{kj}$, we consider that the dominance between the two objects would arise with probability 0.5, and break the tie as follows. We decide either $c_{jk} = 1$ or $c_{jk} = 0$ with probability 0.5. Since $c_{jk}$ and $c_{kj}$ are not independent, $c_{kj} = 0$ should follow from $c_{jk} = 1$, that is, $c_{kj} = 1 - c_{jk}$.

In the fields of sociology and animal psychology, it is of interest to explore linear hierarchy or social hierarchy of objects inherent in dominance matrices (Masure and Allee, 1934; Goldthorpe, 1980). For example, Table 2.2 shows the intergenerational occupation mobility in Japan in 1975. In this case, one is interested in detecting some ordinal structure about the occupation mobility. Table 2.3 gives the dominance matrix derived by (2.1) from Table 2.2, and Table 2.4 is the rearranged dominance matrix.

We now turn to examples of animal psychology. Observations of pecking frequency in two flocks of pigeons are shown in Tables 2.5 and 2.6 (Masure and Allee, 1934). From either set of data, the dominance matrix is provided by (2.1). The pecking behavior indicates attacking. It may be of interest to examine whether a rank order exists or not among individuals. If it does, it may involve dominant–subordinate relationships.

### 2.2.3. Measurement of Inconsistency

Given a dominance matrix for a set of objects, we will examine, by using the coefficient of consistence (Kendall and Babinton Smith, 1940), whether the rank

**TABLE 2.2** Intergenerational Occupation Mobility

| | Son's occupation | | | | |
|---|---|---|---|---|---|
| Father's occupation | Upper nonmanual | Lower nonmanual | Upper manual | Lower manual | Farmer |
| Upper n.m. | 0.4320 | 0.3435 | 0.0816 | 0.1020 | 0.0408 |
| Lower n.m. | 0.1995 | 0.4803 | 0.1485 | 0.1415 | 0.0302 |
| Upper m. | 0.1383 | 0.2347 | 0.3923 | 0.1929 | 0.0418 |
| Lower m. | 0.1556 | 0.2267 | 0.2756 | 0.2933 | 0.0489 |
| Farmer | 0.1012 | 0.1913 | 0.1708 | 0.2349 | 0.3018 |

m. = manual; n.m. = nonmanual.

**TABLE 2.3** Dominance Matrix

| Father's occupation | Son's occupation: Upper nonmanual | Lower nonmanual | Upper manual | Lower manual | Farmer | Row sum |
|---|---|---|---|---|---|---|
| Upper n.m. | — | 1 | 0 | 0 | 0 | 1 |
| Lower n.m. | 0 | — | 0 | 0 | 0 | 0 |
| Upper m. | 1 | 1 | — | 0 | 0 | 2 |
| Lower m. | 1 | 1 | 1 | — | 0 | 3 |
| Farmer | 1 | 1 | 1 | 1 | — | 4 |

m. = manual; n.m. = nonmanual.

order is derived or not. The consistence is based on the measurement of inconsistence in terms of circular triads in $\boldsymbol{C}$. As mentioned above, circular triad $S_iS_jS_k$ arises if $c_{ij} + c_{jk} + c_{kj} = 3$ or 0. The degree of inconsistence in order is measured by counting the number of circular triads. Denoting the number by $d$, it is given by

$$d = \frac{1}{12}m(m-1)(2m-1) - \frac{1}{2}\sum_{j=1}^{m}\left(\sum_{k=1}^{m} c_{jk}\right)^2 \tag{2.2}$$

The coefficient $\zeta$ is defined as

$$\zeta = \begin{cases} 1 - \dfrac{24d}{m^3 - m} & \text{for odd } m \\ 1 - \dfrac{24d}{m^3 - 4m} & \text{for even } m \end{cases} \tag{2.3}$$

**TABLE 2.4** Rearranged Dominance Matrix

| Father's occupation | Son's occupation: Farmer | Lower manual | Upper manual | Upper nonmanual | Lower nonmanual | Row sum |
|---|---|---|---|---|---|---|
| Farmer | — | 1 | 1 | 1 | 1 | 4 |
| Lower m. | 0 | — | 1 | 1 | 1 | 3 |
| Upper m. | 0 | 0 | — | 1 | 1 | 2 |
| Upper n.m. | 0 | 0 | 0 | — | 1 | 1 |
| Lower n.m. | 0 | 0 | 0 | 0 | — | 0 |

m. = manual; n.m. = nonmanual.

**TABLE 2.5** Pecking Frequency Among Seven Female Pigeons

| | BB | BR | BW | BY | GW | RW | RY |
|---|---|---|---|---|---|---|---|
| BB | 0 | 41 | 44 | 68 | 58 | 29 | 45 |
| BR | 6 | 0 | 18 | 40 | 16 | 3 | 11 |
| BW | 10 | 7 | 0 | 11 | 5 | 18 | 12 |
| BY | 62 | 44 | 62 | 0 | 66 | 49 | 31 |
| GW | 36 | 20 | 23 | 28 | 0 | 29 | 11 |
| RW | 0 | 9 | 8 | 22 | 10 | 0 | 12 |
| RY | 10 | 2 | 5 | 2 | 5 | 5 | 0 |

The range of $\zeta$ is $0 \leq \zeta \leq 1$. When $\zeta = 1$, we consider that there exists a rank order in the data. As mentioned above, the existence of a rank order means that one could rank all the objects on some unidimensional scale. However, it does not necessarily mean that they should be positioned on a line in a space derived by a procedure of scaling or graphical representation of data (see Sec. 3.2). When $\zeta$ is thought to be close to unity on a specified criterion, there should be a small number of circular triads and we may regard this situation as where a nearly ordinal structure is involved in the data.

In order to conclude the existence of a rank order (or ordinal scale) in view of the fluctuation in preference judgment, we wish to perform a statistical test of $\zeta$ under the null hypothesis that $\zeta = 0$. However, no procedure for such a test has been proposed to date. Therefore, the investigator should make an appropriate decision in view of the range of $d$ for a specified $m$. According to Kendall (1962), we can derive the distribution of $d$ under the null hypothesis $H_0$ that the pattern of $\{c_{jk}\}$ is random. Then we can test $H_0$ against $H_1$ that the pattern is not random. If the observed $d$ is significantly small, we may

**TABLE 2.6** Pecking Frequency Among Seven Male Pigeons

| | B | BL | G | R | W | Y | YY |
|---|---|---|---|---|---|---|---|
| B | 0 | 0 | 0 | 2 | 0 | 0 | 0 |
| BL | 18 | 0 | 15 | 57 | 22 | 6 | 14 |
| G | 14 | 37 | 0 | 58 | 79 | 25 | 44 |
| R | 10 | 16 | 29 | 0 | 23 | 16 | 13 |
| W | 2 | 7 | 2 | 4 | 0 | 0 | 6 |
| Y | 14 | 48 | 34 | 41 | 38 | 0 | 37 |
| YY | 20 | 49 | 92 | 37 | 49 | 18 | 0 |

conclude at a significant level that there exists some structure in the population. However, even with such a conclusion, we cannot assert that the structure is a rank order (ordinal scale). In passing, it is noted that the coefficient of consistence is equivalent to the hierarchy index proposed by Landau (1951).

## 2.2.4. Example

### Occupation Mobility

For the data in Table 2.1, we have $d = 4$ and $\zeta = 0.500$. So, we may think that no ordinal structure exists. Take up the case of intergenerational occupation mobility. Computing $\zeta$ with the dominance matrix in Table 2.3, we have $\zeta = 1.000$. This indicates that a rank order exists in the mobility. When we compute the sum for each row in $C$, and rearrange the occupations in descending order of the sum, we obtain Table 2.4. When the mobility from $A$ to $B$ is larger than that from $B$ to $A$, denote the situation by $A \rightarrow B$. For Table 2.4, we represent the structure as: farmer $\rightarrow$ lower manual $\rightarrow$ upper manual $\rightarrow$ upper nonmanual $\rightarrow$ lower nonmanual. Moving from farmer to any of the other occupations is more prominent than moving from any of them to farmer. In contrast, moving from lower nonmanual to any of the other occupations is more prominent than moving from any of them to lower nonmanual. It is pointed out that the detection of ordinal structure described here is different from the customary analysis of social mobility in sociology, where researchers transform the asymmetric data to symmetric data by using the odds ratio and apply them to a log-linear model.

### Pecking Frequency

We turn to the case of pecking frequency. For the dominance matrix $\boldsymbol{C}$ derived from the data of Table 2.5, we have $\zeta = 0.929$. This value indicates that nontransitive dominance relations exist between every pair of members in the flock. We may then consider a nearly rank order among the flock. Regarding the data of Table 2.6, we have $\zeta = 1.000$, which reveals the existence of a rank order in terms of the dominant–subordinate relationships among the flock. When individual $P$ dominates $Q$, denote the situation by $P \rightarrow Q$. In descending order of dominance, we represent the structure as: Y $\rightarrow$ YY $\rightarrow$ G $\rightarrow$ BL $\rightarrow$ R $\rightarrow$ W $\rightarrow$ B.

### Morse Code Confusion

Nakajima and Saito (1996) applied the same procedure on the Morse code confusion data of Tables 3.10 to 3.12 (Rothkopf, 1957). A nearly ordinal structure was found for some of the 36 signals: 7, C, 6, J, L, B, 4, V, 5. The structure is traced in this ordered arrangement, which gives $\zeta = 0.938$.

## 2.3. ANALYSIS OF VARIANCE

### 2.3.1. Scheffé's Method of Paired Comparisons

For a combination of stimuli $S_j$ and $S_k$, we distinguish the ordered pairs $(S_j, S_k)$ and $(S_k, S_j)$. Here suppose that stimulus at the left (right) is presented first. No pair of identical stimuli is designed. Then there are $m(m-1)$ pairs in total. The subjects are required to judge the difference in the paired stimuli. When there are $N$ subjects, $n$ subjects are allotted in random for each $(S_j, S_k)$; that is, $N = m(m-1)n$. Let us put forward fundamental assumptions (A1) to (A3) for comparative judgments (Scheffé, 1952).

(A1) For $(S_j, S_k)$, the subject judges the psychological difference of the first stimuli from the second and scores it by $u_{jk}$ on a specified numerical scale. For example of preference judgment, $u_{jk}$ represents the degree of preferring $S_j$ to $S_k$. In the case of aesthetic judgment, $u_{jk}$ represents the degree of aesthetic precedence of $S_j$ over $S_k$.

(A2) The $u_{jk}$ is a random variable, which accords to normal distribution

$$u_{jk} \sim N(\mu_{jk}, \sigma^2) \tag{2.4}$$

(A3) For $(S_k, S_j)$, the subject is required to judge $u_{kj}$, the difference. The difference of $S_j$ from $S_k$ is not judged, but is assumed to be $-u_{kj}$, that is $-u_{kj} \sim N(-\mu_{kj}, \sigma^2)$.

From the assumptions above, the average score of $S_j$ over $S_k$ is given by

$$\lambda_{jk} = \tfrac{1}{2}(\mu_{jk} - \mu_{kj}) \tag{2.5}$$

The average of the difference of $S_j$ to $S_k$ is given by

$$\delta_{jk} = \tfrac{1}{2}(\mu_{jk} - (-\mu_{kj})) = \tfrac{1}{2}(\mu_{jk} + \mu_{kj}) \tag{2.6}$$

It is regarded as the effect due to the order of presentation of stimuli. Apparently we see that

$$\lambda_{jk} = -\lambda_{kj}, \qquad \delta_{jk} = \delta_{kj} \tag{2.7}$$

Define $\mu_{jj} = 0$, $\delta_{jj} = 0$, $\lambda_{jj} = 0$ $(j = 1, 2, \ldots, n)$ and let square matrices

$$\boldsymbol{M} = (\mu_{jk}), \qquad \boldsymbol{\Delta} = (\delta_{jk}), \qquad \boldsymbol{\Lambda} = (\lambda_{jk}) \tag{2.8}$$

Then we have a decomposition asymmetric matrix $\boldsymbol{M}$ into symmetric $\Delta$ and skew-symmetric $\boldsymbol{\Lambda}$ as

$$\boldsymbol{M} = \boldsymbol{\Lambda} + \boldsymbol{\Delta}, \qquad \boldsymbol{\Delta}' = \boldsymbol{\Delta}, \qquad \boldsymbol{\Lambda}' = -\boldsymbol{\Lambda} \tag{2.9}$$

Let us consider in view of the assumptions what $\delta_{jk}$ and $\lambda_{jk}$ mean. We have expectations as

$$\begin{aligned} \mathrm{E}(u_{jk}) &= \mu_{jk} = \lambda_{jk} + \delta_{jk} \\ \mathrm{E}(-u_{kj}) &= -\lambda_{kj} - \delta_{kj} = \lambda_{jk} - \delta_{jk} \end{aligned} \tag{2.10}$$

The expectations of the difference of $S_j$ from $S_k$ consist of two components: $\delta_{jk}$, which depends on the order of presentation, and $\lambda_{jk}$, which does not. The expectation shifts from $\lambda_{jk}$ by $\delta_{jk}$ in the positive or negative direction, corresponding to the order of presentation. In this sense, we call $\delta_{jk}$ the order effect and $\lambda_{jk}$ the combination effect.

### 2.3.2. Structure Model for Skew-Symmetry

For the structure for $\lambda_{jk}$, we set a model such as

$$\lambda_{jk} = \beta_j - \beta_k + \gamma_{jk} \qquad (j \neq k;\ j, k = 1, 2, \ldots, m) \tag{2.11}$$

Given $\lambda_{jk}$, we want to determine $\beta_j$ and $\gamma_{jk}$. At first, we impose a constraint on $\{\beta_j\}$ as

$$\sum_{j=1}^{m} \beta_j = 0 \tag{2.12}$$

From (2.7) we see that

$$\gamma_{jk} = -\gamma_{kj} \tag{2.13}$$

Next setting a constraint,

$$\sum_{k=1}^{m} \gamma_{jk} = 0 \qquad (j = 1, 2, \ldots, m) \tag{2.14}$$

we obtain the solution of $\{\beta_j\}$ as

$$\beta_j = \frac{1}{m}\sum_{j=1}^{m} \lambda_{jk} \tag{2.15}$$

If we do not impose (2.14), setting (2.11), (2.12), and (2.15) yields (2.14). Lastly, we determine $\gamma_{jk}$ from (2.11) and (2.15),

$$\gamma_{jk} = \lambda_{jk} - \frac{1}{m}\sum_{j=1}^{m} \lambda_{jk} + \frac{1}{m}\sum_{k=1}^{m} \lambda_{jk} \tag{2.16}$$

If all $\gamma_{jk} = 0$, then

$$\lambda_{jk} = \beta_j - \beta_k \Leftrightarrow \gamma_{jk} = 0 \qquad \text{for all } (j, k) \tag{2.17}$$

That is, model (2.11) represents the skew-symmetry $\lambda_{jk}$ as the sum of linear component $\beta_j - \beta_k$ and nonlinear component $\gamma_{jk}$ expressed by (2.16).

### 2.3.3. Stochastic Model and Parameter Estimation

Write the score of $(S_j, S_k)$ judged by subject $i$ as $o_{ijk}$ $(i = 1, 2, \ldots, n)$. We note $n$ observations of random variable $u_{jk}$, which accords to normal distribution $N(\mu_{jk}, \sigma^2)$. For $\mu_{jk}$, we retain the structure of (2.10) and (2.11). Denote the random variable of error by $\varepsilon_{ijk}$. Then $o_{ijk}$ is expressed as

$$o_{ijk} = \mu_{jk} + \varepsilon_{ijk} = (\beta_j - \beta_k) + \gamma_{jk} + \delta_{jk} + \varepsilon_{ijk}, \tag{2.18}$$

$$\varepsilon_{ijk} \sim N(0, \sigma^2). \tag{2.19}$$

Given $o_{ijk}$, let us obtain least squares estimates of parameters $\beta_j$, $\gamma_{jk}$, $\delta_{jk}$, and $\sigma^2$. For convenience of estimation, we also estimate $\mu_{jk}$ and $\lambda_{jk}$. Define

$$o_{\cdot jk} = \sum_{i=1}^{n} o_{ijk}, \quad o_{\cdot j \cdot} = \sum_{i=1}^{n}\sum_{k=1}^{m} o_{ijk} \quad \text{and} \quad o_{\cdot\cdot k} = \sum_{i=1}^{n}\sum_{j=1}^{m} o_{ijk} \tag{2.20}$$

First we give the least squares estimate for $\mu_{jk}$ as

$$\hat{\mu}_{jk} = \frac{1}{n} o_{\cdot jk} \tag{2.21}$$

Then we have from (2.5) and (2.6) that

$$\hat{\lambda}_{jk} = \frac{1}{2}(\hat{\mu}_{jk} - \hat{\mu}_{kj}) = \frac{1}{2n}(o_{\cdot jk} - o_{\cdot kj}) \tag{2.22}$$

$$\hat{\delta}_{jk} = \frac{1}{2}(\hat{\mu}_{jk} + \hat{\mu}_{kj}) = \frac{1}{2n}(o_{\cdot jk} + o_{\cdot kj}) \tag{2.23}$$

These estimates meet the following relations:

$$\hat{\lambda}_{jk} = -\hat{\lambda}_{kj}, \qquad \hat{\delta}_{jk} = \hat{\delta}_{kj} \tag{2.24}$$

Substituting $\hat{\lambda}_{jk}$ in (2.11), and using (2.12), (2.13), and (2.14), we obtain least squares estimates as

$$\hat{\beta}_j = \frac{1}{m}\sum_{k=1}^{m} \hat{\lambda}_{jk} = \frac{1}{2mn}(o_{\cdot j\cdot} - o_{\cdot\cdot j}) \tag{2.25}$$

$$\hat{\gamma}_{jk} = \hat{\lambda}_{jk} - \hat{\beta}_j + \hat{\beta}_k \tag{2.26}$$

Using $\mathcal{S}_e$ of (2.29), $\sigma^2$ is estimated as

$$\hat{\sigma}^2 = \frac{1}{m(m-1)(n-1)}\mathcal{S}_e \tag{2.27}$$

### 2.3.4. Analysis of Variance

Let the error term be $e_{ijk} = o_{ijk} - \hat{\mu}_{jk}$. Define sums of squares $\mathcal{S}_t$, $\mathcal{S}_\mu$, $\mathcal{S}_e$ as follows, to obtain the decomposition of (2.31):

$$\mathcal{S}_t = \sum_{i=1}^{n}\sum_{j=1}^{m}\sum_{k=1}^{m} o_{ijk}^2 \tag{2.28}$$

$$\mathcal{S}_e = \sum_{i=1}^{n}\sum_{j=1}^{m}\sum_{k=1}^{m} (o_{ijk} - \hat{\mu}_{jk})^2 \tag{2.29}$$

$$\mathcal{S}_\mu = n\sum_{j=1}^{m}\sum_{k=1}^{m} \hat{\mu}_{jk}^2 = \frac{1}{n}\sum_{j=1}^{m}\sum_{k=1}^{m} o_{\cdot jk}^2 \tag{2.30}$$

$$\mathcal{S}_t = \mathcal{S}_\mu + \mathcal{S}_e \tag{2.31}$$

Next define $\mathcal{S}_\lambda$, $\mathcal{S}_\delta$, and decomposition (2.34) holds:

$$\mathcal{S}_\lambda = 2n\sum_{j=1}^{m}\sum_{k<j}^{m}\hat{\lambda}_{jk}^2 = \frac{1}{2n}\sum_{j=1}^{m}\sum_{k<j}^{m}(o_{\cdot jk} - o_{\cdot kj})^2 \tag{2.32}$$

$$\mathcal{S}_\delta = 2n\sum_{j=1}^{m}\sum_{k<j}^{m}\hat{\delta}_{jk}^2 = \frac{1}{2n}\sum_{j=1}^{m}\sum_{k<j}^{m}(o_{\cdot jk} + o_{\cdot kj})^2 \tag{2.33}$$

$$\mathcal{S}_\mu = \mathcal{S}_\lambda + \mathcal{S}_\delta \tag{2.34}$$

Furthermore, by definition of $\mathcal{S}_\beta$ and $\mathcal{S}_\gamma$, decomposition (2.37) holds:

$$\mathcal{S}_\beta = 2mn\sum_{j=1}^{m}\hat{\beta}_j^2 = \frac{1}{2mn}\sum_{j=1}^{m}(o_{\cdot j\cdot} - o_{\cdot\cdot k})^2 \tag{2.35}$$

$$\mathcal{S}_\gamma = 2n\sum_{j=1}^{m}\sum_{k<j}^{m}\hat{\gamma}_{jk}^2 = \mathcal{S}_\lambda - \mathcal{S}_\beta \tag{2.36}$$

$$\mathcal{S}_\lambda = \mathcal{S}_\beta + \mathcal{S}_\gamma \tag{2.37}$$

Thus $\mathcal{S}_t$ is decomposed as follows,

$$\mathcal{S}_t = \mathcal{S}_\lambda + \mathcal{S}_\delta + \mathcal{S}_e \tag{2.38}$$

$$= \mathcal{S}_\beta + \mathcal{S}_\gamma + \mathcal{S}_\delta + \mathcal{S}_e \tag{2.39}$$

Quantities $\mathcal{S}_\beta$, $\mathcal{S}_\gamma$, $\mathcal{S}_\delta$, and $\mathcal{S}_e$ are statistically independent. Under the assumption of normality of (2.4), $\mathcal{S}_e/\sigma^2$ accords to the $\chi^2$ distribution with the degree of freedom $\phi_e = m(m-1)(n-1)$. By $F(\phi_1, \phi_2)$ we denote the $F$ distribution with first and second degrees of freedom $\phi_1$ and $\phi_2$, respectively.

(a) If $\beta_j = 0$, then $\mathcal{S}_\beta/\sigma^2$ accords to the $\chi^2$ distribution with degree of freedom $\phi_\beta = m - 1$. Then $F_\beta$ accords to $F(\phi_\beta, \phi_e)$, where

$$F_\beta = \frac{\mathcal{S}_\beta/\phi_\beta}{\mathcal{S}_e/\phi_e} = \frac{2m(n-1)\mathcal{S}_\beta}{\mathcal{S}_e} \tag{2.40}$$

(b) If all $\gamma_{jk} = 0$, then $\mathcal{S}_\gamma/\sigma^2$ accords to the $\chi^2$ distribution with degree of freedom $\phi_\gamma = (m-1)(m-2)/2$. Then $F_\gamma$ accords to $F(\phi_\gamma, \phi_e)$, where

$$F_\gamma = \frac{\mathcal{S}_\gamma/\phi_\gamma}{\mathcal{S}_e/\phi_e} = \frac{2m(n-1)\mathcal{S}_\gamma}{(m-2)\mathcal{S}_e} \tag{2.41}$$

(c) If all $\delta_{jk} = 0$, then $\mathcal{S}_\delta/\sigma^2$ accords to the $\chi^2$ distribution with degree of freedom $\phi_\delta = m(m-1)/2$. Then $F_\delta$ accords to

$F(\phi_\delta, \phi_e)$, where

$$F_\delta = \frac{S_\delta/\phi_\delta}{S_e/\phi_e} = \frac{2m(n-1)S_\delta}{S_e} \tag{2.42}$$

Utilizing the properties mentioned above, we can perform analysis of variance (ANOVA). To $S_e$, we will do the $F$-test about the significance of the three kinds of effects, main effects $S_\beta$, order effects $S_\delta$, and combination effects $S_\gamma$. The ANOVA table is provided as Table 2.7.

## Confidence Region

If the main effects are found to be significant by using ANOVA, we can then examine which difference would be significant. The interval with $100 \times (1-\alpha)\%$ is given in terms of the Studentized range $q(m, \phi_e : \alpha)$ and the yardstick $Y(\alpha)$ as

$$Y(\alpha) = q(m, \phi_e : \alpha)\sqrt{\frac{\hat{\sigma}^2}{2nm}} \tag{2.43}$$

$$\hat{\beta}_j - \hat{\beta}_k - Y \leq \beta_j - \beta_k \leq \hat{\beta}_j - \hat{\beta}_k + Y \tag{2.44}$$

When the interval is given in the positive side, we conclude that $\beta_j > \beta_k$, and if given in the negative side we may consider $\beta_j < \beta_k$.

### 2.3.5. Example

Regarding the perfume of four brands of soap, a total of 240 subjects made comparative judgments on a seven-point scale. Twenty subjects are allotted for each pair. The data are summarized in Table 2.8. Table 2.9 shows $\hat{\lambda}_{jk}$ given by (2.22) and $\hat{\beta}_j$ given by (2.25). Figures 2.2 and 2.3 show these estimates. Table 2.10 shows $\hat{\gamma}_{jk}$ given by (2.26), and Table 2.11 shows $\hat{\delta}_{jk}$

**TABLE 2.7** Analysis of Variance

| Source | Sum of squares | Degree of freedom | Variance | $F$-value |
|---|---|---|---|---|
| Main effects | $S_\beta$ | $\phi_\beta = m-1$ | $V_\beta = S_\beta/\phi_\beta$ | $F_\beta = V_\beta/V_e$ |
| Combination effects | $S_\gamma$ | $\phi_\gamma = (m-1)(m-2)/2$ | $V_\gamma = S_\gamma/\phi_\gamma$ | $F_\gamma = V_\gamma/V_e$ |
| Order effects | $S_\delta$ | $\phi_\delta = m(m-1)/2$ | $V_\delta = S_\delta/\phi_\delta$ | $F_\delta = V_\delta/V_e$ |
| Error | $S_e$ | $\phi_e = m(m-1)(n-1)$ | $V_e = S_e/\phi_e$ | |
| Total | $S_T$ | $\phi_T = m(m-1)n$ | | |

**TABLE 2.8** Judgment Frequency in Paired Comparisons

| | Frequency of scores | | | | | | |
|---|---|---|---|---|---|---|---|
| Pair $(j, k)$ | $-3$ | $-2$ | $-1$ | 0 | 1 | 2 | 3 |
| (1, 2) | 0 | 1 | 4 | 5 | 5 | 4 | 1 |
| (1, 3) | 0 | 1 | 3 | 10 | 4 | 2 | 0 |
| (1, 4) | 0 | 0 | 1 | 2 | 5 | 8 | 4 |
| (2, 1) | 0 | 4 | 11 | 4 | 1 | 0 | 0 |
| (2, 3) | 0 | 0 | 2 | 10 | 7 | 0 | 1 |
| (2, 4) | 0 | 0 | 6 | 4 | 5 | 2 | 3 |
| (3, 1) | 3 | 3 | 8 | 4 | 2 | 0 | 0 |
| (3, 2) | 1 | 5 | 2 | 7 | 5 | 0 | 0 |
| (3, 4) | 0 | 1 | 3 | 6 | 6 | 3 | 1 |
| (4, 1) | 5 | 8 | 4 | 3 | 0 | 0 | 0 |
| (4, 2) | 0 | 5 | 1 | 9 | 5 | 0 | 0 |
| (4, 3) | 1 | 4 | 7 | 6 | 2 | 0 | 0 |

given by (2.23). Table 2.12 summarizes ANOVA. By $F$-test, it is found that the main effects are significant at 1% level and the combination effects are significant at 5% level. Yardsticks are computed by (2.43) as $Y(0.05) = 0.3296$ and $Y(0.01) = 0.4022$. The confidence intervals are provided in Table 2.13. We see that the difference between $\beta_2$ and $\beta_3$ is not significant, but the other differences are significant.

## 2.4. PSYCHOLOGICAL SCALING

The dominance matrix represents the result of single judgment for each pair made by a subject. If the subject is required to make multiple judgment for

**TABLE 2.9** Combination Effects and Main Effects

| | $\lambda_{jk}$ | | | | |
|---|---|---|---|---|---|
| | 1 | 2 | 3 | 4 | $\beta_j$ |
| 1 | 0.000 | 0.700 | 0.600 | 1.675 | 0.7437 |
| 2 | $-0.700$ | 0.000 | 0.450 | 0.450 | 0.0500 |
| 3 | $-0.600$ | $-0.450$ | 0.000 | 0.650 | $-0.1000$ |
| 4 | $-1.675$ | $-0.450$ | $-0.650$ | 0.000 | $-0.6938$ |

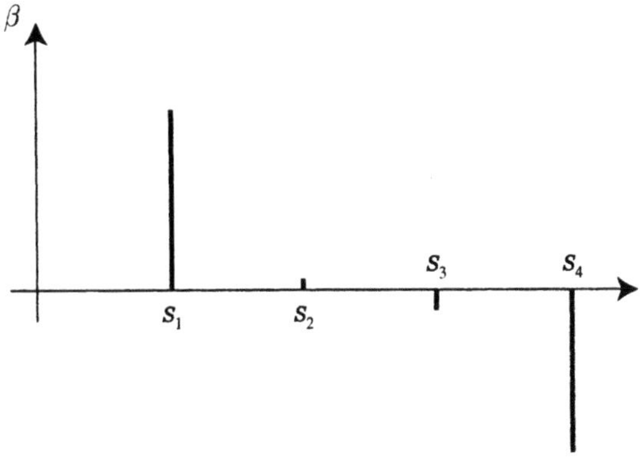


FIGURE 2.2 Main effect

$(S_j, S_k)$, or a group of subjects are required to make a single judgment for the pair, the result would be summarized as proportions $\{\pi_{jk}\}$, which indicate the averaged judgments. Assuming that the judgments are made on a latent axis (psychological continuum) for the paired comparisons (PC), Thurstone's scal-

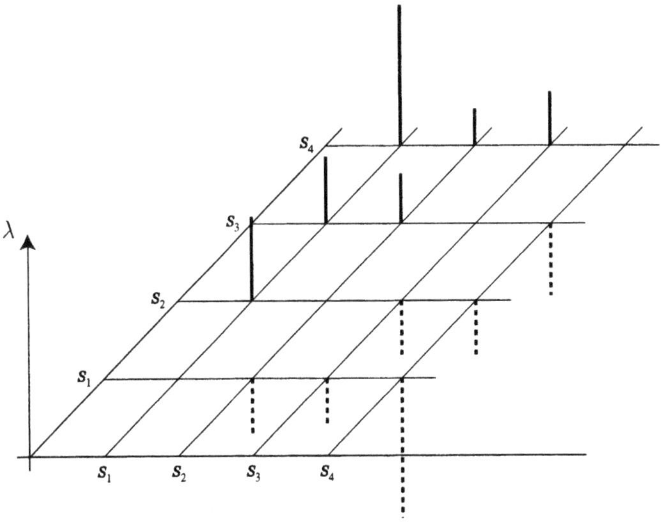


FIGURE 2.3 Combination effect

**TABLE 2.10** Nonlinear Components: $\{\gamma_{jk}\}$

| | 1 | 2 | 3 | 4 |
|---|---|---|---|---|
| 1 | 0.0000 | 0.0063 | −0.2437 | 0.2375 |
| 2 | −0.0063 | 0.0000 | 0.3000 | −0.2938 |
| 3 | 0.2437 | −0.3000 | 0.0000 | 0.0563 |
| 4 | −0.2375 | 0.2938 | −0.0563 | 0.0000 |

**TABLE 2.11** Order Effects: $\{\delta_{jk}\}$

| | 1 | 2 | 3 | 4 |
|---|---|---|---|---|
| 1 | 0.000 | −0.200 | −0.450 | −0.075 |
| 2 | −0.200 | 0.000 | −0.050 | 0.150 |
| 3 | −0.450 | −0.050 | 0.000 | −0.150 |
| 4 | −0.075 | 0.150 | −0.150 | 0.000 |

ing procedure analyzes $\{\pi_{jk}\}$ on the basis of a probabilistic model that is called the discriminal process. Here it is postulated that a random variable $u_j$ represents psychological quantity associated with stimulus $S_j$ and it accords to normal distribution $N(\mu_j, \sigma_j^2)$. Then psychological difference for $(S_j, S_k)$ is stated as $w_{jk} = u_k - u_j$, which is distributed as $N(\mu_k - \mu_j, \sigma_{jk}^2)$. By $\Pi_{jk}$ we denote the probability that $S_k$ is preferred to (or judged to be greater than) $S_j$.

**TABLE 2.12** Analysis of Variance

| Source | Sum of squares | d.f. | Variance | *F*-value |
|---|---|---|---|---|
| Main effects | 167.5125 | 3 | 55.8375 | 43.6216** |
| Combination effects | 11.8125 | 3 | 3.9375 | 3.0761* |
| Order effects | 11.8250 | 6 | 1.9708 | 1.5397 |
| Error | 291.8500 | 228 | 1.2800 | |
| Total | 331.0000 | 192 | | |

d.f., degrees of freedom.
*Significant at 5% level.
**Significant at 1% level.

**TABLE 2.13** Confidence Region

| Difference | Estimate | 95% confidence lb | 95% confidence ub | 99% confidence lb* | 99% confidence ub* |
|---|---|---|---|---|---|
| $\beta_1 - \beta_2$ | 0.6937 | 0.3641 | 1.0233 | 0.2915 | 1.0959 |
| $\beta_1 - \beta_3$ | 0.8437 | 0.5141 | 1.1733 | 0.4415 | 1.2459 |
| $\beta_1 - \beta_4$ | 1.4375 | 1.1079 | 1.7671 | 1.0353 | 1.8397 |
| $\beta_2 - \beta_3$ | 0.1500 | −0.1796 | 0.4796 | −0.2522 | 0.5522 |
| $\beta_2 - \beta_4$ | 0.7438 | 0.4142 | 1.0734 | 0.3416 | 1.1460 |
| $\beta_3 - \beta_4$ | 0.5938 | 0.2642 | 0.9234 | 0.1916 | 0.9960 |

*lb: lower bound; ub: upper bound.

The law of comparative judgment is illustrated in Fig. 2.4 and expressed as

$$\Pi_{jk} = \Pr(S_j < S_k) = \int_0^{\infty} h(w_{jk})\mathrm{d}w_{jk} \tag{2.45}$$

where $h(\cdot)$ is the density function of $w_{jk}$.

According to the procedure, we regard $\pi_{jk}$ as the observed probability of $\Pi_{jk}$. The forced binary response implies that $\pi_{jk} + \pi_{kj} = 1$. Although pairs of identical stimuli are not designed, we usually set $\pi_{jj} = 0.5$. From $\{\pi_{jk}\}$, one derives a unidimensional scale in terms of estimates of $\{\mu_j\}$ and others related to $\{\sigma_j\}$. The data $\{\pi_{jk}\}$ are asymmetric in the form of a matrix. In this viewpoint, Thurstone's scaling procedure is related to such asymmetry. However, it is customarily applied in psychometrics out of interest in asymmetry, and so we do not describe the procedure further here.

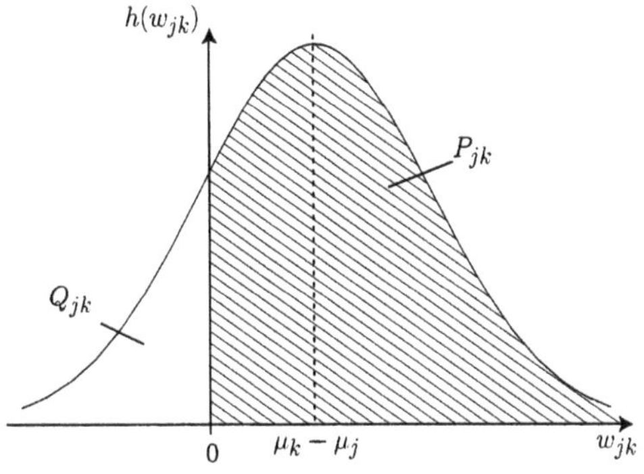


**FIGURE 2.4** The law of comparative judgment

We are concerned with asymmetry observed in judgments in experimental psychology. A typical example of such asymmetry is an order effect, which may be due to spatial or temporal order of stimuli in pairwise presentation. There might be many psychological factors behind the asymmetry observed in PC judgments, such as sensory modality, level of stimulus magnitude, length of interstimulus interval or spatial separation, and other experimental conditions. As the cause of the asymmetry, one would consider response bias, adaptation assimilation, contrast, sequential effect, traces and images, and so on (Hellström, 1978). To construct a psychological process model or theory for the phenomenon in view of those factors and causes, one needs to measure the asymmetry. For the measurement, we will present a generalized procedure of Thurstonean scaling in this section (Saito, 1994).

## 2.4.1. Response in Terms of Three Ordered Categories

As mentioned in the preceding sections, in comparative judgments the subject is usually asked to give the answer in terms of dichotomous categories, for example, "yes, no" or "greater, less." However, there are situations in which for small psychological differences, such a task may be difficult for the subject. Then one would like to give intermediate answers such as "no difference," or "equal." In such situations, it becomes practical to accommodate the use of three ordered categories with the intermediate one.

The judgment task in terms of three ordered categories seems to be particularly meaningful in studies of perceptual processes. Suppose that an experimenter has found any asymmetry between judgments for symmetrically designed pairs consisting of different stimuli, such as $(S_j, S_k)$ and $(S_k, S_j)$, where the stimulus order might be temporal or spatial. Then the experimenter would like to observe judgments for pairs of two identical stimuli $(S_j, S_j)$. In such situations, it is appropriate in view of the accuracy of judgments to require the subject to answer using the three categories.

### Notation

For convenience of exposition, we consider successive presentation of two stimuli. Denote a pair of stimuli by $(S^{(1)}, S^{(2)})$, where $S^{(1)}$ is the first stimulus and $S^{(2)}$ the second in presentation. For each pair of stimuli, the subject is required to make comparative judgments in terms of three ordered categories on the psychological continuum: (G) $S^{(2)}$ is greater than $S^{(1)}$; (L) $S^{(2)}$ is less than $S^{(1)}$; (E) $S^{(2)}$ is equal to $S^{(1)}$. Let frequencies of G, L, and E responses be $l_{jk}$, $m_{jk}$, and $n_{jk}$, respectively. Denote observed probabilities and the corresponding matrices as

$$p_{jk} = \frac{l_{jk}}{N_{jk}}, \qquad q_{jk} = \frac{m_{jk}}{N_{jk}}, \qquad r_{jk} = \frac{n_{jk}}{N_{jk}}$$

$$\boldsymbol{P} = (p_{jk}), \qquad \boldsymbol{Q} = (q_{jk}) \quad \text{and} \quad \boldsymbol{R} = (r_{jk}) \tag{2.46}$$

Here $N_{jk}$ is the number of observations for the pair. The asymmetry dealt with in this section is now defined. It refers to relations on probability matrices such that

$$p_{jk} \neq q_{kj} \quad \text{or} \quad p_{kj} \neq q_{jk} \tag{2.47}$$

This definition implies symmetry that $p_{jk} = q_{kj}$ and $p_{kj} = q_{jk}$. In some cases, not all pairs may be presented because of the experimental design. To indicate this type of missing pair, we define $\delta_{jk} = 1$ if $(S_j^{(1)}, S_k^{(2)})$ is observed, and $\delta_{jk} = 0$ if the pair is not observed. Denote the design matrix by $\boldsymbol{\Delta} = (\delta_{jk})$.

## 2.4.2. Revised Law of Comparative Judgment

When the asymmetry is significantly observed in PC judgments, one may consider that two sets of stimuli, $S_j^{(1)}$ and $S_k^{(2)}$, would play different roles in the judgments. This consideration seems natural, because a pair of physically identical stimuli are never alike (Harris, 1957). That is, they are different in a sense that they are separated spatially (e.g., right and left) or temporally (first and second).

Let us revise Thurstone's model in order to analyze the asymmetry of (2.47) in connection with ternary responses. We consider two discriminal processes for an identical stimulus, a random variable $v_j^{(1)}$ representing $S_j^{(1)}$ and another $v_j^{(2)}$ representing $S_j^{(2)}$. In the sequel, it is assumed that they are normally distributed in such a way that

$$v_j^{(1)} \sim N(\mu_j, \sigma_j^2) \qquad (j = 1, 2, \ldots, m) \tag{2.48}$$

$$v_j^{(2)} \sim N(\nu_j, \sigma_j^2) \qquad (j = 1, 2, \ldots, m) \tag{2.49}$$

Then the perceived difference for $(S_j^{(1)}, S_k^{(2)})$ is represented by

$$u_{jk} = v_k^{(2)} - v_j^{(1)} \tag{2.50}$$

and is distributed as $N(\nu_k - \mu_j, c_{jk}^2)$. Here $c_{jk}^2$ is the so-called comparatal dispersion with correlation $\rho_{jk}$ between two random processes:

$$c_{jk}^2 = \sigma_j^2 + \sigma_k^2 - 2\rho_{jk}\sigma_j\sigma_k \tag{2.51}$$

Now we introduce a threshold parameter $\varepsilon > 0$ to account for the E response. Here it is assumed that the subject will give an E response if $|u_{jk}| < \varepsilon$ and also a G response if $u_{jk} > \varepsilon$ and an L response if $u_{jk} < -\varepsilon$. Then the law of comparative judgment is revised as

$$P_{jk} = \Pr(S_j^{(1)} < S_k^{(2)}) = \int_{\varepsilon}^{\infty} g(u_{jk})\mathrm{d}u_{jk} \tag{2.52}$$

$$Q_{jk} = \Pr(S_j^{(1)} > S_k^{(2)}) = \int_{-\infty}^{-\varepsilon} g(u_{jk})\mathrm{d}u_{jk} \tag{2.53}$$

$$R_{jk} = \Pr(S_j^{(1)} = S_k^{(2)}) = \int_{-\varepsilon}^{\varepsilon} g(u_{jk})\mathrm{d}u_{jk} \tag{2.54}$$

where $g(\cdot)$ is the density function of $u_{jk}$. Figure 2.5 shows the revised law of comparative judgment.

By $\Phi$ we denote an upper probability integral of the density $f(u)$ of the unit normal distribution, that is,

$$\Phi(z) = \int_{z}^{\infty} f(u)\mathrm{d}u \tag{2.55}$$

Write the normal deviates as

$$X_{jk} = \Phi^{-1}(P_{jk}) \quad \text{and} \quad Y_{jk} = \Phi^{-1}(1 - Q_{jk}) \tag{2.56}$$

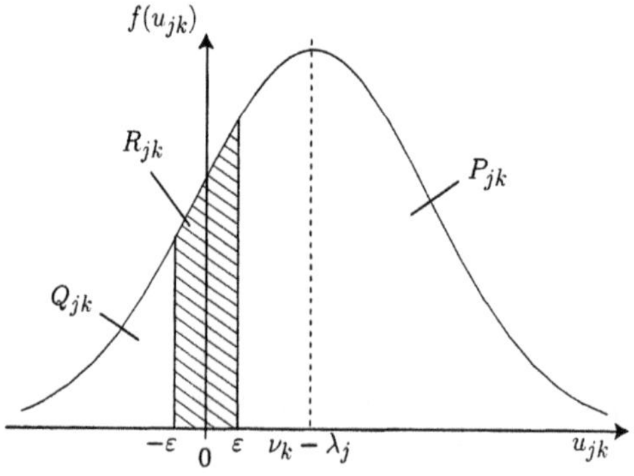


**FIGURE 2.5** The revised law of comparative judgment

Then we have

$$X_{jk} = \frac{1}{c_{jk}}(\mu_j - \nu_k + \varepsilon) \quad \text{and} \quad Y_{jk} = \frac{1}{c_{jk}}(\mu_j - \nu_k - \varepsilon) \tag{2.57}$$

## 2.4.3. Least Squares Estimation

We adopt Thurstone's Case 5 assumption, specifying equal $\sigma_j$ ($j = 1, 2, \ldots, m$) and a null correlation between $v_j^{(1)}$ and $v_k^{(2)}$. Then we may treat all $c_{jk}^2$ as an unknown constant $2\sigma^2$. Setting $2\sigma^2 = 1$ for the measurement unit, we then have

$$X_{jk} = \mu_j - \nu_k + \varepsilon \quad \text{and} \quad Y_{jk} = \mu_j - \nu_k - \varepsilon \tag{2.58}$$

Usually observed probabilities, $p_{jk}$, $q_{jk}$, and $r_{jk}$ are given instead of the population probabilities. Write the observed normal deviates as

$$x_{jk} = \Phi^{-1}(p_{jk}) \quad \text{and} \quad y_{jk} = \Phi^{-1}(1 - q_{jk}) \tag{2.59}$$

We consider that (2.57) approximately holds with the observed deviates. Thus our aim is to determine $\{\mu_j\}$, $\{\nu_j\}$, and $\varepsilon$ with those equations.

Given normal deviates $x_{jk}$, $y_{jk}$ and design indicators $\delta_{jk}$, we wish to estimate parameters $\mu_j$, $\nu_j$, and $\varepsilon$ by using a least squares (LS) criterion. This is defined as

$$S = \sum_{j=1}^{m}\sum_{k=1}^{m} \delta_{jk}(x_{jk} - \mu_j + \nu_k - \varepsilon)^2 + \sum_{j=1}^{m}\sum_{k=1}^{m} \delta_{jk}(y_{jk} - \mu_j + \nu_k + \varepsilon)^2 \tag{2.60}$$

Let

$$L = \sum_{j=1}^{m}\sum_{k=1}^{m} \delta_{jk}, \quad M_j = \sum_{k=1}^{m} \delta_{jk} \quad \text{and} \quad N_k = \sum_{j=1}^{m} \delta_{jk} \tag{2.61}$$

Setting the derivatives of $S$ with respect to the parameters equal to zero gives the following equations:

$$\varepsilon = \frac{1}{2L}\sum_{j=1}^{m}\sum_{k=1}^{m} \delta_{jk}(x_{jk} - y_{jk}) \tag{2.62}$$

$$N_k \nu_k - \sum_{j=1}^{m} \delta_{jk}\mu_j = a_k \tag{2.63}$$

$$M_j \mu_j - \sum_{k=1}^{m} \delta_{jk}\nu_j = b_j \tag{2.64}$$

where

$$a_k = -\frac{1}{2}\sum_{j=1}^{m} \delta_{jk}(x_{jk} + y_{jk}) \qquad (k = 1, 2, \ldots, m) \tag{2.65}$$

$$b_j = \frac{1}{2}\sum_{k=1}^{m} \delta_{jk}(x_{jk} + y_{jk}) \qquad (j = 1, 2, \ldots, m) \tag{2.66}$$

Equations (2.63) and (2.64) may be rewritten in a matrix form as

$$\begin{pmatrix} -\boldsymbol{\Delta}' & \boldsymbol{N} \\ \boldsymbol{M} & -\boldsymbol{\Delta} \end{pmatrix}\begin{pmatrix} \boldsymbol{\mu} \\ \boldsymbol{\nu} \end{pmatrix} = \begin{pmatrix} \boldsymbol{a} \\ \boldsymbol{b} \end{pmatrix} \tag{2.67}$$

where $\boldsymbol{\mu}$, $\boldsymbol{\nu}$, $\boldsymbol{a}$, $\boldsymbol{b}$ are column vectors of the corresponding quantities respectively, and $\boldsymbol{M} = \mathrm{diag}(M_1, M_2, \ldots, M_m)$ and $\boldsymbol{N} = \mathrm{diag}(N_1, N_2, \ldots, N_m)$.

Let $\boldsymbol{\Gamma}$ be the entire matrix on the left-hand side of (2.67). Noting that $\boldsymbol{\Gamma}$ is singular, we impose a constraint on the parameters. Since we are usually interested in relative positioning of stimuli, and noting that (2.58) are invariant with respect to addition of a constant to all the parameters, let us impose that the average of the $\mu_j$ values be the origin,

$$\sum_{j=1}^{m} \mu_j = 0 \tag{2.68}$$

without loss of generality. Although we may replace it by any row in (2.67), we select the last row for simplicity. Define $\boldsymbol{\Gamma}_*$, which differs from $\boldsymbol{\Gamma}$ only in the last row. The $m$th row of $\boldsymbol{\Gamma}_*$ has ones in the first $m$ entries and zeros in the remaining entries. Further, define $\boldsymbol{b}^* = (b_j^*)$ where $b_j^* = b_j$ $(j = 1, 2, \ldots, m -$

1) and $b_m^* = 0$. Then the solution is given by

$$\begin{pmatrix} \boldsymbol{\mu} \\ \boldsymbol{\nu} \end{pmatrix} = \boldsymbol{\Gamma}_*^{-1} \begin{pmatrix} \boldsymbol{a} \\ \boldsymbol{b}^* \end{pmatrix} \tag{2.69}$$

When $\boldsymbol{\Gamma}*$ is still singular, we may use a generalized inverse of $\boldsymbol{\Gamma}*$ in (2.69). Thus we have obtained LS estimates of parameters using (2.62) and (2.69).

For the $\boldsymbol{\Delta}$ matrix of the complete design, solution (2.69) is explicitly described as follows:

$$\mu_j = \sum_{j=1}^{m} \frac{a_j}{m^2} + \frac{b_j}{m} \tag{2.70}$$

$$\nu_j = \frac{a_j}{m} \qquad (j = 1, 2, \ldots, m) \tag{2.71}$$

## 2.4.4. Minimum Normit $\chi^2$ Estimation

Let us regard probabilities $p$, $q$, and $r$ as observed from the trinomial distribution with parameters $P$, $Q$, $R$, and $n$. Let

$$\boldsymbol{u} = (p, q, r)' \quad \text{and} \quad \boldsymbol{\omega} = (P, Q, R)' \tag{2.72}$$

It is known that $\boldsymbol{u}$ accords asymptotically to $N(\boldsymbol{\omega}, \boldsymbol{\Sigma}/n)$, where each element of $\boldsymbol{\Sigma}$ is described in terms of $P$, $Q$, $R$, and $n$. For observed probabilities, some of which might be very small, we treat $p + r$ and $q + r$ instead of $p$ and $q$ in order to avoid unstability that might arise from very small values. Now we consider the distribution of variables transformed by

$$x = \Phi^{-1}(p + r) \quad \text{and} \quad y = \Phi^{-1}(q + r) \tag{2.73}$$

where $\Phi$ is defined by (2.55). Using a theorem from Rao (1964, p. 319), we find that $\boldsymbol{z} = (x, y)'$ accords asymptotically to $N(\boldsymbol{\zeta}, \boldsymbol{S}/n)$, where

$$\boldsymbol{\zeta} = (X, Y)', \qquad X = \Phi^{-1}(P + R), \qquad Y = \Phi^{-1}(Q + R) \tag{2.74}$$

and $S$ is given in terms of $P$, $Q$, $R$, and $n$.

Let us apply this result to our scaling problem (Saito, 1992). Given probabilities $p_{jk}$, $q_{jk}$, $r_{jk}$, and $N_{jk}$ observed for pair $(j, k)$, we define

$$\boldsymbol{z}_{jk} = (x_{jk}, y_{jk})' \quad \text{and} \quad \boldsymbol{\zeta}_{jk} = (X_{jk}, Y_{jk})' \tag{2.75}$$

where $x_{jk}$, $y_{jk}$, $X_{jk}$, and $Y_{jk}$ are given by (2.73) and (2.74), respectively. Define matrix $\boldsymbol{S}_{jk} = (s_{pq}^{jk})$ of size $2 \times 2$, whose entries are given as

$$s_{11}^{jk} = \frac{4Q_{jk}(1 - Q_{jk})}{f(X_{jk})^2} \tag{2.76}$$

$$s_{22}^{jk} = \frac{4P_{jk}(1 - P_{jk})}{f(Y_{jk})^2} \tag{2.77}$$

$$s_{12}^{jk} = s_{21}^{jk} = \frac{-4P_{jk}Q_{jk}}{f(X_{jk})f(Y_{jk})} \tag{2.78}$$

where $f(x_{jk})$ and $f(y_{jk})$ are densities of the unit normal distribution.

Then $z_{jk}$ accords asymptotically to $N(\boldsymbol{\zeta}_{jk}, \boldsymbol{S}_{jk}/N_{jk})$. Putting the Case 5 assumption for the discriminal process, we obtain

$$X_{jk} = \varepsilon + \nu_k - \mu_j \quad \text{and} \quad Y_{jk} = \varepsilon - \nu_k + \mu_j \tag{2.79}$$

These equations may be rewritten as

$$\boldsymbol{\zeta}_{jk}(\boldsymbol{\theta}) = \boldsymbol{A}_{jk}\boldsymbol{\theta} \tag{2.80}$$

where $\boldsymbol{\theta} = (\mu_1, \ldots, \mu_m, \nu_1, \ldots, \nu_m, \varepsilon)'$, and $\boldsymbol{A}_{jk}$ is a matrix of size $2 \times (2m + 1)$, each entry being 1, 0, or $-1$. It follows from the above argument that $z_{jk}$ accords asymptotically to $N(\boldsymbol{\zeta}_{jk}, \boldsymbol{S}_{jk}/N_{jk})$ and statistic $Q_{jk}^2$ defined by

$$Q_{jk}^2 = N_{jk}(z_{jk} - \boldsymbol{A}_{jk}\boldsymbol{\theta})'\boldsymbol{S}_{jk}^{-1}(z_{jk} - \boldsymbol{A}_{jk}\boldsymbol{\theta}) \tag{2.81}$$

accords to the $\chi^2$ distribution with two degrees of freedom. Assuming independence over all $(j, k)$, we find that $Q(\boldsymbol{\theta})$ is asymptotically distributed as the $\chi^2$ with degree of freedom $\phi$, where

$$Q(\boldsymbol{\theta}) = \sum_{j=1}^{m}\sum_{k=1}^{m} \delta_{jk}Q_{jk}^2 \quad \text{and} \quad \phi = 2\left(\sum_{j=1}^{m}\sum_{k=1}^{m} \delta_{jk} - m\right) \tag{2.82}$$

By minimizing $Q$, we like to obtain the estimates of parameters. Setting the derivative of $Q$ with respect to $\boldsymbol{\theta}$ to zero, we have an equation such as

$$\sum_{j=1}^{m}\sum_{k=1}^{m} \boldsymbol{A}_{jk}'\boldsymbol{S}_{jk}^{-1}\boldsymbol{A}_{jk}\boldsymbol{\theta} = \sum_{j=1}^{m}\sum_{k=1}^{m} \boldsymbol{A}_{jk}'\boldsymbol{S}_{jk}^{-1}z_{jk} \tag{2.83}$$

To cope with the linear dependence in each $\boldsymbol{A}_{jk}$, we can impose (2.68) or fix any one of the parameters in view of the indeterminacy of the origin. Then the normal equation is stated with the remaining $2m$ parameters. After this revision, let the coefficient matrix in the left-hand side be $\boldsymbol{B}*$ and the vector in the right-hand side $\boldsymbol{d}*$. Then we obtain the minimum normit $\chi^2$ solution (MNX) by

$$\hat{\boldsymbol{\theta}} = \boldsymbol{B}_*^{-1}\boldsymbol{d}_* \tag{2.84}$$

Here remember that the elements of $S_{jk}$ involve $P_{jk}$ and $Q_{jk}$, which depend on $\boldsymbol{\theta}$. To solve (2.83), we need provisional values for $P_{jk}$ and $Q_{jk}$.

The LS solution serves to provide those values. The resulting solution may further be used as the provisional one for the next step of computation. This iterative process might be repeated until the change in $Q(\boldsymbol{\theta})$ becomes sufficiently small. Define

$$\text{SST} = \sum_{j=1}^{m}\sum_{k=1}^{m} N_{jk}\boldsymbol{z}'_{jk}\boldsymbol{S}_{jk}^{-1}\boldsymbol{z}_{jk} \tag{2.85}$$

The quantity is partitioned into four $\chi^2$ components, each of which corresponds to $\boldsymbol{\mu}$, $\boldsymbol{\nu}$, $\varepsilon$, and the error term respectively. Thus it is possible to test a null hypothesis for each component. Through some manipulation, the covariance matrix of estimates is derived as $\boldsymbol{B}_*^{-1}$, from which the confidence regions of the parameters are constructed.

## 2.4.5. Maximum Likelihood Estimation

### The Likelihood Function

Let us regard $l_{jk}$, $m_{jk}$, and $n_{jk}$ as observations of a random variable $v_{jk}$, which accords to a trinomial distribution with parameters $P_{jk}$, $Q_{jk}$, $R_{jk}$, and $N_{jk}(= l_{jk} + m_{jk} + n_{jk})$. Under the assumption of independence over all pairs, the log-likelihood function is defined for the whole set of observations by

$$L^* = K + \sum_{j=1}^{m}\sum_{k=1}^{m} \delta_{jk}(l_{jk}\log P_{jk} + m_{jk}\log Q_{jk} + n_{jk}\log R_{jk}) \tag{2.86}$$

where $K$ is a constant in terms of observations over all possible pairs. Now let us assume that the distributional parameters $P_{jk}$, $Q_{jk}$, and $R_{jk}$ are determined by the revised law of comparative judgment of (2.52) to (2.54). We use parameters to mean the latent parameters of the discriminal processes hereafter.

For the processes let us adopt Thurstone's Case 2 assumption, which is the most general assumption from a practical point of view; that is,

$$\rho_{jk} = \rho \quad (j \neq k \qquad j, k = 1, \ldots, m) \tag{2.87}$$

Under this assumption the comparatal dispersion becomes

$$c_{jk}^2 = \sigma_j^2 + \sigma_k^2 - 2\rho\sigma_j\sigma_k \tag{2.88}$$

The population probabilities are expressed as

$$P_{jk} = \int_{X_{jk}}^{\infty} f(u)\mathrm{d}u = P(\mu_j, \nu_k, \varepsilon, \sigma_j, \sigma_k, r) \tag{2.89}$$

$$Q_{jk} = \int_{-\infty}^{Y_{jk}} f(u)\mathrm{d}u = Q(\mu_j, \nu_k, \varepsilon, \sigma_j, \sigma_k, r) \tag{2.90}$$

where $X_{jk}$ and $Y_{jk}$ are given by (2.57), and $f(\cdot)$ is the density of the unit normal distribution. For simplicity, we denote a vector of the unknown $3m + 2$ parameters by

$$\boldsymbol{\theta} = (\mu_1, \ldots, \mu_m, \nu_1, \ldots, \nu_m, \varepsilon, \sigma_1, \ldots, \sigma_m, \rho)' \tag{2.91}$$

and the summation term of (2.86) by $L(\boldsymbol{\theta})$. Then,

$$L(\boldsymbol{\theta}) = \sum_{j=1}^{m}\sum_{k=1}^{m} \delta_{jk} T_{jk} \tag{2.92}$$

where

$$T_{jk} = l_{jk} \log P_{jk} + m_{jk} \log Q_{jk} + n_{jk} \log(1 - P_{jk} - Q_{jk}) \tag{2.93}$$

In what follows it is sufficient to deal with $L(\boldsymbol{\theta})$ instead of $L^*(\boldsymbol{\theta})$.

## The Estimation

For maximum likelihood (ML) estimation, the log-likelihood function is maximized with respect to $\boldsymbol{\theta}$. The aim is equivalent to maximizing $L(\boldsymbol{\theta})$. As often with ML estimation, this problem leads to solving a nonlinear equation in terms of $\boldsymbol{\theta}$ in an iterative way. Among numerical procedures applicable to the present purpose, an algorithm of the Newton–Raphson type is provided by

$$\boldsymbol{\theta}^{(t+1)} = \boldsymbol{\theta}^{(t)} + s^{(t)}\boldsymbol{H}(\boldsymbol{\theta}^{(t)})^{-1}\boldsymbol{G}(\boldsymbol{\theta}^{(t)}) \tag{2.94}$$

Here $\boldsymbol{G}$ is the gradient vector, $\boldsymbol{H}$ the Hessian matrix, $s$ is the stepsize with superscript $t$ showing the $t$th iteration. For the initial solution, to start with it is suggested to use the LS solution mentioned above. For the gradient and the Hessian, refer to Saito (1994).

In view of the indeterminacy of parameters as mentioned, we should put constraints to fix one of $\mu_j$ and one of $\sigma_j$ for identifiability. Here we put

$$\mu_1 = \mu_c \text{ (const.)} \quad \text{and} \quad \sigma_1 = \frac{1}{\sqrt{2}} \tag{2.95}$$

Then the number of parameters is reduced to $3m$ for the estimation. Moreover we may restrict some parameters with appropriate assumptions for the discriminal processes. For example, under the Case 3 assumption we put $\rho = 0$, and under the Case 5 assumption add the further constraint $\sigma_j = \sigma$ (const.) $(j = 2, \ldots, m)$.

When some parameters are fixed for the estimation, they are eliminated from the vector $\boldsymbol{\theta}$. Thus $\boldsymbol{G}$ and $\boldsymbol{H}$ contain only those elements that correspond to the free parameters. For the constrained cases, we can apply the treatment suggested by Arbuckle and Nugent (1973).

Assuming a large sample property, we may replace the Hessian with the expected Hessian in the iterative ML estimation. Such a method is called Fisher's scoring method. Denote the expectation of the Hessian matrix $\boldsymbol{H}$ by $\boldsymbol{H}^+ = (h^+_{pq})$, which is given as follows:

$$h^+_{pq} = \sum_{j=1}^{m}\sum_{k=1}^{m} \mathrm{E}\left[\frac{\partial^2 T_{jk}}{\partial x^2_{jk}}\right]\frac{\partial x_{jk}}{\partial \theta_p}\frac{\partial x_{jk}}{\partial \theta_q} + \sum_{j=1}^{m}\sum_{k=1}^{m} \mathrm{E}\left[\frac{\partial^2 T_{jk}}{\partial y^2_{jk}}\right]\frac{\partial y_{jk}}{\partial \theta_p}\frac{\partial y_{jk}}{\partial \theta_q} \tag{2.96}$$

$$(p, q = 1, 2, \ldots, 3m)$$

The first derivatives of $x_{jk}$ and $y_{jk}$ are stated in Saito (1994) and the expectations are computed by

$$\mathrm{E}\left[\frac{\partial^2 T_{jk}}{\partial x^2_{jk}}\right] = -\frac{f(x_{jk})^2 N_{jk}(1 - Q_{jk})}{P_{jk}R_{jk}}$$
$$\mathrm{E}\left[\frac{\partial^2 T_{jk}}{\partial y^2_{jk}}\right] = -\frac{f(y_{jk})^2 N_{jk}(1 - P_{jk})}{Q_{jk}R_{jk}} \tag{2.97}$$

For construction $\boldsymbol{H}^+$ with (2.96) and (2.97) in using the scoring method, one does not need to compute the second derivatives, but only the first derivatives.

## Confidence Intervals and Statistical Tests

Once ML estimates are given, we can derive the confidence interval as follows. Assuming the asymptotic normality of ML estimates for a large sample, the covariance matrix of the estimates is given by the inverse of $-\boldsymbol{H}^{+}$. Let $s^{jk}$ be the $(j, k)$ element of the inverse of the covariance matrix. Then a confidence interval of $\theta_i$ with $100(1 - \alpha)\%$ probability is stated as

$$\hat{\theta}_i - z_\alpha\sqrt{s^{ii}} < \theta_i < \hat{\theta}_i + z_\alpha\sqrt{s^{ii}} \tag{2.98}$$

As mentioned, a class of hypotheses might be set for the discriminal processes, which are constructed hierarchically in the degree of restrictions on the parameters. Let us consider testing a null hypothesis $\mathcal{H}_1$ against an alternative one $\mathcal{H}_2$, where $\mathcal{H}_1$ is associated with likelihood $L_1$ in terms of $\phi_1$ parameters and $\mathcal{H}_2$, the less restricted, with likelihood $L_2$ in terms of $\phi_2$ parameters ($\phi_1 < \phi_2$). The test criterion is provided, under assumptions of asymptotic normality and efficiency, by

$$U = 2(L_2^* - L_1^*) = 2(L_2 - L_1) \tag{2.99}$$

It accords asymptotically to the $\chi^2$ distribution with $\phi_2 - \phi_1$ degrees of freedom. To examine the goodness of fit of the model, the AIC statistic (Akaike, 1974) can be utilized, which is defined for a model with the degree of freedom $\phi$ by

$$\text{AIC} = -2L^* + 2\phi \tag{2.100}$$

### 2.4.6. Example

#### Data

To illustrate the proposed procedures, examples are provided here. The data were collected in an experimental investigation of the time order error (TOE). Five balance weights, $S_1$, $S_2$, $S_3$, $S_4$, and $S_5$, in increasing order of weight with equal steps, were used as stimuli in the experiment. For a successive presentation of each pair of stimuli with an interval of 1.0 seconds, subjects were instructed to make comparative judgments for the stimulus pair about the weight in terms of three ordered categories (G, L, E). Among 25 possible pairs of stimuli, an incomplete set of 17 pairs was designed for the experiment. Table 2.14 shows the set of data.

In the table, row stimuli indicate $S^{(1)}$ and column stimuli $S^{(2)}$, and the observed probabilities $p_{jk}$, $r_{jk}$, and $q_{jk}$ are given for each cell $(S_j^{(1)}, S_k^{(2)})$ designed. The blank cells show the missing pairs. The sample size was $N_{jk} = 72$ for each pair. This size might be rather small to assume the large sample properties. For the present aim, however, it would be permitted to use the data.

**TABLE 2.14** Probabilities Observed in an Experiment with Five Balance Weights

| $S^{(1)} \backslash S^{(2)}$ | | $S_1$ | $S_2$ | $S_3$ | $S_4$ | $S_5$ |
|---|---|---|---|---|---|---|
| $S_1$ | $P$ | 0.333 | | 0.708 | | 0.889 |
| | $R$ | 0.473 | | 0.209 | | 0.097 |
| | $Q$ | 0.194 | | 0.083 | | 0.014 |
| $S_2$ | $P$ | | 0.319 | 0.500 | 0.694 | |
| | $R$ | | 0.389 | 0.319 | 0.223 | |
| | $Q$ | | 0.292 | 0.181 | 0.083 | |
| $S_3$ | $P$ | 0.097 | 0.306 | 0.472 | 0.611 | 0.833 |
| | $R$ | 0.306 | 0.416 | 0.320 | 0.278 | 0.111 |
| | $Q$ | 0.597 | 0.278 | 0.208 | 0.111 | 0.056 |
| $S_4$ | $P$ | | 0.056 | 0.181 | 0.458 | |
| | $R$ | | 0.402 | 0.388 | 0.334 | |
| | $Q$ | | 0.542 | 0.431 | 0.208 | |
| $S_5$ | $P$ | 0.028 | | 0.181 | | 0.347 |
| | $R$ | 0.222 | | 0.319 | | 0.417 |
| | $Q$ | 0.750 | | 0.500 | | 0.236 |

## Scaling Results

We applied the three scaling procedures to the data. We obtained the LS solution first, which was then used as an initial solution for the MNX or the ML estimation. Table 2.15 shows the results. As may be seen, there are small differences in the estimates of scale values across the solutions.

The ML estimation was performed in three ways. In the first case (MLE-1) all $\sigma_i$ were fixed to $1/\sqrt{2}$ so that $c_{jk}$ all be unity. In the second and the third cases (MLE-2, MLE-3), they were free except for $\sigma_1$, which was fixed to $1/\sqrt{2}$. For the three cases $\mu_1$ was set equal to the LS estimate. Because we had no prior information about the degree of correlation $\rho$, it was set to zero for MLE-1 and MLE-2, whereas it was free for MLE-3.

The $L$ (log-likelihood) and AIC values are indicated at the bottom of the table. With those values, we can perform statistical tests for model selection. Regarding MLE-2 and MLE-3, we have a very small value of $U$ that accords to the $\chi^2$ distribution with one degree of freedom. Thus, the MLE-2 solution is accepted against that of MLE-3. Comparing AIC values also, we confirm the

**TABLE 2.15** Parameter Estimates by Three Procedures

| Stimulus | | $\theta$ | LSE est. | MNX est. | MLE-1 est. | MLE-1 std. | MLE-2 est. | MLE-3 est. |
|---|---|---|---|---|---|---|---|---|
| $S^{(1)}$ | $S_1$ | $\mu_1$ | *−0.818 | *−0.818 | *−0.818 | — | *−0.818 | *−0.818 |
| | $S_2$ | $\mu_2$ | −0.241 | −0.212 | −0.220 | 0.108 | −0.136 | −0.161 |
| | $S_3$ | $\mu_3$ | −0.143 | −0.120 | −0.125 | 0.091 | −0.047 | −0.074 |
| | $S_4$ | $\mu_4$ | 0.527 | 0.536 | 0.516 | 0.108 | 0.704 | 0.659 |
| | $S_5$ | $\mu_5$ | 0.674 | 0.729 | 0.707 | 0.099 | 0.937 | 0.884 |
| $S^{(2)}$ | $S_1$ | $\nu_1$ | −0.712 | −0.683 | −0.684 | 0.080 | −0.671 | −0.681 |
| | $S_2$ | $\nu_2$ | −0.208 | −0.153 | −0.157 | 0.108 | −0.070 | −0.100 |
| | $S_3$ | $\nu_3$ | 0.194 | 0.233 | 0.221 | 0.086 | 0.363 | 0.324 |
| | $S_4$ | $\nu_4$ | 0.732 | 0.785 | 0.767 | 0.112 | 0.999 | 0.939 |
| | $S_5$ | $\nu_5$ | 0.955 | 1.029 | 1.002 | 0.102 | 1.306 | 1.241 |
| | | $\varepsilon$ | 0.508 | 0.514 | 0.508 | 0.027 | 0.591 | 0.574 |
| | $S_1$ | $\sigma_1$ | All | All | All | — | *0.707 | 0.707 |
| | $S_2$ | $\sigma_2$ | values | values | values | — | 0.763 | 0.771 |
| | $S_3$ | $\sigma_3$ | fixed | fixed | fixed | — | 0.893 | 0.901 |
| | $S_4$ | $\sigma_4$ | to | to | to | — | 0.813 | 0.820 |
| | $S_5$ | $\sigma_5$ | $1/\sqrt{2}$ | $1/\sqrt{2}$ | $1/\sqrt{2}$ | — | 0.897 | 0.910 |
| | | $r$ | *0.000 | *0.000 | *0.000 | — | *0.000 | 0.076 |
| | | $\phi$ | | | 10 | | 14 | 15 |
| | | $L$ | | | −483.41 | | −482.96 | −482.96 |
| | | $AIC$ | | | 986.82 | | 993.92 | 995.92 |

Note: Estimates shown by (est.) and standard deviations by (std.).
*Indicates fixed values.
LSE, least squares estimation; MNX, minimum normit $\chi^2$ solution; MLE, maximum likelihood estimation.

precedence of MLE-2 over MLE-3. Regarding MLE-1 and MLE-2, we have 0.90 for $U$, which accords to the $\chi^2$ distribution with four degrees of freedom. The corresponding upper probability is between 0.95 and 0.99. Thus we can accept the MLE-1 solution against the MLE-2 solution. On the basis of AIC, we find that MLE-1 fits the data better than MLE-2. This coincides with the result of the likelihood ratio test. In summary, it is concluded that the MLE-1 solution is best. This indicates that the Case 5 assumption is valid, that is, discriminal dispersions are considered constant for the stimulus range and $r$ is null. Figure 2.6 shows the estimates of $\mu$, $\nu$, and $\varepsilon$ of MLE-1.

## Measurement of Time Order Error

The present procedures are usefully applied in research of time order errors (TOE). It is of great concern in perceptual psychology to measure the difference between $\mu_i$ and $\nu_i$, that is,

$$\hat{\tau}_i = \hat{\mu}_i - \hat{\nu}_i \qquad (i = 1, 2, \ldots, m) \tag{2.101}$$

Given those estimates, whether small or large in the absolute value, one would need to examine them on a statistical basis. Let

$$\mathrm{Var}(\hat{\tau}_i) = \mathrm{Var}(\hat{\mu}_i) + \mathrm{Var}(\hat{\nu}_i) - 2\mathrm{Cov}(\hat{\mu}_i, \hat{\nu}_i) \tag{2.102}$$

It is easy to compute the variances and the standard deviations $c_i$ of $\hat{\tau}_i$ from the covariance matrix of the estimates that is obtained with the ML solution. Assuming the large sample property again, we derive the confidence interval of $\tau_i$ with $(1 - \alpha)\%$ probability as

$$\hat{\tau}_i - z_\alpha \hat{c}_i < \tau_i < \hat{\tau}_i + z_\alpha \hat{c}_i \tag{2.103}$$

Figure 2.7 plots estimates of $\tau_j$ (dots) and indicates the two-sided confidence interval with 90% confidence coefficient (bars). In the figure, we see the tendency for TOE to be negative. An impression with only the dots is that the TOE varies with stimulus magnitude to a good extent. Then it seems difficult to recognize a constancy over the stimulus range. In view of the confidence

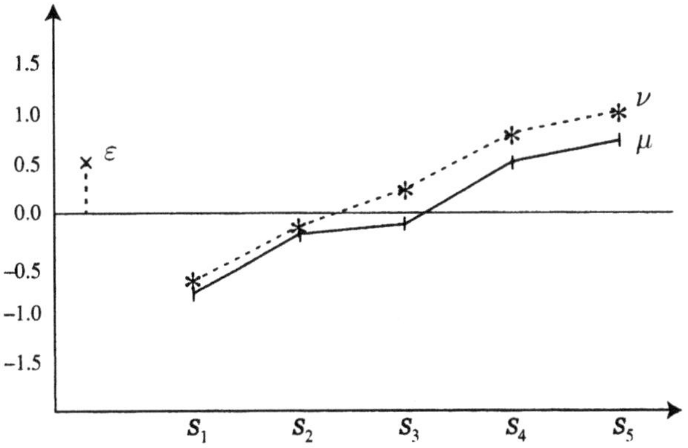


**FIGURE 2.6** Estimates of MLE-1

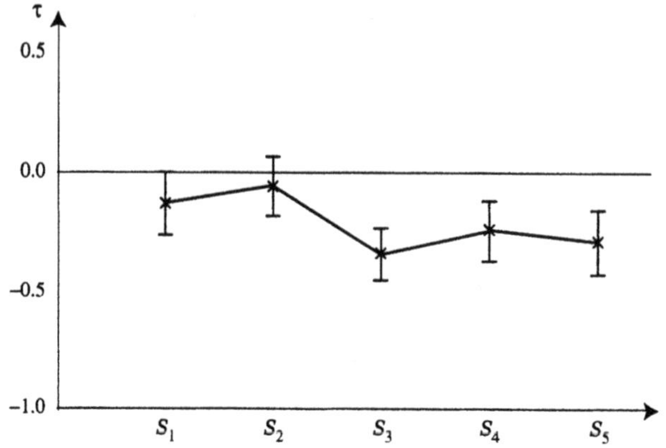


**FIGURE 2.7** Estimates of the time order error with 90% confidence interval

interval, however, we find that the negative tendency of TOE becomes gradually greater with increasing magnitude of stimulus. It would not be revealed so clearly by only inspecting the asymmetry in the probability data, or by using the percent difference (D%) index (Guilford, 1954), which has often been used in studies of TOE. Psychological interpretation of this result is not attempted here.

## 2.4.7. Discussion

### The Basic Assumption

For the revised law of comparative judgment of (2.48) and (2.49), the dispersions with an identical stimulus have been assumed to be equal for the two kinds of discriminal processes. The assumption might be restrictive when one considers a general treatment of asymmetry. According to the ML approach, it is possible to revise the present procedure so as to deal with different dispersions, $\sigma_j^2$ for $v_j^{(1)}$ and $\sigma_j'^{(2)}$ for $v_j^{(2)}$. However, the present assumption seems to be applicable in most practical situations.

### Related Work on Asymmetry in Unidimensional Scaling

Bock and Jones (1968) considered an asymmetry due to the order effect that should be constant over the range of stimuli. It is expressed as

$$\nu_j = \mu_j + \lambda \qquad (j = 1, 2, \ldots, m) \tag{2.104}$$

They treated PC judgments with binary responses, and then $Q_{jk} = 1 - P_{jk}$. Thus the analysis was confined only to $P_{jk}$. Instead of the normal response function, they adopted the angular response function. As a result, $\{c_{jk}\}$ becomes constant, which leads to equations of a simple form such as

$$X_{jk} = \mu_k + \lambda - \mu_j \qquad (j \neq k) \tag{2.105}$$

where $X_{jk}$ denotes the angular deviate. They proposed a procedure by the minimum normit $\chi^2$ approach for the PC judgment with the complete design. If desired, the ML procedure can incorporate such a constant order effect by imposing (2.104) on parameters. The constraint requires, under the normal response function, that $v_j^{(2)}$ accords to $N(\mu_j + \lambda, \sigma_j^2)$ $(j = 1, 2, \ldots, m)$.

## Binary Response

Let us examine the applicability of the procedures to comparative judgments with binary responses, that is, G and L responses only, and $R_{jk} = 0$ for all $(j, k)$. First we take up the LS solution. By definition (2.59) $x_{jk} - y_{jk} = 0$, which then gives $\varepsilon = 0$ in (2.62), and the solution (2.69) is still tractable even in this case. For the complete design, the solution given by (2.70) and (2.71) is identical to the solution given by Harris (1957). Turning to the ML and the MNX procedures, we see that they are based on the trinomial distribution and thus the formulations should be revised.

## Comparison of MNX with MLE

Let us examine the MNX approach in comparison with the ML approach. Both approaches need the assumption of statistical independence across the pairwise judgments, but they differ in the requirement of the large sample property. The formulation of MNX starts with assuming the property. In the ML approach, in contrast, the property becomes necessary at derivation of the covariance matrix of the estimates and at statistical testing. Even setting aside those differences, the ML approach is more advantageous than the MNX one. As the best point of it, we can implement a class of hypotheses into the algorithm and perform statistical tests about them for model selection on the bases of likelihood ratio and AIC criteria. In summary, the MNX solution may serve theoretical interests rather than practical purposes.

## Treatment in Thurstone's Cases 1 to 5

Let us examine those cases in our approach. As is known, Case 1 is specified with a condition of multiple judgment given by a single subject (Thurstone, 1927). This is a rather special and impractical case for usual experiments, and then we may omit it and consider the other four cases for conditions of multiple judgment given by a group of subjects. It should be remembered that as we move from Case 2 to Case 5 the assumptions involved become stronger. In other words, they form a hierarchy of constraints on parameters in the degree of restriction. Now remember that our ML procedure has been implemented for Case 2, the most general one. As a prominent advantage of the ML approach, we note its computational convenience, in that we can easily incorporate constraints on parameters in the estimation process without changing the framework of an algorithm. Suffice to say that such an incorporation is handled by a simple matrix operation of reparametrization (Arbucke and Nugent, 1973; Bock and Jones, 1968; among others). Hence the ML procedure can handle those cases easily.

For the numerical examples, the three executions of the ML procedure were carried out with such reparametrization. In view of constraints and free parameters, we see that Case 2 corresponds to MLE-3, Case 3 to MLE-2, Case 5 to MLE-1. Turning to the LS procedure that is based on Case 5, we find that it benefits from an easy computation due to its linear form. A great deal of revision of the present LS algorithm is required to deal with Cases 2 to 4 expressed in the nonlinear forms. However, such a revision is not useful in a practical sense. Rather we should utilize the LS procedure to provide initial solutions for the nonlinear computations of the ML estimation in those cases.

## 2.5. OPERATIONAL SCALING

Saaty (1980) proposed a method to construct a scale based on a paired comparison (PC) matrix, as a means for a decision-making process that is performed at multilevels in a hierarchical structure. At each level, there is a set of objects for which there is a relative importance or priority that a decision maker wants to determine. Then one aims to assign weights to the objects based on PC judgments among them. We call this set of weights scale. Thus such scale construction is carried out at each level of the hierarchical process to reach the final decision. The whole process of decision making is called analytic hierarchy process (AHP), which is mainly applied in the field of operations research and management science. In the approach, no model is assumed for subjective judgment of paired comparisons. We like to call this type of scaling *operational scaling*, differentiating it from the approach of the preceding section.

Consider a situation in which a subject (decision maker) provides relative evaluation among $n$ objects through PC judgments. Suppose that the judgment is performed on the basis of a latent ratio scale and is expressed in terms of a positive ratio. We denote the paired comparison matrix by $\boldsymbol{A} = (a_{ij})$ where $a_{ij}$ represents the ratio score of object $i$ to object $j$, and

$$a_{ij} > 0 \qquad (i, j = 1, 2, \ldots, n) \tag{2.106}$$

A square matrix whose elements are all positive is often called *positive matrix.* For PC judgments used in AHP, the subject is required to meet a strong condition such that

$$a_{ij} = 1/a_{ji} \quad \text{and} \quad a_{ii} = 1 \tag{2.107}$$

When $\boldsymbol{A}$ satisfies Eqs. (2.106) and (2.107), it is called a *positive reciprocal matrix.* Among those matrices, we are interested in matrices that satisfy the following condition,

$$a_{ij}a_{jk} = a_{ik} \qquad \text{for all} \quad (i, j, k) \tag{2.108}$$

This condition is called *consistency.* In practical cases, it is often found that a positive reciprocal matrix through PC judgments does not satisfy consistency. For example, the PC matrix of Table 2.16 satisfies (2.108), but the matrix of Table 2.17 does not.

### 2.5.1. A Typical Case

Let us set out a typical case in which one makes PC judgments the exact way using weights known to objects, such as $w_1, w_2, \ldots, w_n$. Then one would give $a_{ij}$ by

$$a_{ij} = \frac{w_i}{w_j} \qquad (i, j = 1, 2, \ldots, n) \tag{2.109}$$

Thus consistency obviously holds as

$$a_{ij}a_{jk} = \frac{w_i}{w_j} \cdot \frac{w_j}{w_k} = \frac{w_i}{w_k} = a_{ik} \tag{2.110}$$

**TABLE 2.16** Positive Reciprocal Matrix (1)

| | 1 | 2 | 3 | 4 |
|---|---|---|---|---|
| 1 | 1 | 1/3 | 1/2 | 1/4 |
| 2 | 3 | 1 | 3/2 | 3/4 |
| 3 | 2 | 2/3 | 1 | 2/4 |
| 4 | 4 | 4/3 | 4/2 | 1 |

**TABLE 2.17** Positive Reciprocal Matrix (2)

| | 1 | 2 | 3 | 4 |
|---|---|---|---|---|
| 1 | 1 | 1/3 | 1/2 | 4 |
| 2 | 3 | 1 | 5 | 2 |
| 3 | 2 | 1/5 | 1 | 3 |
| 4 | 1/4 | 1/2 | 1/3 | 1 |

and also does (2.107). Let

$$\boldsymbol{w} = (w_1, w_2, \ldots, w_n)' \quad \text{and} \quad \boldsymbol{w}_* = \left(\frac{1}{w_1}, \frac{1}{w_2}, \ldots, \frac{1}{w_n}\right)' \tag{2.111}$$

We observe that $\boldsymbol{A} = (a_{ij})$ is of rank 1 and the following equations hold:

$$\boldsymbol{A}\boldsymbol{w} = n\boldsymbol{w} \tag{2.112}$$

$$\boldsymbol{A}'\boldsymbol{w}_* = n\boldsymbol{w}_* \tag{2.113}$$

Then $n$ is the largest eigenvalue of $\boldsymbol{A}$ or $\boldsymbol{A}'$, which is a simple root because the remaining eigenvalues are all zero. For simplicity, let $\lambda_{\max} = n$. Vector $\boldsymbol{w}$ is the right eigenvector associated with $\lambda_{\max}$, and $\boldsymbol{w}_*$ is the left eigenvector associated with $\lambda_{\max}$. Note that any column of $\boldsymbol{A}$ is also a right eigenvector associated with $\lambda_{\max}$, and any row of $\boldsymbol{A}$ is also a left eigenvector associated with it. It is pointed out that ordering of objects in terms of $\boldsymbol{w}$ is inversely related to that in terms of $\boldsymbol{w}_*$. In this sense, $\boldsymbol{w}$ and $\boldsymbol{w}_*$ are intrinsically the same scales.

## 2.5.2. Positive Matrix with Some Structure

We consider a positive matrix $\boldsymbol{P}$ of order $n$ before proceeding to general cases in which conditions (2.107) and/or (2.108) are not necessarily satisfied. Note that $\boldsymbol{P}$ is asymmetric, hence some of the eigenvalues and the associated eigenvectors may be given in complex numbers. Writing an eigenvalue as $\lambda = \mu + i\nu$ where $i^2 = -1$, we have its absolute value (or modulus) as $|\lambda| = (\mu^2 + \nu^2)^{1/2}$. Denote those eigenvalues in descending order of absolute value as $\lambda_1, \lambda_2, \ldots, \lambda_n$.

It is useful to refer to a theorem for positive matrices due to Perron (Bellman, 1970; Saaty, 1980).

*Theorem 2.1.*

1. Matrix $\boldsymbol{P}$ has real positive eigenvalues. The largest one, $\lambda_{\max}$, is simple, which is also the greatest in absolute value among all $\lambda_i$ values; that is, $\lambda_{\max} = \lambda_1$.
2. Denote the right eigenvector and the left eigenvector associated with $\lambda_{\max}$ by $\boldsymbol{x} = (x_1, x_2, \ldots, x_n)'$ and $\boldsymbol{y} = (y_1, y_2, \ldots, y_n)'$, respectively. Either vector has positive components; that is, all $x_j > 0$ and all $y_j > 0$.
3. A positive right (or left) eigenvector is a multiple of $\boldsymbol{x}$ (or $\boldsymbol{y}$).

It is noted that $\lambda_{\max}$ is called the Perron root (or Perron–Frobenius root), and $\boldsymbol{x}$ and $\boldsymbol{y}$ are called principal eigenvectors. In view of the positive components, it is convenient for utilization to normalize $\boldsymbol{x}$ and $\boldsymbol{y}$ so that the component sum be unity. Now we focus on a positive reciprocal matrix that is provided through the ratio judgments for paired objects. Of principal concern is the consistency condition, which a positive reciprocal matrix may satisfy. Regarding the theoretical relationship between positive reciprocity and consistency, Saaty (1980) presented the following theorem.

*Theorem 2.2.*

1. For a positive reciprocal matrix $\boldsymbol{A}$, it holds that $\lambda_{\max} \geq n$. Matrix $\boldsymbol{A}$ is consistent if and only if $\lambda_{\max} = n$.
2. If $\boldsymbol{A}$ is positive and consistent, then it is a positive reciprocal matrix.
3. A positive matrix $\boldsymbol{A}$ is consistent if and only if it has unit rank and its diagonal entries are equal to unity.
4. A positive reciprocal matrix $\boldsymbol{A} = (a_{ij})$ is consistent if and only if $a_{ij} = \alpha_i/\alpha_j$ for all $(i, j)$, where $\alpha_i$ are all positive.

Given the fourth statement of the theorem, we like to express relation (2.111) clearly as a theorem.

*Theorem 2.3.* For a positive reciprocal consistent matrix, the componentwise reciprocal of the left principal eigenvector is equal to the right principal eigenvector.

For a positive reciprocal matrix $\boldsymbol{A}$ of order $n$, the following relations hold:

$$\sum_{i=1}^{n} \lambda_i = \mathrm{tr}\,(\boldsymbol{A}) = n \tag{2.114}$$

$$\sum_{i=1}^{n} \lambda_i^2 = \mathrm{tr}\,(\boldsymbol{A}^2) = n^2 \tag{2.115}$$

To measure the degree of inconsistency of $\boldsymbol{A}$, Saaty proposed an index such as

$$\theta = \frac{\lambda_{\max} - n}{n-1} = -\frac{1}{n-1}\sum_{k=2}^{n} \lambda_k \tag{2.116}$$

As a somewhat misleading matter, it is commonly called the index of consistency. Regarding the third term, it is noted that there may be some pairs of complex conjugate eigenvalues. To show $\theta$ from another point of view, let us express $\boldsymbol{A}$ that is inconsistent as

$$a_{ij} = \frac{\alpha_i}{\alpha_j}(1 + \varepsilon_{ij}), \qquad \text{where} \quad \varepsilon_{ij} > -1 \tag{2.117}$$

Then it is seen that $\theta$ indicates the variance of $\{\varepsilon_{ij}\}$.

### 2.5.3. Scaling with a Positive Reciprocal Matrix

In practical cases, it is commnon for us to observe that the consistency condition is violated to a small or large extent. The violation might be attributed to some psychological process in PC judgments rather than the random fluctuation. If the extent of inconsistency is supposed to be small, one would expect a kind of continuity that a small perturbation to a positive reciprocal and consistent matrix results in a small change of $\lambda_{\max}$ and its associated eigenvectors, $\boldsymbol{x}$ and $\boldsymbol{y}$. Saaty (1980) investigated sensitivity of eigenvectors from this point of view. In business applications of AHP, that sort of continuity seems to be assumed tacitly in treating the eigenvalue problem (2.122) given below.

When we deal with an $n \times n$ positive reciprocal matrix $\boldsymbol{A}$ provided by PC judgments, derivation of the eigen equation can be supported without considering the continuity as follows. At first it is noticed that if $\boldsymbol{A}$ is inconsistent, relation (2.111) does not hold any longer in the case for $n > 3$. Let us derive two scales from a given $\boldsymbol{A}$, one representing weights to row objects and the other representing weights to column objects. In view of Theorem 2.1, we then set an objective function such as

$$f(\boldsymbol{u}, \boldsymbol{v}) = \boldsymbol{v}'\boldsymbol{A}\boldsymbol{u} \tag{2.118}$$

It may be regarded as a kind of total utility for a set of objects. We aim to maximize it under constraints:

$$u'v = 1 \tag{2.119}$$

$$u > 0 \quad \text{and} \quad v > 0 \tag{2.120}$$

To deal with the constrained maximization problem, we define a Lagrangean function by

$$\phi(u, v, \lambda) = v'Au - \lambda(u'v - 1) \tag{2.121}$$

Then we derive eigen equations that $u$ and $v$ should satisfy as follows:

$$Au = \lambda u \tag{2.122}$$

$$A'v = \lambda v \tag{2.123}$$

Using (2.119), we have $f(u, v) = \lambda$. Applying Theorem 2.1, we find that the solution $u$ and $v$ to maximize $f(u, v)$ is given by the left eigenvector and the right eigenvector associated with $\lambda_{max}$ of $A$, respectively. It is noted that constraint (2.120) is satisfied by Theorem 2.1, without incorporating it in the Lagrangean function. Hence we obtain two scales $u$ and $v$ from $A$.

### 2.5.4. Related Methods

For a positive reciprocal matrix with some violation of consistency, Saaty and Vargas (1984a) studied scaling problems in a different approach from that mentioned above, suggesting two methods. In the first problem, they dealt with a logarithmic least squares method (LLSM). Here one aims to find $\alpha_i$ to minimize

$$\sum_{i=1}^{n}\sum_{j=1}^{n}(\log a_{ij} - (\log \alpha_i - \log \alpha_j))^2 \tag{2.124}$$

under a constraint that

$$\sum_{i=1}^{n} \alpha_i = 1 \tag{2.125}$$

The solution is given by

$$\alpha_i = \left(\prod_{j=1}^{n} a_{ij}\right)^{1/n} \sum_{h=1}^{n}\left(\prod_{j=1}^{n} a_{hj}\right)^{1/n} \qquad (i = 1, 2, \ldots, n) \tag{2.126}$$

It is seen that to obtain $\alpha_i$, elements in other rows of $\boldsymbol{A}$ make no contribution. Thus inconsistent relations involved in $\boldsymbol{A}$ are not reflected in the solution.

In the second problem, one aims to approximate $\boldsymbol{A}$ by a consistent matrix of rank 1. The purpose leads to a least squares method (LSM), finding $\alpha_i$ to minimize

$$\sum_{i=1}^{n}\sum_{j=1}^{n}\left(a_{ij}-\frac{\alpha_i}{\alpha_j}\right)^2 \tag{2.127}$$

The solution is given in terms of the eigenvector of $\boldsymbol{A}'\boldsymbol{A}$ and the eigenvector of $\boldsymbol{A}\boldsymbol{A}'$ associated with the largest eigenvalue.

Let EM stand for the method of treating eigenequations (2.122) and (2.123). Saaty and Vargas (1984a) showed that if $\boldsymbol{A}$ is consistent, then the three kinds of solutions derived by EM, LLSM, and LSM coincide. The proof is easy to follow in view of the definition of consistency.

When one regards $\boldsymbol{A} = (a_{ij})$ as representing PC judgments on the basis of row objects, it is of interest to argue the rank order of a set of objects derived from $\boldsymbol{A}$. Then Saaty and Vargas (1984b) showed, for an inconsistent matrix $\boldsymbol{A}$, that the rank order is determined in terms of powers of $\boldsymbol{A}$. The argument leads to eigen equation (2.122), indicating that the relative dominance of objects is given by the right principal eigenvector. Furthermore, they investigated conditions for rank preservation in a positive reciprocal matrix that is inconsistent. For other methodological studies in connection with this sort of positive reciprocal matrix, see references in Saaty and Vargas (1984b).

Furthermore, they investigated conditions for rank preservation in a positive reciprocal matrix that is inconsistent. For other methodological studies in connection with that sort of matrix, also see references in Saaty and Vargas (1984b).

**Remark** Let $\boldsymbol{A}$ be a positive reciprocal matrix, and define

$$\boldsymbol{B} = (b_{ij}) \qquad \text{where} \quad b_{ij} = \log a_{ij} \tag{2.128}$$

Equation (2.107) is rewritten as

$$b_{ij} + b_{ji} = 0 \quad \text{and} \quad b_{ii} = 0 \tag{2.129}$$

Thus $\boldsymbol{B}$ is a skew-symmetric matrix and (2.108) is stated as a relation of transitivity or additivity as

$$b_{ij} + b_{jk} = b_{ik} \qquad \text{for all} \quad (i, j, k) \tag{2.130}$$

In connection with the condition, it may be of interest to refer to Theorem 3.2 described in Sec. 3.2.4, which shows that a line pattern or unidimensional scale is involved in $\boldsymbol{B}$. In this viewpoint, LLSM is regarded as a method to approximate $\boldsymbol{B}$ by a matrix of rank 2.

## 2.5.5. Numerical Example

### Example 1

Here we would like to provide a numerical illustration of two types of positive reciprocal matrices regarding the description of Sec. 2.5.2. For the consistent matrix of Table 2.16, we have $\lambda_{\max} = \lambda_1 = 4$, with the remaining eigenvalues being zero. Then the index of inconsistency shows that $\theta = 0$. It is easily seen that the right and the left principal eigenvectors are expressed $\boldsymbol{x} = (1, 3, 2, 4)'$ and $\boldsymbol{y} = (1, \frac{1}{3}, \frac{1}{2}, \frac{1}{4})'$, respectively. Normalizing them to be the component sum of unity gives $\tilde{\boldsymbol{x}} = (0.1, 0.3, 0.2, 0.4)'$ and $\tilde{\boldsymbol{y}} = (0.48, 0.16, 0.24, 0.12)'$. For the inconsistent matrix of Table 2.17, we have

$$\lambda_1 = 4.6421, \qquad \lambda_2 = -0.0863 + 1.7006i,$$

$$\lambda_3 = -0.0863 - 1.7006i, \qquad \lambda_4 = -0.4695$$

As the degree of inconsistency, we see that $\theta = 0.6421/3 = 0.2140$. The normalized principal eigenvectors are expressed as

$$\tilde{\boldsymbol{x}} = (0.1859, 0.5005, 0.2127, 0.1009)'$$
$$\tilde{\boldsymbol{y}} = (0.2255, 0.0967, 0.2068, 0.4711)'$$

It may be of interest to construct dominance matrices from the matrices of both tables according to (2.1). Then we compute the coefficient of consistence $\zeta$ defined by (2.3). It is found that $\zeta = 1.0$ for both cases, indicating existence of a rank order of objects involved in pairwise dominance relations. This is a natural outcome, since the consistency of $\zeta$ is defined in a weak sense in comparison with the consistency defined by (2.108).

### Example 2

Consider a case in which the president of a company has to decide on an executive member from several candidates. Their profiles may be described in terms of five criteria as factors for consideration. The criteria would be leadership, character, ability, experience, and knowledge. Then the president may have a relative evaluation of the criteria and construct a PC

**TABLE 2.18** Paired Comparison of Five Criteria

| | A | B | C | D | E |
|---|---|---|---|---|---|
| A: leadership | 1 | 5 | 2 | 4 | 3 |
| B: character | 1/5 | 1 | 1/3 | 1/3 | 1/2 |
| C: ability | 1/2 | 3 | 1 | 1/2 | 1/3 |
| D: experience | 1/4 | 3 | 2 | 1 | 2 |
| E: knowledge | 1/3 | 2 | 3 | 1/2 | 1 |

matrix $\boldsymbol{A} = (a_{ij})$ as shown in Table 2.18. The element $a_{ij}$ indicates that criterion $i$ is $a_{ij}$ times as important as criterion $j$.

The largest eigenvalue of the PC matrix is $\lambda_{max} = \lambda_1 = 5.4182$. The index of inconsistency is expressed by $\theta = 0.1045$, indicating that the degree of inconsistency is small. The normalized principal eigenvectors are given as

$$\tilde{\boldsymbol{x}} = (0.4319, 0.0641, 0.1284, 0.2038, 0.1718)'$$
$$\tilde{\boldsymbol{y}} = (0.0648, 0.3993, 0.2340, 0.1342, 0.1676)'$$

From a candidate profile in terms of five scores of the criteria, one can compute the candidate's total score as a weighted sum of them by using the components of an eigenvector. Then one would have a priority scale of the candidates according to the total score.

## 2.6. SUMMARY

We now summarize this section. Sec. 2.2 deals with the dominance matrix and exploring ordinal structure inherent in the data. The subject is apparently different from those of the last three sections. Let us discuss the differences in the three, which deal with seemingly similar subjects in PC judgments. We call the subject of Sec. 2.3 the *ANOVA approach*, that of Sec. 2.4 the *scaling approach*, and that of Sec. 2.5 the *operational approach* in short.

First, let us take up the design of stimulus pairs and the requirement of judgment response. In the ANOVA approach, no pair of identical stimuli is designed and a complete design is required. Score rating in terms of a numerical or a graded scale is assumed for the judgment response. In the scaling approach, pairs of identical stimuli may be considered and an incomplete design is allowed. The response in terms of three ordered categories is required by incorporating indifferent judgment. In the operational approach, the subject is required to make PC judgments on a ratio scale, and unlike observing probability, no multiple judgment is imposed for each pair. According to the design of the PC matrix, the diagonal entries are forced to be unity.

For comparison of the models, we consider a psychological continuum for the ANOVA and the scaling approaches, and rewrite the psychological difference (2.50) as $u_{jk} = \mu_j^{(1)} - \nu_k^{(2)}$. In the ANOVA approach, it is assumed that $u_{jk}$ is distributed as $N(\mu_{jk}, \sigma^2)$. The term $\mu_{jk}$ represents the sum of symmetric $\delta_{jk}$ and skew-symmetric $\lambda_{jk}$; the latter consists of linear and interaction terms in (2.11), respectively. Then

$$\mu_{jk} = \beta_j - \beta_k + \gamma_{jk} + \delta_{jk} \tag{2.131}$$

In the scaling approach, it is assumed that $u_{jk}$ is distributed as $N(\mu_j - \nu_k, \sigma_{jk}^2)$, thus the expectation represents the skew-symmetric component of two linear terms. In contrast to the two approaches, no model is assumed for the operational scaling.

Turn to the treatment of the order effect in the three approaches. In the ANOVA approach, the order effect $\delta_{jk}$ is considered in connection with each pair, not with each stimulus. The model may be simplified with such a modification as $\delta_{jk} = 0$, or there might be cases in which a null hypothesis for $\delta_{jk}$ is accepted. In contrast, in the scaling approach, the formulation starts with postulating order effect. It is incorporated in (2.48) and (2.49); that is, the assumption of two discriminal processes. In the framework of the MLE procedure, it is easy to deal with some model versions by incorporating constraints on parameters and perform model selection on the basis of AIC. In the operational approach, the order effect is incorporated in a positive reciprocal matrix by imposing (2.107) in the judgments.

# 3

# Graphical Representation of Asymmetric Data

## 3.1. OVERVIEW AND PRELIMINARIES

Suppose that we are given a square asymmetric data matrix $\boldsymbol{O} = (o_{jk})$ for a set of objects where $o_{jk}$ represents the measure of relationship between a pair of objects. In order to explore some structure inherent in the data or derive some information from them, we can utilize procedures that are provided without referring to the meaning of the measure, in other words, whether the measure represents similarity or dissimilarity. In this chapter we describe those procedures independent of whatever the measure means. At the observation of a phenomenon, the diagonal entries may or may not be defined for the matrix $\boldsymbol{O} = (o_{jk})$, or some of the diagonal entries may be missing. In such cases, we fill zero values in those diagonal elements and will treat $\boldsymbol{O}$ with all entries defined in what follows.

### 3.1.1. Decomposition of Asymmetric Matrix

Define a symmetric matrix $\boldsymbol{S} = (s_{jk})$ and a skew-symmetric matrix $\boldsymbol{A} = (a_{jk})$ as follows:

$$\boldsymbol{S} = \tfrac{1}{2}(\boldsymbol{O} + \boldsymbol{O}') \tag{3.1}$$

$$\boldsymbol{A} = \tfrac{1}{2}(\boldsymbol{O} - \boldsymbol{O}') \tag{3.2}$$

Then we have a unique decomposition of $\boldsymbol{O}$ as (3.3):

$$\boldsymbol{O} = \boldsymbol{S} + \boldsymbol{A} \tag{3.3}$$

Noting that

$$\operatorname{tr}(\boldsymbol{SA}) = \sum_{i=1}^{n}\sum_{j=1}^{n} s_{ij}a_{ij} = 0 \tag{3.4}$$

we derive (3.5), which is alternatively expressed as (3.6) in terms of the matrix norm:

$$\sum_{i=1}^{n}\sum_{j=1}^{n} o_{ij}^2 = \sum_{i=1}^{n}\sum_{j=1}^{n} s_{ij}^2 + \sum_{i=1}^{n}\sum_{j=1}^{n} a_{ij}^2 \tag{3.5}$$

$$\|\boldsymbol{O}\|^2 = \|\boldsymbol{S}\|^2 + \|\boldsymbol{A}\|^2 \tag{3.6}$$

Noting that the average of $\{o_{jk}\}$ is equal to the average of $\{s_{jk}\}$ and the average of $\{a_{jk}\}$ is zero, it follows from (3.5) that

$$\text{Var}(o_{jk}) = \text{Var}(s_{jk}) + \text{Var}(a_{jk}) \tag{3.7}$$

There are two approaches. In the first approach, we deal with $\boldsymbol{S}$ and $\boldsymbol{A}$ separately. Then we apply a method to analyze symmetric data (for example, MDS) and another to analyze asymmetric data. The latter analysis becomes the focus of analysis of skew-symmetry. In the second approach, we deal with $\boldsymbol{S}$ and $\boldsymbol{A}$ simultaneously. We describe the first approach in Sec. 3.2 and the second in Sec. 3.3.

Given some asymmetric data, descriptive models serve for graphical representation. In Sec. 3.4, we treat a vector model, which is differentiated from the vector model in MDS. In Sec. 3.5, we consider illustrating properties of asymmetric data in terms of a vector field model.

## 3.2. GOWER'S PROCEDURE

### 3.2.1. Singular Value Decomposition of a Skew-Symmetric Matrix

A skew-symmetric matrix is characterized as

$$\boldsymbol{A}' = -\boldsymbol{A}, \qquad \text{that is} \quad a_{jk} + a_{kj} = 0 \tag{3.8}$$

Then $\boldsymbol{A}'\boldsymbol{A} = \boldsymbol{A}\boldsymbol{A}'$, and so $\boldsymbol{A}$ has properties of a normal matrix. Let $\boldsymbol{u}$ and $\boldsymbol{v}$ be the pair of singular vectors associated with singular value $\mu$. The following relations hold:

$$\boldsymbol{A}\boldsymbol{u} = -\mu\boldsymbol{v} \tag{3.9}$$

$$\boldsymbol{A}\boldsymbol{v} = \mu\boldsymbol{u} \tag{3.10}$$

We see that $\boldsymbol{u}$ and $\boldsymbol{v}$ are also eigenvectors associated with an eigenvalue $\mu^2$ (double root) of symmetric matrix $\boldsymbol{A}'\boldsymbol{A}(=\boldsymbol{A}\boldsymbol{A}')$:

$$\boldsymbol{A}'\boldsymbol{A}\boldsymbol{u} = \mu^2\boldsymbol{u} \tag{3.11}$$

$$\boldsymbol{A}'\boldsymbol{A}\boldsymbol{v} = \mu^2\boldsymbol{v} \tag{3.12}$$

It is known that the rank of a skew-symmetric matrix is generally an even number. Let $m = [n/2]$ and $r = \text{rank}(\boldsymbol{A})/2$. Then $\text{rank}(\boldsymbol{A}) \leq 2m$. Denote singular values in descending order of magnitude as $\mu_1 \geq \mu_2 \geq \cdots \geq \mu_r > 0$. Let the pair of singular vectors associated with $\mu_t$ be $\boldsymbol{u}_t$ and $\boldsymbol{v}_t$. They are orthogonal to each other and can be normalized with the norm 1, such that

$$\boldsymbol{u}_p'\boldsymbol{u}_q = \boldsymbol{u}_p'\boldsymbol{v}_q = \boldsymbol{v}_p'\boldsymbol{u}_q = \boldsymbol{v}_p'\boldsymbol{v}_q = \delta_{pq} \tag{3.13}$$

where $\delta_{pq}$ represents Kronecker's delta. Hence the singular value decomposition (SVD) of $\boldsymbol{A}$ is given as:

$$\boldsymbol{A} = \sum_{t=1}^{r} \mu_t(\boldsymbol{u}_t\boldsymbol{v}_t' - \boldsymbol{v}_t\boldsymbol{u}_t') \tag{3.14}$$

For convenience of description in subsequent chapters, we provide the SVD in a matrix form. Denote eigenvectors associated with null eigenvalues of (3.12) by $\boldsymbol{w}_1, \boldsymbol{w}_2, \ldots, \boldsymbol{w}_{n-2r}$. Let

$$\boldsymbol{Z} = (\boldsymbol{u}_1, \boldsymbol{v}_1, \ldots, \boldsymbol{u}_r, \boldsymbol{v}_r, \boldsymbol{w}_1, \boldsymbol{w}_2, \ldots, \boldsymbol{w}_{n-2r}) \tag{3.15}$$

Define $\boldsymbol{M}$ of order $n \times n$ as a block-diagonal matrix with $2 \times 2$ matrices $\boldsymbol{M}_t$ $(t = 1, 2, \ldots, r)$ along the diagonal positions, and a square matrix $\boldsymbol{E}$ at the last diagonal position. Also define $\boldsymbol{J}$ of order $n \times n$ to be of the same form as $\boldsymbol{M}$ by replacing $\boldsymbol{M}_t$ by $\boldsymbol{J}_t$ along the diagonal. Here $\boldsymbol{E}$ is a square matrix of order $n - 2r$ with all null entries, and matrices $\boldsymbol{M}_t$ and $\boldsymbol{J}_t$ $(t = 1, 2, \ldots, r)$

are written as

$$M_t = \begin{pmatrix} 0 & \mu_t \\ -\mu_t & 0 \end{pmatrix}, \quad J_t = \begin{pmatrix} 0 & 1 \\ -1 & 0 \end{pmatrix} \tag{3.16}$$

$$M = \begin{pmatrix} M_1 & & & & O \\ & M_2 & & & \\ & & \ddots & & \\ & & & M_r & \\ O & & & & E \end{pmatrix}, \quad J = \begin{pmatrix} J_1 & & & & O \\ & J_2 & & & \\ & & \ddots & & \\ & & & J_r & \\ O & & & & E \end{pmatrix} \tag{3.17}$$

Note that $Z$ and $J$ are orthogonal matrices. Define a diagonal matrix of order $n$ as

$$M_\mu = \mathrm{diag}(\mu_1, \mu_1, \mu_2, \mu_2, \ldots, \mu_r, \mu_r, 0, \ldots, 0) \tag{3.18}$$

Using the matrices above, we have the SVD of $A$ in a matrix form, such as

$$A = ZM_\mu JZ' = ZMZ' \tag{3.19}$$

Define a symmetric data matrix by

$$B = (b_{jk}) = AA' = A'A \tag{3.20}$$

From (3.12) and (3.13), the spectral decomposition and the trace of $B$ are given as:

$$B = A'A = \sum_{t=1}^{r} \mu_t^2(u_t u_t' + v_t v_t') \tag{3.21}$$

$$\mathrm{tr}(B) = \mathrm{tr}(A'A) = \|A\|^2 = 2\sum_{t=1}^{r} \mu_t^2 \tag{3.22}$$

### 3.2.2. Skew-Symmetry Analysis (SSA)

Gower (1977) proposed applying SVD in data analysis of the skew-symmetric matrix $A$. This procedure is called skew-symmetry analysis (SSA). For each pair of ($u_t$, $v_t$), one has a plot of $n$ points in two dimensions, which represents a configuration of objects in terms of ($u_{jt}$, $v_{jt}$). The graphical representation in this way is sometimes called a Gower diagram. In view of (3.22), we set an index on the basis of $\|A\|^2$ as

$$\theta(p) = \frac{\sum_{t=1}^{p} \mu_t^2}{\sum_{t=1}^{r} \mu_t^2} \tag{3.23}$$

This indicates the degree to which $A$ of rank $2r$ is approximated by a matrix of rank $2p$. In the SVD of (3.14), we see that $u_t v_t'$ is of rank 1 and $u_t v_t' - v_t u_t'$ is of rank 2. In view of this point, ten Berge (1997) studied a problem in which approximation of $A$ by using those rank 1 matrices was considered. He suggested two kinds of indices to measure the degree of asymmetry, one of which is equivalent to $\|A\|$.

When $\theta(p)$ is close to unity, one may perform data analysis of $A$ by referring to $p$ pairs of singular vectors. It should be noted that the two-dimensional configuration is provided with rotational indeterminacy. Let us state the reason. Define an $n \times 2$ matrix $W = (u, v)$. Let $I_2$ be an identity matrix of order 2, and denote $J_t$ of (3.16) simply by $J$. A matrix expression of (3.9) and (3.10) is given as

$$AW = \mu WJ \tag{3.24}$$

Let $T$ be a $2 \times 2$ orthogonal matrix, that is, $TT' = T'T = I_2$. Postmultiplication of $T$ on both sides of (3.24) leads to

$$AWT = \mu WJT = \mu WT \cdot T'JT \tag{3.25}$$

Note that $T'JT = J$ and let $W_* = (u_*, v_*) = WT$. Then (3.25) becomes

$$AW_* = \mu W_* J \tag{3.26}$$

showing that $u_*$ and $v_*$ are also singular vectors associated with the singular value $\mu$. One way to cope with the indeterminacy is suggested (Gower, 1977), is to put a restriction that

$$\sum_{i=1}^{n} v_i = 0 \tag{3.27}$$

We can interpret the configuration in terms of each pair of the vectors from geometrical points of view in what follows.

## (I) The Case of $r = 1$

For the sake of exposition we take up a case of $r = 1$ i.e., rank $(A) = 2$ at first. In such a case $A$ is expressed as

$$A = \mu(uv' - vu'), \qquad \text{that is,} \quad a_{jk} = \mu(u_j v_k - u_k v_j) \tag{3.28}$$

Using $\boldsymbol{u} = (u_j)$ and $\boldsymbol{v} = (v_j)$, we define a vector $\boldsymbol{g}_j = (u_j, v_j)'$. For the two-dimensional configuration of $n$ points with coordinates $(u_j, v_j)$, we note the following properties.

*(a) Triangle.* Let $\triangle_{ojk}$ be the quantity determined by a triangle formed by the origin, points $j$ and $k$, which are oriented counterclockwise, such that

$$\triangle_{ojk} = \tfrac{1}{2}(u_j v_k - u_k v_j) \tag{3.29}$$

We call $\triangle_{ojk}$ the oriented triangle. Then

$$\triangle_{okj} = -\triangle_{ojk} \tag{3.30}$$

These two oriented triangles are in a skew-symmetric relation, leading to the following relation such that

$$a_{jk} = 2\mu\triangle_{ojk}, \qquad a_{kj} = 2\mu\triangle_{okj} \tag{3.31}$$

That is, the absolute value of $\triangle_{ojk}$ is proportional to $\|a_{jk}\|$ and the direction of the triangle corresponds to the skew-symmetry of $\boldsymbol{A}$. Even if the measure $o_{jk}$ represents dissimilarity, $a_{jk}$ does not correspond to interpoint distance in the plane. However, this property does not exclude considering Euclidean distance in the derived configuration, as shown in what follows. Regarding this property, there seems to be misunderstanding even among researchers.

*(b) Interpoint Distance.* Define interpoint distance $d_{jk}$ by

$$d_{jk}^2 = (u_j - u_k)^2 + (v_j - v_k)^2 \tag{3.32}$$

$$= \|\boldsymbol{g}_j\|^2 + \|\boldsymbol{g}_k\|^2 - 2\boldsymbol{g}_j'\boldsymbol{g}_k \tag{3.33}$$

and consider what in terms of observations it indicates.

Since $r = 1$, $\boldsymbol{B}$ of (3.21) is described as

$$\boldsymbol{B} = \mu^2(\boldsymbol{uu}' + \boldsymbol{vv}') \tag{3.34}$$

$$b_{jk} = \mu^2(u_j u_k + v_j v_k) = \mu^2 \boldsymbol{g}_j'\boldsymbol{g}_k \tag{3.35}$$

Rewriting (3.33) in terms of $b_{jk}$ gives

$$\mu^2 d_{jk}^2 = b_{jj} + b_{kk} - 2b_{jk} \tag{3.36}$$

Let the right-hand side be $c_{jk}$. Then

$$\begin{aligned} c_{jk} = b_{jj} + b_{kk} - 2b_{jk} &= \sum_{i=1}^{n} (a_{ij} - a_{ik})^2 \\ &= \frac{1}{4}\sum_{i=1}^{n} (o_{ij} - o_{ik} + o_{ki} - o_{ji})^2 \end{aligned} \tag{3.37}$$

Hence (3.36) is stated as

$$\mu^2 d_{jk}^2 = c_{jk} \tag{3.38}$$

Thus $d_{jk}^2$ corresponds to the transformation of data (3.37). It shows that $d_{jk}^2$ is the squared norm of the difference of two data vectors, the $j$th and the $k$th columns in $\boldsymbol{A}$, or equivalently, one combining $j$th column and the $k$th row and the other combining the $k$th column and the $j$th row in $\boldsymbol{O}$. Write the distance from the origin to point $j$ as $d_{oj} = \|\boldsymbol{g}_j\|$. From (3.35) it follows that

$$\mu^2 d_{oj}^2 = b_{jj} = \frac{1}{4}\sum_{i=1}^{n} (o_{ij} - o_{ji})^2 \tag{3.39}$$

## (II) General Case of $r > 1$

Now we are concerned with a general case of $r$. In a two-dimensional configuration in terms of singular vectors $(\boldsymbol{u}_t, \boldsymbol{v}_t)$, we denote the oriented triangle by $\Delta_{ojk}(t)$ and distances by $d_{jk}(t)$ and $d_{oj}(t)$. According to similar arguments above, we have extensions of (3.31), (3.38), and (3.39) as follows:

$$a_{jk} = 2\sum_{t=1}^{r} \mu_t \Delta_{ojk}(t) \tag{3.40}$$

$$c_{jk} = \sum_{t=1}^{r} \mu_t^2 d_{jk}^2(t) \tag{3.41}$$

$$b_{jj} = \sum_{t=1}^{r} \mu_t^2 d_{oj}^2(t) \tag{3.42}$$

It should be noted that the left-hand sides are stated in terms of transformed data and the right-hand sides are expressed by singular values and singular vectors (Saito, 2003a).

The $c_{jk}$ is equal to the squared difference between two vectors, one is the sum of the $j$th column and the $k$th row of $\boldsymbol{O}$, the other is the sum of the $j$th row and the $k$th column of $\boldsymbol{O}$. Relation (3.41) indicates that the symmetric $c_{jk}$ represents the weighted sum of the squared distances. Suppose a case in which entries of the $j$th row are all equal to those of the $j$th column in the asymmetric matrix $\boldsymbol{O}$, that is,

$$o_{ij} = o_{ji} \qquad (i = 1, 2, \ldots, n) \tag{3.43}$$

Then $b_{jj} = 0$, leading to $d_{oj}^2(t) = 0$ because $\mu_t^2 \neq 0\ (t = 1, 2, \ldots, r)$ by assumption. Accordingly, object $j$ is located at the origin on each plane.

### 3.2.3. General Applicability of SSA

Given an asymmetric matrix $\boldsymbol{O}$, one can construct skew-symmetric matrices in various ways, differently from (3.2). Then, apart from Gower's original suggestion of treating $\boldsymbol{A}$, which is paired with a symmetric matrix $\boldsymbol{S}$, we may perform SSA with any skew-symmetric matrix, for which a symmetric counterpart may not necessarily exist. For example, we construct a skew-symmetric matrix $\boldsymbol{A}^+$ from $\boldsymbol{O}$ in such a way that

$$\boldsymbol{S}^+ = \tfrac{1}{2}(\boldsymbol{OO} + \boldsymbol{O}'\boldsymbol{O}') \tag{3.44}$$

$$\boldsymbol{A}^+ = \tfrac{1}{2}(\boldsymbol{OO} - \boldsymbol{O}'\boldsymbol{O}') \tag{3.45}$$

$$\boldsymbol{OO} = \boldsymbol{S}^+ + \boldsymbol{A}^+ \tag{3.46}$$

Given $\boldsymbol{O}$, whose diagonal entries are not all null, we may construct skew-symmetric matrices as:

$$\boldsymbol{A}^* = (a_{jk}^*) \qquad \text{where} \quad a_{jk}^* = \tfrac{1}{2}(o_{jk}^2 - o_{kj}^2 + o_{jj}^2 - o_{kk}^2) \tag{3.47}$$

$$\tilde{\boldsymbol{A}} = (\tilde{a}_{jk}) \qquad \text{where} \quad \tilde{a}_{jk} = \tfrac{1}{2}(o_{jk} - o_{kj} + o_{jj} - o_{kk}) \tag{3.48}$$

Note that we have no expression of the counterpart of symmetry for definitions (3.47) and (3.48).

Let us consider a transformation of data such as

$$p_{jk} = \alpha' o_{jk} + \beta' \tag{3.49}$$

and let $\boldsymbol{P} = (p_{jk})$. Applying SSA to skew-symmetric matrices constructed from $\boldsymbol{P}$, we derive the same configuration as that from $\boldsymbol{O}$ except for the scale unit. Therefore SSA is applicable even to data that are measured on an interval scale.

### 3.2.4. Special Pattern Derived by SSA

Utilizing SSA, one obtains a two-dimensional representation of points in terms of coordinates. If the data reveal a noticeable pattern such as a line or circle in the representation, one would interpret it as a meaningful structure or significant information involved in the data. From this viewpoint, Saito (1997, 2000, 2002) studied conditions to obtain a line or circle patterns from the asymmetric data matrix. In both cases, the interpoint distance $d_{jk}$ is proportional to $|a_{jk}|$ in the two-dimensional configuration.

#### Line Pattern

If all the points are positioned on a line $l$ in the plot, we call this sort of configuration a *line pattern*. In such a case, $v_j = \alpha + \beta u_j$ ($\beta \neq 0$; $j = 1, 2, \ldots, n$), all the triangles formed by the origin and two points have the same heights (Figs. 3.1 and 3.2). Furthermore, since

$$a_{jk} = 2\mu\Delta_{ojk} = \tfrac{1}{2}\alpha(u_j - u_k) \tag{3.50}$$

the absolute value of $\Delta_{ojk}$ is proportional to $|u_j - u_k|$. In this sense, data matrix $\boldsymbol{A}$ involves one-dimensional structure. If $\boldsymbol{v} = a\mathbf{1}$, it holds that

$$\sum_{j=1}^{n} u_j = 0 \tag{3.51}$$

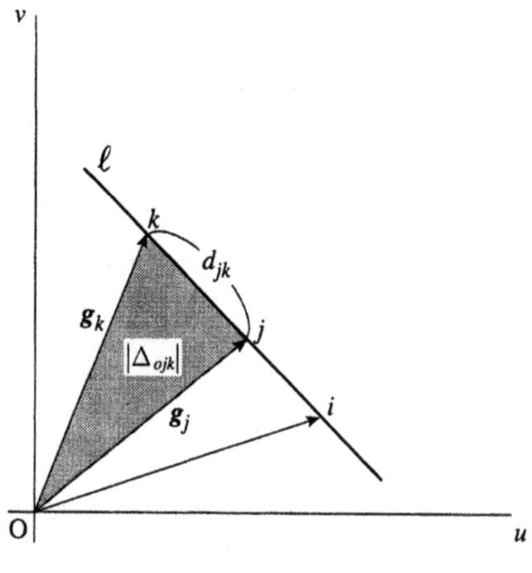


**FIGURE 3.1** Line pattern ($a$)

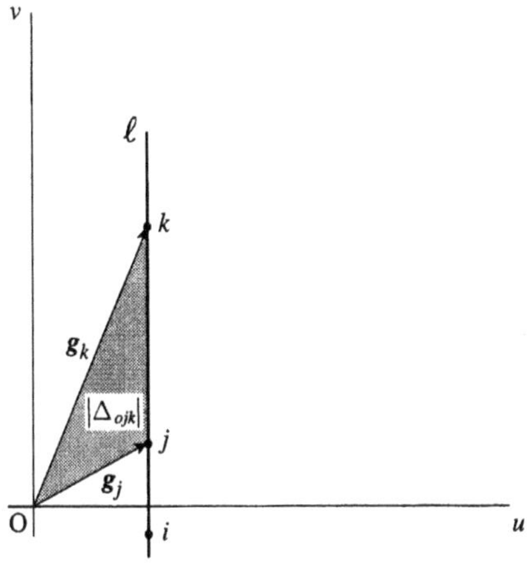


**FIGURE 3.2** Line pattern (*b*)

from the orthogonality (3.13). This corresponds to the case of Fig 3.2. Following are theorems regarding line patterns (Saito, 1997).

*Theorem 3.1.* Let $\boldsymbol{u}$ and $\boldsymbol{v}$ be a pair of singular vectors to satisfy (3.9) and (3.10) where $\mu \neq 0$. A necessary and sufficient condition that either $\boldsymbol{v} = a\mathbf{1} + b\boldsymbol{u}$ or $\boldsymbol{u} = a\mathbf{1} + b\boldsymbol{v}$ with $a \neq 0$ is that

$$\mathbf{A}^2\mathbf{1} = -\mu^2\mathbf{1} \tag{3.52}$$

where $\mathbf{1}$ is a vector of unities.

*Theorem 3.2.* If the following additivity holds for all the triads $(i, j, k)$,

$$a_{ik} = a_{ij} + a_{jk} \tag{3.53}$$

then there is a line on the plane spanned by a pair of singular vectors of $\boldsymbol{A}$. The line is perpendicular to one of the axes and its centroid is on (corresponds to the origin of) another axis. In this sense (3.53) indicates unidimensionality.

*Theorem 3.3.* Define $\boldsymbol{A}$ by (3.1) if diagonal entries are zeros in $\boldsymbol{O}$, otherwise by (3.48) if they are not. When (3.53) holds for every triad $(i, j, k)$, then for every pair $(j, k)$ holds

$$o^*_{jk} = o^*_{kj} \qquad \text{where} \quad o^*_{jk} = o_{jk} - o_{j\cdot} - o_{\cdot k} + o_{\cdot\cdot} \tag{3.54}$$

and the dot notation indicates the mean value over the specified suffix. Conversely, if (3.54) holds then (3.53) holds.

Supposing a case for which (3.53) holds, we let

$$c_j = a_{j\cdot} \qquad \text{and} \qquad \|\boldsymbol{c}\| = \sum_{j=1}^{n} c_j^2 \tag{3.55}$$

Define

$$\mu = (n\|\boldsymbol{c}\|)^{1/2}, \qquad \boldsymbol{u} = \frac{\boldsymbol{c}}{\|\boldsymbol{c}\|} \qquad \text{and} \qquad \boldsymbol{v} = \frac{\mathbf{1}}{\sqrt{n}} \tag{3.56}$$

Then $\boldsymbol{u}$ and $\boldsymbol{v}$ are singular vectors associated with singular value $\mu$. The configuration in terms of those vectors shows a line perpendicular to axis $v$. When (3.53) holds, however, we may not always obtain such a line but the pattern such as $v = a\mathbf{1} + b\boldsymbol{u}$, because of the rotational indeterminacy stated by (3.24) and (3.25).

## Circle Pattern

When the vectors $\boldsymbol{g}_j$ are of the same length, that is, $\|\boldsymbol{g}_j\| = \rho\ (j = 1, 2, \ldots, n)$, all the points are located on a circle. Then the degree of angle spanned by $\boldsymbol{g}_j$ and $\boldsymbol{g}_k$ is proportional to $|a_{jk}|$. We are interested in some conditions under which the sort of circle pattern appears in the plot. When such a pattern is found in the plot, we will say that the pair of singular vectors yields a circle structure. For convenience of exposition, we state properties of a cyclic (or circulant) matrix, and then consider conditions to derive circle patterns.

An $n \times n$ asymmetric matrix $\boldsymbol{C} = (c_{jk})$ is defined to be a cyclic matrix if it satisfies

$$c_{jk} = q_{n-j+k} \qquad \text{where} \qquad q_0 = q_n \qquad \text{and} \qquad q_{n+k} = q_k \tag{3.57}$$

It is explicitly stated that

$$\boldsymbol{C} = \begin{pmatrix} q_0 & q_1 & q_2 & \cdots & q_{n-1} \\ q_{n-1} & q_0 & q_1 & \cdots & q_{n-2} \\ \vdots & \vdots & \vdots & \ddots & \vdots \\ q_2 & q_3 & q_4 & \cdots & q_1 \\ q_1 & q_2 & q_3 & \cdots & q_0 \end{pmatrix} \tag{3.58}$$

Write the eigenproblem of $\boldsymbol{C}$ as

$$\boldsymbol{C}\boldsymbol{z} = \lambda\boldsymbol{z} \tag{3.59}$$

The eigenvalue and eigenvector of this equation were provided, probably for the first time, by Berlin and Kac (1952) in connection with a physical problem. We will refer to their result as Theorem 3.4 hereafter.

*Theorem 3.4.* The eigenvalue and eigenvector of (3.59) are provided as:

$$\lambda = q_0 + q_1\omega + q_2\omega^2 + \cdots + q_{n-1}\omega^{n-1} \tag{3.60}$$

$$z = (1, \omega, \omega^2, \ldots, \omega^{n-1}) \tag{3.61}$$

Here $\omega = e^{i\theta}$ ($i^2 = 1$) with $\theta = 2\pi/n$, which is a root of unity, that is, $\omega^n = 1$.

Write the identical value of the row sum of $\boldsymbol{C}$ as $c$, where

$$c = q_0 + q_1 + q_2 + \cdots + q_{n-1} \tag{3.62}$$

Matrix $\boldsymbol{C}$ has an eigenvector $\mathbf{1} = (1, 1, \ldots, 1)'$ associated with the elgenvalue $c$. Setting aside such an eigenvector, we are interested only in the complex eigenvector $z = \boldsymbol{x} + i\boldsymbol{y}$. Let $\tilde{z}_j = (x_j, y_j)'$. Consider making a plot of $n$ points in terms of $z$ (or $z^*$), which means plotting them on a real $x$–$y$ plane.

According to Theorem 3.4, we know that in each of these plots, the $n$ points are equally spaced on a circle (i.e., with *equal arcs*), but the points may be positioned differently in order on each circle. Except for the difference in order, the configurations are therefore all the same. This result holds with whatever entries are in the original matrix $\boldsymbol{O}$.

We describe some theorems about circle patterns (Saito, 2002). As mentioned above, if a skew-symmetric matrix is cyclic, then applying SVD to it we have the circle structure with equal arcs for each pair of singular vectors. Conversely, if the SVD of a skew-symmetric matrix yields the circle structure with equal arcs for each pair of singular vectors, then it is cyclic as shown by Theorem 3.5. However, there can be an asymmetric and noncyclic $\boldsymbol{O}$ from which construction of $\boldsymbol{A}$ by (3.2) results in a cyclic matrix, as indicated by Theorem 3.6. Table 3.1 shows an example, and the noncyclic asymmetric matrix gives a skew-symmetric and cyclic matrix.

*Theorem 3.5.* If every pair of singular vectors of a skew-symmetric matrix $\boldsymbol{A}$ yields the circle structure with equal arcs, then $\boldsymbol{A}$ is a cyclic matrix.

*Theorem 3.6.* Let $\boldsymbol{O}$ be an asymmetric matrix. If the skew-symmetric matrix $\boldsymbol{A}$ defined by (3.2) is cyclic, then it is necessary that

$$\sum_{k=1}^{n} o_{jk} = \sum_{k=1}^{n} o_{kj} \qquad (j = 1, 2, \ldots, n) \tag{3.63}$$

**TABLE 3.1** Noncyclic Asymmetric Matrix

| | 1 | 2 | 3 | 4 | 5 |
|---|---|---|---|---|---|
| 1 | 1 | 2 | 3 | 4 | 2 |
| 2 | 1 | 2 | 4 | 1 | 2 |
| 3 | 5 | 3 | 3 | 2 | 0 |
| 4 | 2 | 3 | 1 | 4 | 2 |
| 5 | 3 | 0 | 2 | 1 | 5 |

We have so far dealt with cyclic matrices, which may be provided as original data like (3.57), otherwise be constructed from them by (3.2). From the cyclic form, we know that those matrices involve circle structures with equal arcs. Then this type of structure may not be of principal concern for data analysis. Here we are interested in circle structures that are not determined by the matrix pattern but by the entries. The following theorem shows some conditions for skew-symmetric matrices that yield circle structures on which points may or may not be equally spaced.

*Theorem 3.7.* If every pair of singular vectors of $\boldsymbol{A}$ yields a circle structure, then

$$\sum_{j=1}^{n} a_{ij}^2 = K \text{ (const.)} \qquad (i = 1, 2, \ldots, n) \tag{3.64}$$

where

$$K = -\frac{1}{n}\text{tr}(\boldsymbol{A}^2) = \frac{2}{n}\sum_{k=1}^{m} \mu_k^2 \tag{3.65}$$

and their radii take an identical value of $\sqrt{2/n}$.

Thus condition (3.64) is necessary for all the points to be located on a circle in all the two-dimensional plots. It indicates that symmetric $\boldsymbol{A}^{(2)} = (a_{ij}^2)$ has an eigenvector of $\mathbf{1}$ associated with an eigenvalue of the constant $K$. From (3.13) and (3.14), it follows generally that

$$\sum_{j=1}^{n} a_{ij}^2 = \mu_1^2(u_{i1}^2 + v_{i1}^2) + \mu_2^2(u_{i2}^2 + v_{i2}^2) + \cdots + \mu_m^2(u_{im}^2 + v_{im}^2) \tag{3.66}$$

$$(i = 1, 2, \ldots, n)$$

If rank($\boldsymbol{A}$) = 2 in particular, we have only one pair of singular vectors. Then (3.66) becomes

$$\sum_{j=1}^{n} a_{ij}^2 = \mu^2(u_i^2 + v_i^2) \qquad (i = 1, 2, \ldots, n) \tag{3.67}$$

Therefore (3.64) is a necessary and sufficient condition for $\boldsymbol{A}$ to yield the circle pattern in the case of rank($\boldsymbol{A}$) = 2.

### 3.2.5. Example

#### Artificial Data

We will now illustrate a circle pattern derived by SSA, using artificial data (Saito, 2002). Table 3.2 shows a noncyclic skew-symmetric matrix. It is a noncyclic matrix that satisfies (3.64), a necessary condition to yield a circle pattern. Figure 3.3 shows a configuration of six points in terms of the first pair of singular vectors, representing a circle pattern with unequal arcs. Figure 3.4 shows another circle pattern given by the second pair of singular vectors. The configuration of points given by the third pair also shows a circle pattern, for which a plot is omitted here. The radii take an identical value of $1/\sqrt{3}$ across the three circle patterns.

#### Soft Drinks Brand Switching Data

Now we present examples of SSA, using real data. First we provide analysis of brand switching data, which are taken from DeSarbo and De Soete (1984). The original data are based on the work of marketing research conducted by Bass et al. (1972). Table 3.3 shows the data; its entries indicate 1000 times the original values.

**TABLE 3.2** Noncyclic Skew-Symmetric Matrix

| | 1 | 2 | 3 | 4 | 5 | 6 | sq. sum |
|---|---|---|---|---|---|---|---|
| 1 | 0 | $-1/2$ | $-\sqrt{3}/2$ | 1 | $\sqrt{2}/2$ | $\sqrt{2}/2$ | 3 |
| 2 | $1/2$ | 0 | $-1$ | $\sqrt{3}/2$ | $\sqrt{2}/2$ | $\sqrt{2}/2$ | 3 |
| 3 | $\sqrt{3}/2$ | 1 | 0 | $-1/2$ | $\sqrt{2}/2$ | $-\sqrt{2}/2$ | 3 |
| 4 | $-1$ | $-\sqrt{3}/2$ | $1/2$ | 0 | $-\sqrt{2}/2$ | $\sqrt{2}/2$ | 3 |
| 5 | $-\sqrt{2}/2$ | $-\sqrt{2}/2$ | $-\sqrt{2}/2$ | $\sqrt{2}/2$ | 0 | 1 | 3 |
| 6 | $-\sqrt{2}/2$ | $-\sqrt{2}/2$ | $\sqrt{2}/2$ | $-\sqrt{2}/2$ | $-1$ | 0 | 3 |

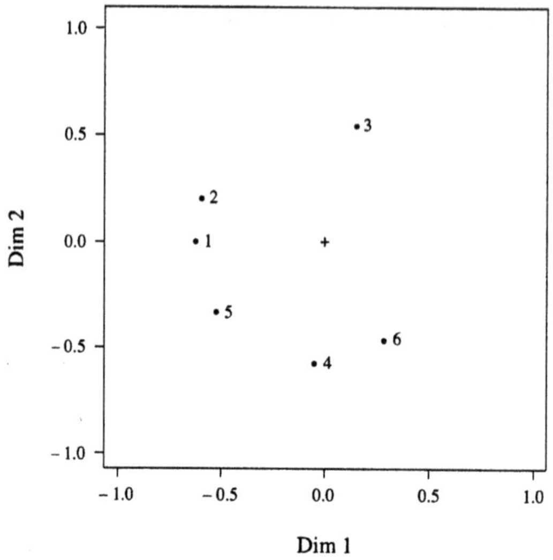


**FIGURE 3.3** Representation of the first singular vectors

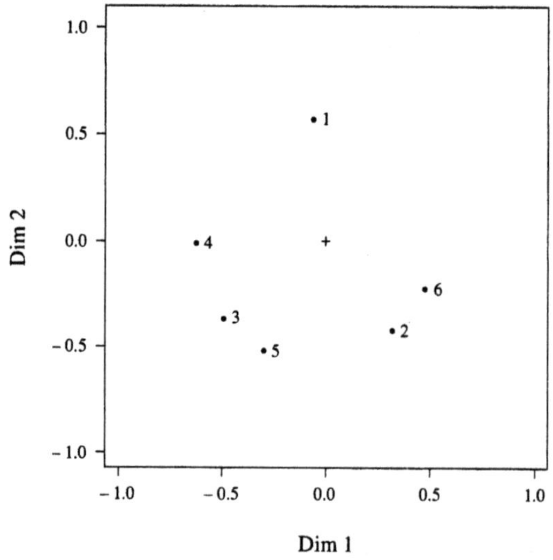


**FIGURE 3.4** Representation of the second singular vectors

**TABLE 3.3** Soft Drinks Brand Switching

| | Period $[t+1]$ | | | | | | | |
|---|---|---|---|---|---|---|---|---|
| Period $[t]$ | A | B | C | D | E | F | G | H |
| A: Coke (Ck) | 612 | 107 | 10 | 33 | 134 | 55 | 13 | 36 |
| B: 7-Up (7Up) | 186 | 448 | 5 | 64 | 140 | 99 | 12 | 46 |
| C: Tab (Tb) | 80 | 120 | 160 | 360 | 80 | 40 | 80 | 80 |
| D: Like (Lk) | 87 | 152 | 87 | 152 | 239 | 43 | 131 | 109 |
| E: Pepsi (Pp) | 177 | 132 | 8 | 30 | 515 | 76 | 26 | 37 |
| F: Sprite (Sp) | 114 | 185 | 29 | 71 | 157 | 329 | 29 | 86 |
| G: Diet Pepsi (DP) | 93 | 47 | 186 | 93 | 116 | 93 | 256 | 116 |
| H: Fresca (Fr) | 226 | 93 | 53 | 107 | 147 | 107 | 67 | 200 |

From the data matrix $\boldsymbol{O}$, we constructed three kinds of skew-symmetric matrices: $\boldsymbol{A}$ by (3.2), $\boldsymbol{A}^*$ by (3.47), and $\tilde{\boldsymbol{A}}$ by (3.48). Note that diagonal entries were filled with zeros in constructing $\boldsymbol{A}$ whereas they were used in constructing $\tilde{\boldsymbol{A}}$ or $\boldsymbol{A}^*$. Skew-symmetric analysis was applied to each matrix, of which the rank is 8 and $r = 4$. Table 3.4 summarizes the results of three cases: singular values $\mu_t$, the index $\theta(t)$ indicating the proportion of $\mu_t$, and its cumulated sum (cum) for $t = 1, 2, 3$, and 4. From the table, it is seen that the results of analysis are influenced to a good extent by the transformation of data.

Figures 3.5, 3.6, and 3.7 represent a two-dimensional configuration of brands in terms of the first pair of singular vectors derived from $\boldsymbol{A}$, $\tilde{\boldsymbol{A}}$, and $\boldsymbol{A}^*$, respectively. Comparing $\theta$-values across the three cases, the contribution

**TABLE 3.4** Singular Values (Soft Drinks Data)

| | | 1 | 2 | 3 | 4 |
|---|---|---|---|---|---|
| $\boldsymbol{A}$ | $\mu$ | 0.2201 | 0.1251 | 0.0465 | 0.0079 |
| | $\theta$ | 0.5508 | 0.3130 | 0.1163 | 0.0198 |
| | cum | 0.5508 | 0.8638 | 0.9801 | 1.0000 |
| $\tilde{\boldsymbol{A}}$ | $\mu$ | 0.4935 | 0.1204 | 0.0461 | 0.0022 |
| | $\theta$ | 0.7452 | 0.1818 | 0.0696 | 0.0034 |
| | cum | 0.7452 | 0.9270 | 0.9966 | 1.0000 |
| $\boldsymbol{A}^*$ | $\mu$ | 0.9068 | 0.1056 | 0.0268 | 0.0068 |
| | $\theta$ | 0.8669 | 0.1010 | 0.0256 | 0.0065 |
| | cum | 0.8669 | 0.9679 | 0.9935 | 1.0000 |

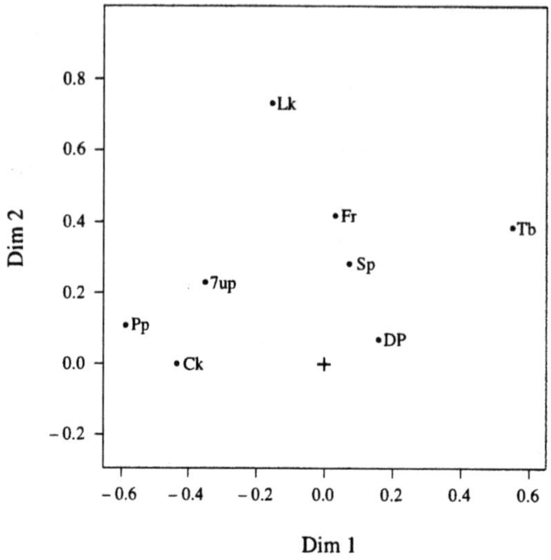


**FIGURE 3.5** Two-dimensional configuration of eight soft drinks (*A*, SSA)

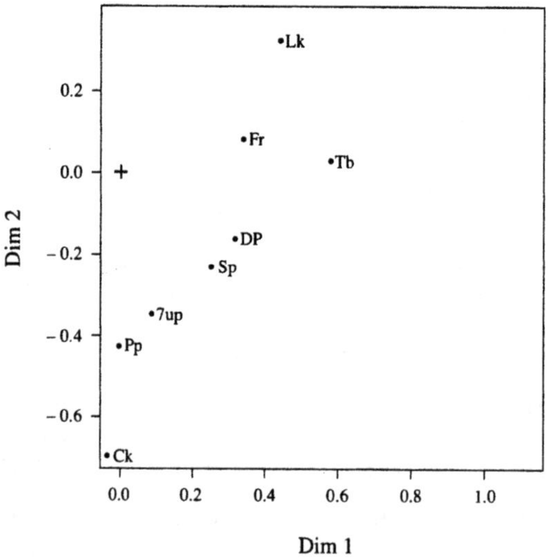


**FIGURE 3.6** Two-dimensional configuration of eight soft drinks ($\tilde{A}$, SSA)

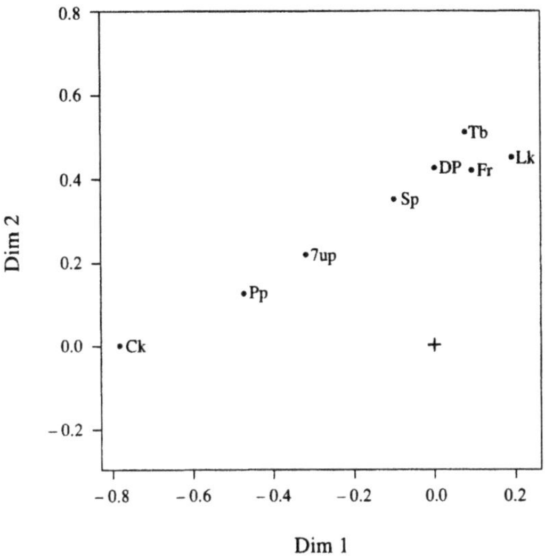


**FIGURE 3.7** Two-dimensional configuration of eight soft drinks ($\tilde{A}^*$, SSA)

of the first pair of singular vectors indicated by $\theta(1)$ becomes greater from the top to the bottom cases. The configuration in Fig. 3.5 looks different from the other two. Inspecting the oriented triangles in the figure, we find a large degree of asymmetry for pairs (Ck, Lk), (Ck, Tb) (Pp, Lk), and (Pp, Tb). Figure 3.6 shows a pattern of line shape (cor = 0.875) and such a pattern appears prominent in Fig. 3.7 (cor = 0.978). In both figures, we find large asymmetry between Ck and each of three brands Lk, Tb, and Fr.

## Pecking Data

Next we show analysis of the pecking data $\{o_{jk}\}$ of Table 2.6. The entries of the table were transformed so as to give pairwise proportions $\boldsymbol{P} = (p_{ij})$ where $p_{ij} = o_{ij}/(o_{ij} + o_{ji})$. Then SSA was applied to the skew-symmetric $\boldsymbol{A}$ constructed from $\boldsymbol{P}$. Table 3.5 gives the result. Figure 3.8 illustrates the result

**TABLE 3.5** Singular Values (Pecking Data)

| $t$ | 1 | 2 | 3 |
|---|---|---|---|
| $\mu$ | 1.6107 | 0.3395 | 0.1532 |
| $\theta$ | 0.7658 | 0.1614 | 0.0728 |
| cum | 0.7658 | 0.9272 | 1.0000 |

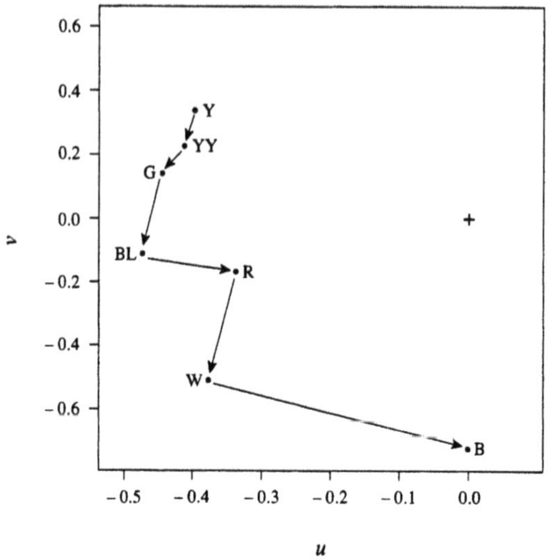


**FIGURE 3.8** Two-dimensional configuration of seven male pigeons ($A$, SSA)

by a plot in terms of the first pair of singular vectors. We note the degree of approximating the data by those vectors, referring to (3.23), as $\theta(1) = 0.7658$.

It is remembered that in the example based on the data in Sec. 2.1.4, we detected a rank order with $\zeta = 1.000$ and it revealed the dominant–subordinate relationships among those pigeons. In the same way as Digby and Kempton (1987), it is of interest to superpose the derived order of dominance on the two-dimensional configuration. In Fig. 3.8 arrows indicate the direction of the dominance. It is noticeable that the order agrees with positioning of pigeons in the configuration.

## 3.3. ESCOUFIER AND GRORUD'S PROCEDURE

In this section we describe the procedure that analyzes $S$ and $A$ simultaneously, which are given by (3.1) and (3.2), respectively. In view of (3.6), we define an index by

$$\kappa_1 = \frac{\|A\|^2}{\|O\|^2} \tag{3.68}$$

Then it shows the proportion of the skew-symmetric part in the norm, and it takes a range such that $0 \leq \kappa_1 \leq 1$. On the basis of variance (3.7), we define another index by

$$\kappa_2 = \frac{\mathrm{Var}(a_{jk})}{\mathrm{Var}(o_{jk})} \tag{3.69}$$

which takes the range $0 \leq \kappa_2 \leq 1$. When $\kappa_1$ or $\kappa_2$ are not small, the procedure will be applied in a practical data analysis.

### 3.3.1. Spectral Decomposition of a Hermitian Matrix

Let us consider the eigenvalue problem of an $n \times n$ Hermitian matrix $\boldsymbol{H}$. It is stated as

$$\boldsymbol{H}z_t = \lambda_t z_t \tag{3.70}$$

The eigenvalue $\lambda_t$ is real, whereas the eigenvector $z_t = (z_{jt})$ is generally complex ($t = 1, 2, \ldots, n$). Without loss of generality, we assume that $r = \mathrm{rank}(\boldsymbol{H})$. Denote the eigenvalues of $\lambda \neq 0$ by $\lambda_1 \geq \lambda_2 \geq \cdots \geq \lambda_r$, and also let $\lambda_{r+1} = \cdots = \lambda_n = 0$. Let $z_t = \boldsymbol{x}_t + i\boldsymbol{y}_t$, $z_t = (z_{jt})$, $\boldsymbol{x}_t = (x_{jt})$, and $\boldsymbol{y}_t = (y_{jt})$. Then elements are expressed as $z_{jt} = x_{jt} + iy_{jt}$ ($t = 1, 2, \ldots, n$). Define the matrice as:

$$\begin{array}{ll} \boldsymbol{Z} = (z_1, z_2, \ldots, z_n), & \boldsymbol{\Lambda} = \mathrm{diag}(\lambda_1, \ldots, \lambda_r, 0, \ldots, 0) \\ \boldsymbol{Z}_r = (z_1, z_2, \ldots, z_r), & \boldsymbol{\Lambda}_r = \mathrm{diag}(\lambda_1, \ldots, \lambda_r) \end{array} \tag{3.71}$$

The eigenequation is expressed in terms of matrices as

$$\boldsymbol{H}\boldsymbol{Z} = \boldsymbol{Z}\boldsymbol{\Lambda} \tag{3.72}$$

Denote complex conjugates of corresponding terms as $\bar{z}$, $\bar{z}$, $\bar{\boldsymbol{Z}}$, and the transpose and conjugate as $z^*$, $\boldsymbol{Z}^*$. Then $\boldsymbol{Z}^{-1} = \bar{\boldsymbol{Z}}' = \boldsymbol{Z}^*$, and $\boldsymbol{Z}$ is a unitary matrix.

The spectral decomposition of $\boldsymbol{H}$ is given by

$$\begin{aligned} \boldsymbol{H} &= \boldsymbol{Z}\boldsymbol{\Lambda}\boldsymbol{Z}^* = \boldsymbol{Z}_r\boldsymbol{\Lambda}_r\boldsymbol{Z}_r^* = \sum_{t=1}^{r} \lambda_t z_t z_t^* \\ &= \sum_{t=1}^{r} \lambda_t((\boldsymbol{x}_t\boldsymbol{x}_t' + \boldsymbol{y}_t\boldsymbol{y}_t') + i(\boldsymbol{y}_t\boldsymbol{x}_t' - \boldsymbol{x}_t\boldsymbol{y}_t')) \end{aligned} \tag{3.73}$$

Write the $j$th row of $\boldsymbol{Z}$ as $\boldsymbol{w}_j = (z_{j1}, \ldots, z_{jr})$. Equation (3.73) is rewritten elementwise as:

$$\begin{aligned} h_{jk} &= \boldsymbol{w}_j \boldsymbol{\Lambda}_r \boldsymbol{w}_k^* \\ &= \sum_{t=1}^{r} \lambda_t z_{jt} \bar{z}_{kt} \\ &= \sum_{t=1}^{r} \lambda_t (x_{jt} x_{kt} + y_{jt} y_{kt}) + i \sum_{t=1}^{r} \lambda_t (y_{jt} x_{kt} - x_{jt} y_{kt}) \end{aligned} \tag{3.74}$$

Using the spectral decomposition, we derive the following relation about the norm of $\boldsymbol{H}$,

$$\|\boldsymbol{H}\|^2 = \mathrm{tr}(\boldsymbol{H}\boldsymbol{H}^*) = \sum_{t=1}^{n} \lambda_t^2 = \sum_{t=1}^{r} \lambda_t^2 \tag{3.75}$$

### 3.3.2. Canonical Analysis

Escoufier and Grorud (1980) proposed applying spectral decomposition of a Hermitian matrix to analyze asymmetric data. From an asymmetric data matrix $\boldsymbol{O}$, we construct a symmetric matrix $\boldsymbol{S}$ and a skew-symmetric matrix $\boldsymbol{A}$ by (3.1) and (3.2), respectively. Define a Hermitian matrix $\boldsymbol{H} = (h_{jk})$ by

$$\boldsymbol{H} = \boldsymbol{S} + i\boldsymbol{A}, \qquad \text{where} \quad h_{jk} = s_{jk} + i a_{jk} \tag{3.76}$$

Here we assume that $\mathrm{rank}(\boldsymbol{H}) = r$. From the eigenvalue problem (3.70) it follows that

$$\sum_{j=1}^{n} o_{jj} = \mathrm{tr}(\boldsymbol{H}) = \sum_{t=1}^{r} \lambda_t \tag{3.77}$$

$$\sum_{j=1}^{n} \sum_{k=1}^{n} o_{jk}^2 = \|\boldsymbol{H}\|^2 = \sum_{t=1}^{r} \lambda_t^2 \tag{3.78}$$

Now applying the theory of spectral decomposition and considering the correspondence of real and imaginary parts for (3.73) and (3.76), we obtain expressions for $\boldsymbol{S}$ and $\boldsymbol{A}$:

$$\boldsymbol{S} = \sum_{t=1}^{r} \lambda_t (\boldsymbol{x}_t \boldsymbol{x}_t' + \boldsymbol{y}_t \boldsymbol{y}_t'), \qquad s_{jk} = \sum_{t=1}^{r} \lambda_t (x_{jt} x_{kt} + y_{jt} y_{kt}) \tag{3.79}$$

$$\boldsymbol{A} = \sum_{t=1}^{r} \lambda_t (\boldsymbol{y}_t \boldsymbol{x}_t' - \boldsymbol{x}_t \boldsymbol{y}_t'), \qquad a_{jk} = \sum_{t=1}^{r} \lambda_t (y_{jt} x_{kt} - x_{jt} y_{kt}) \tag{3.80}$$

Noting that $o_{jk} = s_{jk} + a_{jk}$, we derive in view of (3.79) and (3.80)

$$o_{jk} = \sum_{t=1}^{r} \lambda_t((x_{jt}x_{kt} + y_{jt}y_{kt}) + (y_{jt}x_{kt} - x_{jt}y_{kt})) \tag{3.81}$$

Escoufier and Grorud suggested treating (3.70) in a simpler way in real numbers than the way in complex numbers. Equation (3.70) is represented as an eigenvalue problem of a symmetric matrix of order $2n \times 2n$ as follows:

$$\begin{pmatrix} \boldsymbol{S} & \boldsymbol{A} \\ -\boldsymbol{A} & \boldsymbol{S} \end{pmatrix} \begin{pmatrix} \boldsymbol{x}_t \\ \boldsymbol{y}_t \end{pmatrix} = \lambda_t \begin{pmatrix} \boldsymbol{x}_t \\ \boldsymbol{y}_t \end{pmatrix} \tag{3.82}$$

Corresponding to $\lambda_1, \lambda_2, \ldots, \lambda_n$ in (3.70), we see that $\lambda_1, \lambda_1, \lambda_2, \lambda_2, \ldots, \lambda_n, \lambda_n$ are roots of multiplicity 2 in (3.82). Write the eigenvectors associated with $\lambda_t$ by $z_{1t}, z_{2t}$. Then we find that

$$z_{1t} = \begin{pmatrix} \boldsymbol{x}_t \\ \boldsymbol{y}_t \end{pmatrix} \quad \text{and} \quad z_{2t} = \begin{pmatrix} -\boldsymbol{y}_t \\ \boldsymbol{x}_t \end{pmatrix} \tag{3.83}$$

Write the left-hand side of (3.82) as $\boldsymbol{E}$, so we then have

$$\boldsymbol{E} = \sum_{t=1}^{n} \lambda_t (z_{1t}z'_{1t} + z_{2t}z'_{2t}) \tag{3.84}$$

For each vector $z_t = \boldsymbol{x}_t + i\boldsymbol{y}_t$, we plot $n$ objects in terms of $(x_{jt}, y_{jt})$ on an $x$–$y$ plane. Setting an index

$$\phi(p) = \frac{\sum_{t=1}^{p} \lambda_t^2}{\sum_{t=1}^{r} \lambda_t^2} \tag{3.85}$$

on the basis of (3.78), we find that it indicates the degree to which $\boldsymbol{H}$ is approximated by a matrix of rank $p$. When $\phi(p)$ is close to 1, we can perform data analysis of $\boldsymbol{O}$ by referring to $p$ eigenvectors. We can interpret the two-dimensional configurations from a geometrical point of view as follows.

## (I) The Case of $r = 1$

For simplicity, we describe the case of $r = 1$. Using $z = \boldsymbol{x} + i\boldsymbol{y}$, $\boldsymbol{x} = (x_j)$, $\boldsymbol{y} = (y_j)$, we rewrite $\boldsymbol{H} = \lambda z z^*$ as

$$\boldsymbol{H} = \lambda((\boldsymbol{x}\boldsymbol{x}' + \boldsymbol{y}\boldsymbol{y}') + i(\boldsymbol{y}\boldsymbol{x}' - \boldsymbol{x}\boldsymbol{y}')) \tag{3.86}$$

$$h_{jk} = \lambda((x_j x_k + y_j y_k) + i(y_j x_k - x_j y_k)) \tag{3.87}$$

Define vectors $f_j = (x_j, y_j)'$. Then we have expressions for symmetric $s_{jk}$ and skew-symmetric $a_{jk}$ as follows:

$$s_{jk} = \lambda(x_j x_k + y_j y_k) = \lambda f_j' f_k \tag{3.88}$$

$$a_{jk} = \lambda(y_j x_k - x_j y_k) = \lambda f_j \times f_k \tag{3.89}$$

Thus $s_{jk}$ corresponds to the inner product, and $a_{jk}$ to the outer product or equivalently the oriented paralleotope with sign of $a_{jk}$ (Fig. 3.9).

The interpoint distance $d_{jk}$ is expressed by

$$\begin{aligned} d_{jk}^2 &= (x_j - x_k)^2 + (y_i - y_k)^2 \\ &= \|f_j\|^2 + \|f_k\|^2 - 2f_j' f_k \end{aligned} \tag{3.90}$$

$$\lambda d_{jk}^2 = s_{jj} + s_{kk} - 2s_{jk} \tag{3.91}$$

Now put the right-hand side as $q_{jk}$,

$$\begin{aligned} q_{jk} &= s_{jj} + s_{kk} - 2s_{jk} \\ &= o_{jj} + o_{kk} - o_{jk} - o_{kj} \end{aligned} \tag{3.92}$$

Then it is found that the interpoint distance represents the transformation of data such that

$$q_{jk} = \lambda d_{jk}^2 \tag{3.93}$$

Note that the sign of $q_{jk}$ is not always positive, showing the sign of $\lambda$.

## (II) General Case of $r > 1$

Now we consider the case of rank $r$. In the configuration in terms of $(\boldsymbol{x}_t, \boldsymbol{y}_t)$, we define two-dimensional vectors $f_j(t) = (x_{jt}, y_{jt})'$. Denote the interpoint distance

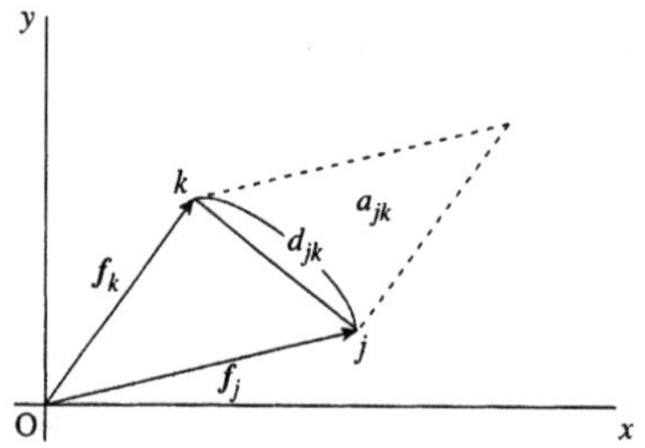


**FIGURE 3.9** Vectors $f_j$ and $f_k$

by $d_{jk}(t)$ and also the distance of point $j$ from the origin by $d_{oj}(t)$,

$$d_{jk}^2(t) = (x_{jt} - x_{kt})^2 + (y_{jt} - y_{kt})^2 \tag{3.94}$$

$$d_{oj}^2(t) = x_{jt}^2 + y_{jt}^2 \tag{3.95}$$

Then we have

$$s_{jk} = \sum_{t=1}^{r} \lambda_t \boldsymbol{f}_j(t)' \boldsymbol{f}_k(t) \tag{3.96}$$

$$a_{jk} = \sum_{t=1}^{r} \lambda_t \boldsymbol{f}_j(t) \times \boldsymbol{f}_k(t) \tag{3.97}$$

$$o_{jk} = \sum_{t=1}^{r} \lambda_t \boldsymbol{f}_j(t)' \boldsymbol{f}_k(t) + \sum_{t=1}^{r} \lambda_t \boldsymbol{f}_j(t) \times \boldsymbol{f}_k(t) \tag{3.98}$$

$$q_{jk} = \sum_{t=1}^{r} \lambda_t d_{jk}^2(t) \tag{3.99}$$

$$o_{jj} = \sum_{t=1}^{r} \lambda_t d_{oj}^2(t) \tag{3.100}$$

In the last two expressions, the transformed data correspond to the weighted sum of squared distances (Saito, 2003a).

## Line Pattern

An easy interpretable pattern will be such that $n$ points show a pattern of line. For example, the following theorem gives a condition to derive such a pattern (Saito, 1997).

*Theorem 3.8.* For (3.70) to give a solution $\boldsymbol{y} = b\boldsymbol{x}$ $(b \neq 0)$, it is necessary and sufficient that there exists a real eigenvector $\boldsymbol{u}$ satisfying

$$\boldsymbol{O}\boldsymbol{u} = \nu \boldsymbol{u} = \boldsymbol{O}'\boldsymbol{u} \tag{3.101}$$

If (3.101) holds, then $\lambda = \nu$ and $\boldsymbol{x} = \boldsymbol{u}$, which leads to

$$\boldsymbol{S}\boldsymbol{x} = \lambda \boldsymbol{x}, \qquad \boldsymbol{A}\boldsymbol{x} = \boldsymbol{0} \tag{3.102}$$

### 3.3.3. Representation with a Hermitian Form

Suppose that $\boldsymbol{H}$ defined by (3.73) is positive semidefinite with rank $r$, that is,

$$\lambda_1 \geq \cdots \geq \lambda_r > 0, \qquad \lambda_{r+1} = \cdots = \lambda_n = 0 \tag{3.103}$$

Let $\mathbf{\Lambda}_r = \text{diag}(\lambda_1, \lambda_2, \ldots, \lambda_r)$. We retain definition of $\{\mathbf{w}_j\}$ in Sec. 3.3.1. Now let us consider the complex vector space $\mathcal{W}_r$ $(r \leq n)$ that is spanned by $\{\mathbf{w}_j\}$. We define an inner product $(\mathbf{w}_j, \mathbf{w}_k)$ by

$$(\mathbf{w}_j, \mathbf{w}_k) = \mathbf{w}_j \mathbf{\Lambda}_r \mathbf{w}_k^* \tag{3.104}$$

The norm and the metric are defined in usual way, that is,

$$\|\mathbf{w}_j\| = (\mathbf{w}_j \mathbf{\Lambda}_r \mathbf{w}_j^*)^{1/2} \tag{3.105}$$

$$\begin{aligned} D_{jk} &= \|\mathbf{w}_j - \mathbf{w}_k\| \\ &= \left( \sum_{t=1}^{r} \lambda_t \{ (x_{jt} - x_{kt})^2 + (y_{jt} - y_{kt})^2 \} \right)^{1/2} \end{aligned} \tag{3.106}$$

On the basis of the norm defined above, we derive identities, through simple manipulations, as (3.107) and (3.108):

$$\|\mathbf{w}_j\|^2 + \|\mathbf{w}_k\|^2 - \|\mathbf{w}_j - \mathbf{w}_k\|^2 = \mathbf{w}_j \Lambda_r \mathbf{w}_k^* + \mathbf{w}_k \Lambda_r \mathbf{w}_j^* \tag{3.107}$$

$$i(\|\mathbf{w}_j\|^2 + \|\mathbf{w}_k\|^2 - \|\mathbf{w}_j - i\mathbf{w}_k\|^2) = \mathbf{w}_j \Lambda_r \mathbf{w}_k^* - \mathbf{w}_k \Lambda_r \mathbf{w}_j^* \tag{3.108}$$

From (3.74), (3.107), and (3.108), it follows that

$$\begin{aligned} h_{jk} &= \tfrac{1}{2}(\|\mathbf{w}_j\|^2 + \|\mathbf{w}_k\|^2 - \|\mathbf{w}_j - \mathbf{w}_k\|^2) \\ &\quad + \tfrac{1}{2} i (\|\mathbf{w}_j\|^2 + \|\mathbf{w}_k\|^2 - \|\mathbf{w}_j - i\mathbf{w}_k\|^2) \end{aligned} \tag{3.109}$$

In addition to $D_{jk}$, we define

$$\bar{D}_{jk} = \|\mathbf{w}_j - i\mathbf{w}_k\| \qquad \text{and} \qquad D_{0j} = \|\mathbf{w}_j\| \tag{3.110}$$

Then (3.109) is alternatively expressed as

$$h_{jk} = \tfrac{1}{2}(D_{0j}^2 + D_{0k}^2 - D_{jk}^2) + \tfrac{1}{2} i (D_{0j}^2 + \mathbf{D}_{0k}^2 - \bar{D}_{jk}^2) \tag{3.111}$$

Rewrite (3.76) as

$$h_{jk} = \tfrac{1}{2}(o_{jk} + o_{kj}) + i \tfrac{1}{2}(o_{jk} - o_{kj}) \tag{3.112}$$

By the correspondence of (3.109) and (3.112) in view of real and imaginary parts, we have two expressions:

$$o_{jk} + o_{kj} = \|\boldsymbol{w}_j\|^2 + \|\boldsymbol{w}_k\|^2 - \|\boldsymbol{w}_j - \boldsymbol{w}_k\|^2 \tag{3.113}$$

$$o_{jk} - o_{kj} = \|\boldsymbol{w}_j\|^2 + \|\boldsymbol{w}_k\|^2 - \|\boldsymbol{w}_j - i\boldsymbol{w}_k\|^2 \tag{3.114}$$

Then we derive (3.115), which is rewritten as (3.116):

$$o_{jk} = \|\boldsymbol{w}_j\|^2 + \|\boldsymbol{w}_k\|^2 - \tfrac{1}{2}\|\boldsymbol{w}_j - \boldsymbol{w}_k\|^2 - \tfrac{1}{2}\|\boldsymbol{w}_j - i\boldsymbol{w}_k\|^2 \tag{3.115}$$

$$= D_{0j}^2 + D_{0k}^2 - \tfrac{1}{2}D_{jk}^2 - \tfrac{1}{2}\bar{D}_{jk}^2 \tag{3.116}$$

These two expressions were first presented by Chino and Shiraiwa (1993). Thus we find that expression (3.81) of observation $o_{jk}$ is alternatively given by (3.115) or (3.116). It is natural that asymmetric $o_{jk}$ is represented by the expression including an asymmetric term. Note that the three expressions are stated in terms of eigenvalues and eigenvectors of $\boldsymbol{H}$. Combining (3.99) and (3.106), we obtain

$$q_{jk} = D_{jk}^2 \tag{3.117}$$

Thus it is realized that the squared distance $D_{jk}^2$ coincides with the transformed observation $q_{jk}$ that is defined in (3.92) (Saito, 2003a). In (3.106) one does not know any explicit quantity in terms of data that $D_{jk}$ might represent, because $D_{jk}$ is given by eigenvalues and eigenvectors of $\boldsymbol{H}$. In (3.117) we find a clear correspondence of $D_{jk}$ to $q_{jk}$. The relation (3.117) was found by Saito (2003a). From the description so far given from (3.106) to (3.117), we will develop arguments about the possibility of MDS of asymmetric data in connection with Escoufier and Grorud's (E–G) procedure in Sec. 4.2.

### 3.3.4. Discussion

#### Remark on the Distance

It should be noted that distance (3.106) is defined in the $r$-dimensional vector space $\mathcal{W}_r$ under assumption (3.103). It is stressed that treatment of distance in a descriptive data analysis using Gower's (SSA) or Escoufier and Grorud's (E–G) procedure differs fundamentally from the treatment of distance in the context of $\mathcal{W}_r$. To show this point clearly, let us consider an asymmetric $\boldsymbol{O}$ that is *cyclic*.

Suppose that we make two-dimensional plots of objects in terms of eigenvectors derived from any of matrices $\boldsymbol{O}'$, $\boldsymbol{S}, \boldsymbol{A}$, and $\boldsymbol{H}$ that are constructed from $\boldsymbol{O}$. According to Theorem 3.4, the configurations are all the same. Writing the eigenvalue problem of $\boldsymbol{O}$ as (3.118), we see that (3.119) holds by the theorem:

$$\boldsymbol{O}z = \alpha z \tag{3.118}$$

$$\boldsymbol{O}'\bar{z} = \alpha\bar{z} \tag{3.119}$$

From them it follows that

$$\boldsymbol{S}z = \beta z, \qquad \text{that is,} \qquad \boldsymbol{S}\boldsymbol{x} = \beta\boldsymbol{x} \quad \text{and} \quad \boldsymbol{S}\boldsymbol{y} = \beta\boldsymbol{y} \tag{3.120}$$

$$\boldsymbol{A}z = i\gamma z, \qquad \text{that is,} \qquad \boldsymbol{A}\boldsymbol{x} = -\gamma\boldsymbol{y} \quad \text{and} \quad \boldsymbol{A}\boldsymbol{y} = \gamma\boldsymbol{x} \tag{3.121}$$

where $\beta = \mathrm{Re}(\alpha)$, $\gamma = \mathrm{Im}(\alpha)$ and $z = \boldsymbol{x} + i\boldsymbol{y}$. Then we find from (3.120) and (3.121) that

$$\boldsymbol{H}z = \lambda z \qquad \text{where} \qquad \lambda = \beta - \gamma \tag{3.122}$$

As described above, we can refer to Euclidean distance in the two-dimensional configuration derived by E–G for any $\boldsymbol{O}$. For a cyclic matrix $\boldsymbol{O}$, the sign of $\lambda$ may be positive, negative, or even zero. It means that $\boldsymbol{H}$ can generally be indefinite even if $\boldsymbol{S}$ is positive definite. Then we cannot define distance, as an entire property implied in $\boldsymbol{H}$ with definitions (3.104) and (3.106), even for the same $\boldsymbol{O}$.

## Properties of the Procedure

The computational steps of the E–G procedure hold without respect to the direction of the measure of relationship (similarity or dissimilarity). Let us examine the properties of the procedure. When $o_{jk}$ indicates similarity, it is allowable to regard (3.93) or (3.99) as a model representation. In the case of similarity, we may consider that $o_{jj} \geq o_{jk}$ usually, then

$$o_{jj} + o_{kk} \geq o_{jk} + o_{kj} \tag{3.123}$$

leading to $q_{jk} \geq 0$ with a natural treatment for fitting. When $o_{jk}$ indicates dissimilarity, in contrast, one would apply E–G formally, but there is no reasonable coincidence between both sides in (3.93) or (3.99). Hence one would not derive practically meaningful information from the analysis.

## Null Diagonal Entries

There can be situations in which some of the diagonal entries are missing for the data matrix. Now we focus on a different situation in which the diagonal element is undefined in the original observations due to the property of the phenomenon concerned (e.g., pecking frequency), so that all the diagonal entries should be given zero values prior to data analysis using the E–G procedure. Let us examine such a case in the context of representation with a Hermitian form. From (3.74), (3.76), and (3.106) it follows that

$$o_{jj} = h_{jj} = \|\boldsymbol{w}_j\|^2 = \sum_{t=1}^{r} \lambda_t (x_{jt}^2 + y_{jt}^2) = D_{oj}^2 \tag{3.124}$$

Thus, for object $j$ such that $o_{jj} = 0$, the norm (3.106) is not defined. If all the elements are null, we see from (3.77) that

$$\sum_{j=1}^{n} o_{jj} = \sum_{t=1}^{n} \lambda_t = 0 \tag{3.125}$$

leading to an indefinite $\boldsymbol{H}$. Therefore, in order that the $D_{jk}$ defined by (3.106) should be valid, all the diagonals must be defined and positive, in addition to $\boldsymbol{H}$ being positive semidefinite. In passing the following point is noted. When $\text{rank}(\boldsymbol{H}) = r = 1$, we see by (3.124) that

$$o_{jj} = \lambda(x_j^2 + y_j^2) \qquad (j = 1, 2, \ldots, n) \tag{3.126}$$

Accordingly, if all the diagonals are null, $r$ cannot be 1.

Consider a transformation of data, giving a constant $\alpha$ to all the diagonal elements in such a way that

$$\begin{aligned} \tilde{o}_{jj} &= o_{jj} + \alpha \qquad (j = 1, 2, \ldots, n) \\ \tilde{o}_{jk} &= o_{jk} \qquad (j \neq k) \end{aligned} \tag{3.127}$$

This treatment is not meaningful as we consider the distance $D_{jk}$. Construct $\tilde{\boldsymbol{H}}$ from $\tilde{\boldsymbol{O}} = (\tilde{o}_{jk})$ by (3.76). Since $\tilde{\boldsymbol{H}} = \boldsymbol{H} + \alpha\boldsymbol{I}$, its eigenproblem is written as

$$\tilde{\boldsymbol{H}}\tilde{z} = \tilde{\lambda}\tilde{z}, \qquad \text{that is,} \quad (\boldsymbol{H} + \alpha\boldsymbol{I})z = (\lambda + \alpha)z \tag{3.128}$$

showing that

$$\tilde{z} = z \qquad \text{and} \qquad \tilde{\lambda} = \lambda + \alpha \tag{3.129}$$

Thus it is seen that $\boldsymbol{H}$ and $\tilde{\boldsymbol{H}}$ have eigenvectors in common. Take a large $\alpha$ so as to make all $\tilde{\lambda}$ positive. Then, with invariant eigenvectors, distance is defined for $\tilde{\boldsymbol{H}}$ whereas it is not for $\boldsymbol{H}$. Therefore, it does not make sense to argue the validity of distance (3.106) under data transformation (3.127). Again it should be remembered that the distance $D_{jk}$ is suggested as the entire property associated with the $\{z_1, z_2, \ldots, z_r\}$, not with a single $z_t$. The E–G procedure is always applied to $\boldsymbol{H}$ regardless of its definiteness.

### 3.3.5. Example

### Soft Drinks Brand Switching Data

We present an analysis of the soft drinks data of Table 3.3, using the E–G procedure. The degree of asymmetry was indicated by $\kappa_1 = 0.072$ and $\kappa_2 = 0.156$. Computation of the eigenvalue problem of $\boldsymbol{H}$ yielded positive and negative eigenvalues. Table 3.6 shows the largest four eigenvalues and their corresponding quantities. The degree of approximation of $\boldsymbol{H}$ by the first eigenvector $z_1 = \boldsymbol{x}_1 + i\boldsymbol{y}_1$ is shown by $\phi(1) = 0.6711$.

Figure 3.10 shows the two-dimensional configuration in terms of $\boldsymbol{x}_1$ and $\boldsymbol{y}_1$. In the figure the horizontal axis indicates $(-y_1)$ and the vertical axis $x_1$, for comparison with figures in Sec. 3.2.5 derived from the same data. Comparing those figures, it is interesting to find that a pattern of line shape appears more prominently in Fig. 3.10 (cor = 0.900) than in Figs. 3.6 and 3.7. It should be noted that Figs. 3.6 and 3.7 were derived from skew-symmetric matrices $\tilde{\boldsymbol{A}}$ and $\boldsymbol{A}^*$, respectively, whereas Fig. 3.10 was derived by using symmetric $\boldsymbol{S}$ and skew-symmetric $\boldsymbol{A}$.

### Pecking Data

For comparison, we analyzed the pecking data of Table 2.6 by the E–G procedure. The degree of asymmetry was indicated by $\kappa_1 = 0.342$ and $\kappa_2 = 0.785$. Table 3.7 summarizes the result, with the largest four eigenvalues and related quantities. As is known theoretically from the matrix with all null diagonal entries, the sum of eigenvalues is zero, and so $\boldsymbol{H}$ is indefinite.

Figure 3.11 represents the configuration in terms of the first eigenvector. The degree of approximation of $\boldsymbol{H}$ is given by $\phi(1) = 0.8103$. On the figure we superposed the dominance relations detected in Sec. 2.2.4. Comparing the

**TABLE 3.6** Eigenvalues (Soft Drinks Data)

| $t$ | 1 | 2 | 3 | 4 |
|---|---|---|---|---|
| $\lambda$ | 1.1144 | 0.4759 | 0.3699 | 0.2597 |
| $\lambda^2$ | 1.2419 | 0.2265 | 0.1368 | 0.1255 |
| $\phi$ | 0.6711 | 0.1224 | 0.0739 | 0.0678 |
| cum | 0.6711 | 0.8674 | 0.9352 | 0.9716 |

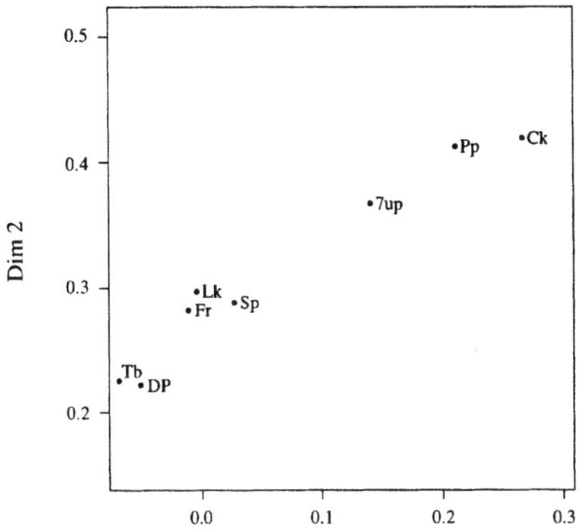


**FIGURE 3.10** Two-dimensional configuration of eight soft drinks (***H***, E–G)

figure with Fig. 3.8, we find that the linear hierarchy mentioned in Sec. 2.2 is revealed clearly by the pattern of line shape in Fig. 3.11.

Considering the property of pairwise proportions that $p_{ij} + p_{ji} = 1$, all $s_{ij} = 0.5$ $(i \neq j)$ and $s_{ii} = 0$, we thus find that the symmetric part has little meaningful information. This is reflected by the values of $\kappa_1$ and $\kappa_2$ mentioned above. For this reason, analysis of SSA based on $\boldsymbol{A}$ yielded nearly the same configuration as that of E–G based on $\boldsymbol{S}$ and $\boldsymbol{A}$.

## 3.4. VECTOR MODEL

In this section, we describe a vector model proposed by Yadohisa and Niki (2000). Given an $n \times n$ asymmetric matrix $\boldsymbol{O} = (o_{ij})$ where $o_{ij}$ represents (dis)similarity for a pair of objects $(i, j)$, we apply decomposition (3.3) to $\boldsymbol{O}$ in such a way as $\boldsymbol{O} = \boldsymbol{S} + \boldsymbol{A}$. We assume that the locations of objects have

**TABLE 3.7** Eigenvalues (Pecking Data)

| $t$ | 1 | 2 | 3 | 4 |
|---|---|---|---|---|
| $\lambda$ | 3.5969 | −0.1563 | −0.3056 | −0.3596 |
| $\lambda^2$ | 12.9377 | 0.0244 | 0.0934 | 0.1293 |
| $\phi$ | 0.8103 | 0.0015 | 0.0058 | 0.0081 |
| cum | 0.8103 | 0.8118 | 0.8176 | 0.8257 |

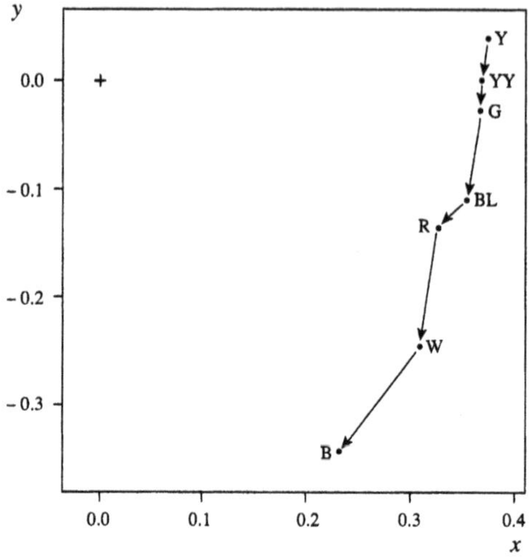


**FIGURE 3.11** Two-dimensional configuration of seven male pigeons (***H***, E–G)

already been derived from the symmetric $\boldsymbol{S}$ through some suitable multidimensional scaling procedure, then we aim to determine from $\boldsymbol{A}$ a set of vectors that represent the latent structure of asymmetry in $\boldsymbol{A}$. We discuss the asymmetric structure, suggesting several indices to measure asymmetry.

### 3.4.1. Model

Let us begin with a simple case in which there are two objects 1 and 2 with dissimilarity ratings $o_{12}$ and $o_{21}$, which satisfy $o_{12} \geq o_{21}$. These objects may properly be positioned by coordinates $\boldsymbol{p}_1$ and $\boldsymbol{p}_2$, respectively. The distance $\|\boldsymbol{p}_1 - \boldsymbol{p}_2\|$ should coincide with the symmetrized dissimilarity $s_{12} = (o_{12} + o_{21})/2$. This arrangement may not be congruent, because it might be too near for the first object as well as too far for the second, $a_{12} = -a_{21} = o_{12} - s_{12} = s_{12} - o_{21}$. We then consider a situation in which the objects float in the direction indicated by $\mathrm{sgn}(\boldsymbol{p}_1 - \boldsymbol{p}_2)$ with a strength proportional to $a_{12}$.

Given $n(\geq 3)$ objects with dissimilarity matrix $\boldsymbol{O}$, the above conceptual model can be extended in several ways. Here we are interested in the simplest and possibly most useful way.

We consider a dissimilarity data matrix $\boldsymbol{O}$, with diagonal elements all being zero. It is decomposed into the symmetric matrix $\boldsymbol{S}$ and skew-symmetric matrix $\boldsymbol{A}$. From $\boldsymbol{S}$, a set $P = \{\boldsymbol{p}_i | i = 1, 2, \ldots, n\}$ of the coordinates of $n$ objects are derived by applying an appropriate symmetric MDS procedure.

Another set of vectors $V = \{\boldsymbol{v}_{i/j} | i = 1, 2, \ldots, n\}$, which is called the "set of vectors at $P$," is derived from $\boldsymbol{A}$ as follows. Each element $a_{ij}$ is compared with the projection of $\boldsymbol{v}_i$ on the directed line $\boldsymbol{p}_i - \boldsymbol{p}_j$ in such a way that

$$a_{ij}\lambda(\boldsymbol{p}_i - \boldsymbol{p}_j) = \boldsymbol{v}_{i/j} + \boldsymbol{e}_{ij} \tag{3.130}$$

Here $\lambda(\boldsymbol{x})$ denotes the direction vector

$$\lambda(\boldsymbol{x}) = \begin{cases} 0, & \text{if } \boldsymbol{x} = 0, \\ \dfrac{\boldsymbol{x}}{\|\boldsymbol{x}\|}, & \text{otherwise,} \end{cases} \tag{3.131}$$

and $\boldsymbol{e}_{ij}$ is an error vector. The least squares estimate of $V$, for which

$$e_i^2 = \sum_{j \neq i} \boldsymbol{e}'_{ij}\boldsymbol{e}_{ij} \qquad (i = 1, 2, \ldots, n) \tag{3.132}$$

attains its minimum, is given by $\hat{V} = \{\hat{\boldsymbol{v}}_{i/j} | i = 1, 2, \ldots, n\}$, where

$$\hat{\boldsymbol{v}}_{i/j} = \frac{1}{n-1} \sum_{j \neq i} a_{ij}\lambda(\boldsymbol{p}_i - \boldsymbol{p}_j) \tag{3.133}$$

This model is characterized by the following points: (i) there is no requirement of prior information or assumptions about the asymmetric structure of the data; (ii) the asymmetry is not represented by a numerical value but graphically by arrows with lengths proportional to the strength of asymmetry; (iii) the asymmetry is associated with objects in space. From (ii) and (iii), we can interpret the results intuitively. In view of vector properties we can extend the model for a more complicated analysis in what follows.

The vector $\boldsymbol{v}_{i/j}$ is in proportion to $(o_{ij} - o_{ji})$ in the same manner as "flow" in Tobler (1976–77). The $\boldsymbol{v}_{i/j}$ and the flow differ in that the $\boldsymbol{v}_{i/j}$ has the weight $\lambda(\boldsymbol{p}_i - \boldsymbol{p}_j)$. Accordingly, the vector depends on the object coordinates, which have been obtained from $\boldsymbol{S}$ through a suitable MDS procedure.

The residual from the symmetric part is not formulated by this model. However, it can easily be extended for representing this residual with the asymmetric component $\boldsymbol{A}$ proposed by Yadohisa (1998). Asymmetric MDS models in view of this sort of residual have been proposed by Saito and Takeda (1990) and Saito (1991). Hereafter, we denote $\boldsymbol{v}_{i/j}$ by $\boldsymbol{v}_i$ for the sake of simplicity.

### 3.4.2. Cluster Properties in Terms of Dynamics

Let us consider the above results according to dynamical interpretation. Let $N$ be a set of $n$ objects, that is, $N = \{1, 2, \ldots, n\}$. Hereafter, a nonempty subset

$C$ of $N$, where $C \in 2^N \setminus \{\phi\}$, is called a "cluster," and the entire cluster $N$ is called the "system." The number of elements of set $C$ is denoted by $n(C)$. An arbitrarily selected location $\boldsymbol{g}_c$ to represent the cluster $C$ is called the "center" of $C$. Typically, $\boldsymbol{g}_c$ is chosen as

$$\boldsymbol{g}_c = \frac{1}{n(C)} \sum_{i \in C} \boldsymbol{p}_i \tag{3.134}$$

which is the equiweighted center $\boldsymbol{g}_c$ of the objects of $C$. For applications, it is also possible to take $\boldsymbol{g}_c$ as the position of a representative object, otherwise the "center" of the subcluster in $C$ and so forth.

A cluster is characterized by a list of properties such as "linear flow," "rotation," and "divergence." Hereafter, we consider that the objects vectors $P$ and their associated vectors $V$ span a three-dimensional space or a two-dimensional subspace in three dimensions.

## Linear Flow

Linear flow $\boldsymbol{l}_c$ of the cluster $C$ shows the direction in which $C$ is moving toward or away, and the magnitude of its speed or acceleration. The least squares estimate $\hat{\boldsymbol{l}}_c$ is given by

$$\hat{\boldsymbol{l}}_c = \frac{1}{n(C)} \sum_{i \in C} \boldsymbol{v}_i \tag{3.135}$$

for which

$$\sum_{i \in C} \|\boldsymbol{l}_c - \boldsymbol{v}_i\|^2 \tag{3.136}$$

attains its minimum. Note that $\hat{\boldsymbol{l}}_c$ is regarded as the wind velocity in the "Jet-stream model" due to Gower (1977).

## Rotation

The rotation vector $\boldsymbol{r}_c$ of the cluster $C$ is defined as

$$\boldsymbol{v}_i \times \frac{\boldsymbol{p}_i - \boldsymbol{g}_c}{\|\boldsymbol{p}_i - \boldsymbol{g}_c\|^\rho} = \boldsymbol{r}_c + \boldsymbol{\varepsilon}_i \qquad (i = 1, 2, \ldots, n) \tag{3.137}$$

with a specified constant $\rho(\leq 2)$ and a noise vector $\boldsymbol{\varepsilon}_i$. Here we denote the outer product by "$\times$."

For $\rho = 0$, (3.137) indicates the "Galaxy model," where each object takes its trajectory conforming to the second Kepler's law. The "Roulette model" is obtained by setting $\rho = 2$, where the cluster rotates like a solid mass. For any other values of $\rho$, even if negative, we have the "Viscous-liquid model" with some level of viscosity. Setting $\rho = 1$ leads to the "Homogeneous model," which is further reduced to the "Cyclone model" (Gower, 1977).

Instead of (3.134), the center $\boldsymbol{g}_c$ is chosen appropriately. We can choose $\rho < 1$ so as to induce angular velocity about the origin in the neighborhood of $\boldsymbol{g}_c$, unless specified otherwise. However, it is allowed and even recommended to set $\rho = 1$ and use $\lambda(\boldsymbol{p}_i - \boldsymbol{g}_c)$ in place of $(\boldsymbol{p}_i - \boldsymbol{g}_c)/\|\boldsymbol{p}_i - \boldsymbol{g}_c\|$ for stability in estimation.

The least squares estimate

$$\hat{\boldsymbol{r}}_c = \frac{1}{n(C)} \sum_{i \in C} \boldsymbol{v}_i \times \lambda(\boldsymbol{p}_i - \boldsymbol{g}_c) \tag{3.138}$$

is derived by minimizing

$$R^2 = \sum_{i \in C} \left\| \boldsymbol{r}_c - \frac{\boldsymbol{v}_i \times (\boldsymbol{p}_i - \boldsymbol{g}_c)}{\|\boldsymbol{p}_i - \boldsymbol{g}_c\|^\rho} \right\|^2 \tag{3.139}$$

If the magnitude of $\boldsymbol{r}_c$ is large to some extent, this type of asymmetry should be called "rotationally asymmetric."

## Divergence

Divergence characterizes another aspect of a cluster, which grows bigger in size and thinner in density otherwise shrinks in size and becomes thicker in density.

The general divergence model, like rotation, is formulated with a scalar constant $d_c$ in such a way that

$$\boldsymbol{v}_i' \frac{\boldsymbol{p}_i - \boldsymbol{g}_c}{\|\boldsymbol{p}_i - \boldsymbol{g}_c\|^\rho} = d_c + \varepsilon_i \qquad (i = 1, 2, \ldots, n) \tag{3.140}$$

where $\rho$ is a specified constant and $\varepsilon_i$ is a scalar noise.

For $\rho = 0$, the model is called "Constant source/sink model." On the other hand, setting $\rho = 1$ in (3.140) gives "Stable model," where little motion will arise in the neighborhood of the center when the noise is negligible.

To avoid singularity at the center, it is required that $\rho > 1$. In practice however, setting $\rho = 1$ may not cause substantial problems, and usually it stabilizes the numerical computation for estimating $d_c$. Then the least squares estimate of $d_c$, for which

$$\sum_{i \in C} \{\boldsymbol{v}_i' \lambda(\boldsymbol{p}_i - \boldsymbol{g}_c) - d_c\}^2 \tag{3.141}$$

attains its minimum, is given by

$$\hat{d}_c = \frac{1}{n(C)} \sum_{i \in C} \boldsymbol{v}_i' \Lambda(\boldsymbol{p}_i - \boldsymbol{g}_c) \tag{3.142}$$

If $\hat{d}_c$ is positive, the cluster expands away from the source point $\boldsymbol{g}_c$, while a negative value indicates shrinking towards $\boldsymbol{g}_c$. Note that if $\boldsymbol{g}_c$ is the center of the mass and $C = N$, the value of $\hat{d}_c$ nearly equals zero.

### 3.4.3. System Property in Terms of Dynamics

Let us consider that the system is divided into $m$ disjoint clusters, that is,

$$N = \bigcup_{j=1}^{m} C_j \tag{3.143}$$

$$C_j \cap C_k = \phi \qquad (j \neq k; j, k = 1, 2, \ldots, m) \tag{3.144}$$

$$C_j \neq \phi \qquad (j = 1, 2, \ldots, m) \tag{3.145}$$

For $m$ clusters with weights $n(C_1), n(C_2), \ldots, n(C_m)$, we regard the sets of centers $G = \{\boldsymbol{g}_j | j = 1, 2, \ldots, m\}$ as the set of position vectors $P$, and linear flows $L = \{\boldsymbol{l}_j | j = 1, 2, \ldots, m\}$ as asymmetric vectors $V$, respectively. Then we suggest intercluster indices, called "System properties" in terms of the cluster properties. When $m = n$ with $C_i = \{i\}$ $(i = 1, 2, \ldots, n)$, the system properties are called "plain system properties." When $m < n$, these should be called "clustered system properties," if desired.

For a clustered system, system linear flow $\boldsymbol{l}^*$, system rotation $\boldsymbol{r}^*$, and system divergence $d^*$ are defined as:

$$\boldsymbol{l}^* = \frac{1}{m} \sum_{j=1}^{m} n(C_j) \boldsymbol{l}_j \tag{3.146}$$

$$\boldsymbol{r}^* = \frac{1}{m} \sum_{j=1}^{m} n(C_j) \boldsymbol{l}_j \times \Lambda(\boldsymbol{g}_j - \boldsymbol{g}) \tag{3.147}$$

$$d^* = \frac{1}{m} \sum_{j=1}^{m} n(C_j) \boldsymbol{l}_j' \Lambda(\boldsymbol{g}_j - \boldsymbol{g}) \tag{3.148}$$

Here $\boldsymbol{g}$ is the center of the system, which is arbitrarily given, or typically chosen as

$$\boldsymbol{g} = \frac{1}{n} \sum_{i=1}^{n} \boldsymbol{p}_i \tag{3.149}$$

### 3.4.4. Numerical Example

Let $\boldsymbol{O}_1$, $\boldsymbol{O}_2$, and $\boldsymbol{O}_3$ be dissimilarity data matrices. Each matrix is decomposed by (3.3) with respective skew-symmetric component as given in the second term in each of the equations:

$$\boldsymbol{O}_1 = \begin{pmatrix} 0 & 2 & 1+2\sqrt{2} & 2 \\ 2 & 0 & 2 & 1+2\sqrt{2} \\ -1+2\sqrt{2} & 2 & 0 & 2 \\ 2 & =1+2\sqrt{2} & 2 & 0 \end{pmatrix}$$

$$= \begin{pmatrix} 0 & 2 & 2\sqrt{2} & 2 \\ 2 & 0 & 2 & 2\sqrt{2} \\ 2\sqrt{2} & 2 & 0 & 2 \\ 2 & 2\sqrt{2} & 2 & 0 \end{pmatrix} + \begin{pmatrix} 0 & 0 & 1 & 0 \\ 0 & 0 & 0 & 1 \\ -1 & 0 & 0 & 0 \\ 0 & -1 & 0 & 0 \end{pmatrix} \tag{3.150}$$

$$\boldsymbol{O}_2 = \begin{pmatrix} 0 & 1 & 2\sqrt{2} & 3 \\ 3 & 0 & 1 & 2\sqrt{2} \\ 2\sqrt{2} & 3 & 0 & 1 \\ 1 & 2\sqrt{2} & 3 & 0 \end{pmatrix}$$

$$= \begin{pmatrix} 0 & 2 & 2\sqrt{2} & 2 \\ 2 & 0 & 2 & 2\sqrt{2} \\ 2\sqrt{2} & 2 & 0 & 2 \\ 2 & 2\sqrt{2} & 2 & 0 \end{pmatrix} + \begin{pmatrix} 0 & -1 & 0 & 1 \\ 1 & 0 & -1 & 0 \\ 0 & 1 & 0 & -1 \\ -1 & 0 & 1 & 0 \end{pmatrix} \tag{3.151}$$

$$\boldsymbol{O}_3 = \begin{pmatrix} 0 & 0 & \sqrt{2} & 2 & 1+\sqrt{2} & 1+\sqrt{13} & 1+\sqrt{13} \\ 2 & 0 & 0 & \sqrt{2} & \sqrt{13} & 2\sqrt{5} & 3\sqrt{2} \\ \sqrt{2} & 2 & 0 & 0 & 3\sqrt{2} & 5 & 5 \\ 0 & \sqrt{2} & 2 & 0 & \sqrt{13} & 3\sqrt{2} & 2\sqrt{5} \\ -1+2\sqrt{2} & \sqrt{13} & 3\sqrt{2} & \sqrt{13} & 0 & 1 & 1 \\ -1+2\sqrt{13} & 2\sqrt{5} & 5 & 3\sqrt{2} & 1 & 0 & \sqrt{2} \\ -1+\sqrt{13} & 3\sqrt{2} & 5 & 2\sqrt{5} & 1 & \sqrt{2} & 0 \end{pmatrix}$$

$$
= \begin{pmatrix}
0 & 1 & \sqrt{2} & 1 & 2\sqrt{2} & \sqrt{13} & \sqrt{13} \\
1 & 0 & 0 & \sqrt{2} & \sqrt{13} & 2\sqrt{5} & 3\sqrt{2} \\
\sqrt{2} & 1 & 0 & 1 & 3\sqrt{2} & 5 & 5 \\
1 & \sqrt{2} & 1 & 0 & \sqrt{13} & 3\sqrt{2} & 2\sqrt{5} \\
2\sqrt{2} & \sqrt{13} & 3\sqrt{2} & \sqrt{13} & 0 & 1 & 1 \\
\sqrt{13} & 2\sqrt{5} & 5 & 3\sqrt{2} & 1 & 0 & \sqrt{2} \\
\sqrt{13} & 3\sqrt{2} & 5 & 2\sqrt{5} & 1 & \sqrt{2} & 0
\end{pmatrix}
+ \begin{pmatrix}
0 & -1 & 0 & 1 & 1 & 1 & 1 \\
1 & 0 & -1 & 0 & 0 & 0 & 0 \\
0 & 1 & 0 & -1 & 0 & 0 & 0 \\
-1 & 0 & 1 & 0 & 0 & 0 & 0 \\
-1 & 0 & 0 & 0 & 0 & 0 & 0 \\
-1 & 0 & 0 & 0 & 0 & 0 & 0 \\
-1 & 0 & 0 & 0 & 0 & 0 & 0
\end{pmatrix} \qquad (3.152)
$$

Applying the model to $\boldsymbol{O}_1$, $\boldsymbol{O}_2$, and $\boldsymbol{O}_3$, we obtain the results in Figs. 3.12(a), (b), and (c), respectively. In the figures, the dots indicate the object locations $P$ and the arrows the vectors $V$. In Figs. 3.12(a) and (b), the locations of the objects show a square shape configuration commonly, but the tendency of asymmetry looks quite different. Figure 3.12(a) reveals that two objects direct equally towards the remaining two objects, which look like moving away from the other two objects; Fig. 3.12(b) shows a circular structure with asymmetries of equal strength. In Fig. 3.12(c) we find a more complicated feature where there are two clusters with different structures of asymmetry. The first cluster, which contains objects $1, 2, 3$, and 4, indicates rotation in a counterclockwise direction. The second cluster contains the remaining objects, and indicates moving towards the first cluster.

The values of the three indices are provided in Table 3.8. Referring to cluster properties, one can interpret the results in view of numerical values in the table. Interpretation of the results is omitted for the present demonstration of our model.

## 3.4.5. Application

### World Trade Data

We analyzed international trade data. The source of Table 3.9 is based on the table of "World exports by commodities and by regions" in the *1992*

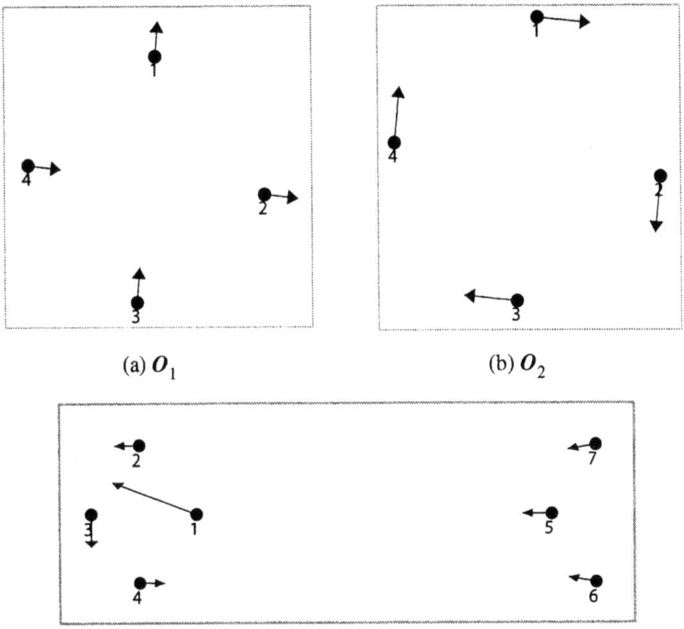


(a) $O_1$ (b) $O_2$

(c) $O_3$

**FIGURE 3.12** Two-dimensional representation of $O_1$, $O_2$, and $O_3$

*International Trade Statistics Yearbook*. Here we have chosen ten countries or regions from the original table. The definition of OPEC, EEC, EFTA, and LAIA are "Organization of Petroleum Exporting Countries," the "European Economic Community," the "European Free Trade Association," and the "Latin American Integration Association," respectively.

**TABLE 3.8** Cluster Properties in Terms of Dynamics

| | System (a) | System (b) | System (c) |
|---|---|---|---|
| Linear flow | (0.52, 0.43, 0.00) | (0.00, 0.00, 0.00) | (−0.38, −0.01, 0.00) |
| Rotation | (0.00, 0.00, 0.00) | (−0.79, 0.00, 0.00) | (0.04, 0.00, 0.00) |
| Divergence | 0.00 | 0.00 | −0.04 |
| | Cluster (c)−1 | Cluster (c)−2 | |
| Linear flow | (−0.33, −0.02, 0.00) | (−0.44, 0.00, 0.00) | |
| Rotation | (0.25, 0.00, 0.00) | (0.00, 0.00, 0.00) | |
| Divergence | −0.10 | −0.05 | |

**TABLE 3.9** World Exports by Commodities and by Regions in 1990

| | A | B | C | D | E |
|---|---|---|---|---|---|
| A: OPEC | 0 | 860 | 44586 | 2651 | 1413 |
| B: Former USSR | 946 | 0 | 32252 | 6300 | 111 |
| C: EEC | 43351 | 13640 | 0 | 138967 | 11832 |
| D: EFTA | 5330 | 5625 | 130477 | 0 | 2806 |
| E: Canada | 1630 | 964 | 10269 | 1927 | 0 |
| F: USA | 13418 | 3072 | 93049 | 10666 | 78212 |
| G: Japan | 13575 | 2563 | 53846 | 8437 | 6726 |
| H: Australia & New Zealand | 2651 | 500 | 6816 | 1041 | 763 |
| I: Africa | 1666 | 1072 | 39200 | 917 | 705 |
| J: LAIA | 3685 | 844 | 24378 | 1433 | 1083 |

| | F | G | H | I | J |
|---|---|---|---|---|---|
| A: OPEC | 23211 | 31861 | 1348 | 5823 | 1350 |
| B: Former USSR | 1189 | 3114 | 62 | 1515 | 305 |
| C: EEC | 95908 | 28486 | 10149 | 42727 | 16942 |
| D: EFTA | 15310 | 5954 | 2078 | 3203 | 3008 |
| E: Canada | 95217 | 7038 | 909 | 903 | 1675 |
| F: USA | 0 | 46130 | 9406 | 6068 | 41969 |
| G: Japan | 90893 | 0 | 8106 | 3835 | 5354 |
| H: Australia & New Zealand | 5483 | 11877 | 0 | 600 | 610 |
| I: Africa | 12278 | 1444 | 60 | 0 | 458 |
| J: LAIA | 36211 | 6782 | 607 | 1605 | 0 |

OPEC, Organization of Petroleum Exporting Countries; EEC, European Economic Community; EFTA, European Free Trade Association; LAIA, Latin American Integration Association.

Figures 3.13 and 3.14 show two- and three-dimensional representations of the object locations $P$ and the vectors $V$, respectively. In addition to information about the object locations, we can examine the relationship among countries in terms of arrows. For example, it is noted that Japan and the United States have relatively large asymmetry to other countries and regions, and also that the largest asymmetry directs from Japan to the United States.

## 3.5. VECTOR FIELD MODEL

In this section, we describe a vector field model proposed by Yadohisa and Niki (1999). As in the case of the preceding section, we start with an asymmetric

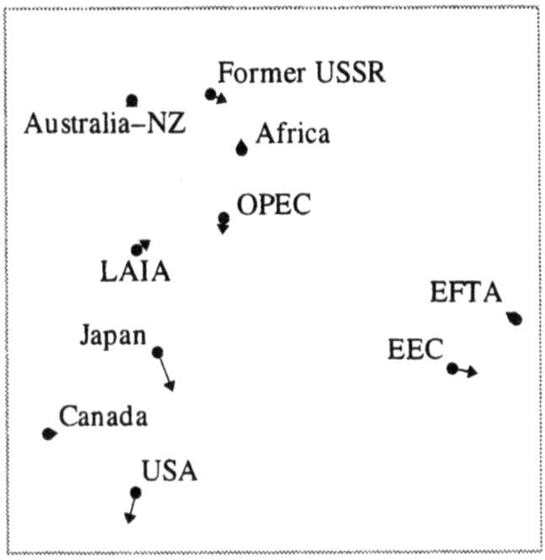


**FIGURE 3.13** Two-dimensional representation of the world trade data

dissimilarity data matrix $\boldsymbol{O}$ with diagonal elements all being zero, and its decomposition as (3.3). We assume that a set $P = \{\boldsymbol{p}_i | i = 1, 2, \ldots, n\}$ of the location coordinates of $n$ objects and a set, $V = \{\boldsymbol{v}_i | i = 1, 2, \ldots, n\}$ of the vectors at $P$ have been derived from $\boldsymbol{S}$ and $\boldsymbol{A}$, respectively, according to the vector model.

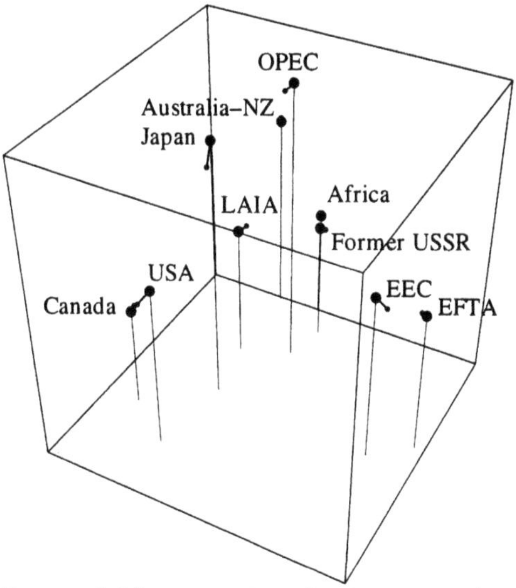


**FIGURE 3.14** Three-dimensional representation of the world trade data

### 3.5.1. Scalar Potential

Let us consider a vector field $\boldsymbol{U}$ in terms of $P$ with scalar potential $h$,

$$\boldsymbol{U}(\boldsymbol{p}_i) = \text{grad } h(\boldsymbol{p}_i) \tag{3.153}$$

We present a model to account for $\boldsymbol{v}_i$ in the vector field such that

$$\begin{aligned} \boldsymbol{v}_i &= -\boldsymbol{U}(\boldsymbol{p}_i) + \boldsymbol{e}_i & (3.154)\\ &= -\text{grad } h(\boldsymbol{p}_i) + \boldsymbol{e}_i & (3.155)\end{aligned}$$

where $\boldsymbol{e}_i$ is a noise vector.

Given $P$ and $V$, we aim to determine the scalar potential $h$. For this purpose, we propose a method of estimation so that the values of $h$ for lattice points are determined to satisfy a kind of *smoothness.*

We begin with an assumption, for simplicity, that the vectors $\boldsymbol{v}_i$ and $\boldsymbol{p}_i$ are two-dimensional. Let $\Delta_1$ and $\Delta_2$ denote the difference operators, such that

$$\begin{aligned} &\Delta_1 z(x, y) = z(x + c, y) - z(x, y) \qquad \text{and}\\ &\Delta_2 z(x, y) = z(x, y + c) - z(x, y) \end{aligned} \tag{3.156}$$

which are defined on a bounded lattice $L$ with unit interval $c$. On the same lattice $L$, we define the operator $\Delta$ as

$$\Delta z(x, y) = (\tfrac{1}{2}\{\Delta_1 z(x, y) + \Delta_1 z(x, y + c)\}, \tfrac{1}{2}\{\Delta_2 z(x, y) + \Delta_2 z(x + c, y)\}) \tag{3.157}$$

and also operator $\Upsilon z(x, y)$ as

$$\Upsilon z(x, y) = (\Delta_1^2 z(x, y), \Delta_2^2 z(x, y)) \tag{3.158}$$

We approximate grad $h$ by $\Delta h(x, y)/c$. First, we define

$$\delta_i(x, y) = \begin{cases} 1, & \text{if } \boldsymbol{p}_i \in [x, x + c) \times [y, y + c)\\ 0, & \text{otherwise} \end{cases} \tag{3.159}$$

Then we have

$$\text{grad } h(\boldsymbol{p}_i) \approx \sum_{(x,y)\in L} \delta_i(x, y)\frac{1}{c}\Delta h(x, y) \tag{3.160}$$

From (3.155), we estimate $h(x, y)$ for every lattice point $(x, y) \in L$ by minimizing

$$Q = \sum_{i}^{n} \sum_{(x,y)\in L} \delta_i(x, y) \left\| \boldsymbol{v}_i + \frac{1}{c} \Delta h(x, y) \right\|^2 + \frac{u^2}{c^2} \sum_{(x,y)\in L} \|\Upsilon h(x, y)\|^2 \tag{3.161}$$

where $u$ is a parameter to control how *smooth* the function $h$ should be in the sense that neighboring differences are nearly equal to each other.

It is allowed to consider that $Q$ is distributed as

$$\frac{1}{\sqrt{2\pi}} \exp\left(-\frac{1}{2\sigma^2} Q\right) \propto \prod_{\substack{(x,y)\in L \\ \delta_i(x,y)=1}} \frac{1}{\sqrt{2\pi}} \exp\left(-\frac{1}{2\sigma^2} \left\| \boldsymbol{v}_i + \frac{1}{c} \Delta h(x, y) \right\|^2\right) \cdot K \exp\left(-\frac{1}{2\sigma^2} \frac{u^2}{c^2} \sum_{(x,y)\in L} \|\Upsilon h(x, y)\|^2\right) \tag{3.162}$$

with some constant $K$. We regard the noise vectors $\{\boldsymbol{e}_i\}$ as random vectors, which should satisfy

$$E(\boldsymbol{e}_i) = 0 \qquad \text{and} \qquad E(\boldsymbol{e}_i \boldsymbol{e}_j') = \sigma^2 \boldsymbol{I} \qquad (\boldsymbol{I}\text{: identity matrix}) \tag{3.163}$$

The first term in the right-hand side of (3.162) is regarded as the joint density of $\boldsymbol{v}_i$ when the parameters $h(x, y)$ are given. The second term is interpreted as the prior distribution of the same parameters. In this Bayesian point of view (Akaike, 1980), minimizing $Q$ in (3.161) is equivalent to maximizing the *posterior* likelihood of the parameters. The parameter $u$ is often called *hyperparameter*, because it determines the *prior* distribution.

Without loss of generality, we assume that $c = 1$ and that

$$\boldsymbol{p}_i \in [0, m_x - 1) \times [0, m_y - 1) \; (i = 1, 2, \ldots, n; m_x, m_y > 3) \tag{3.164}$$

For the lattice $L = \{(j, k) | j = 0, 1, \ldots, m_x - 1; k = 0, 1, \ldots, m_y - 1\}$, we write the column vector $\boldsymbol{\theta}$ of parameters as

$$\boldsymbol{\theta} = (\theta_{jm_x+k}) = (h(j, k)) \qquad (0 \le j < m_x, 0 \le k < m_y) \tag{3.165}$$

and set the following natural boundary conditions:

$$\begin{aligned} \Delta_1^2 h(j, k) &= 0, \quad \text{if} \quad j \le 0 \quad \text{or} \quad j \ge m_x - 1 \\ \Delta_2^2 h(j, k) &= 0, \quad \text{if} \quad k \le 0 \quad \text{or} \quad k \ge m_y - 1 \end{aligned} \tag{3.166}$$

Differentiating $Q$ with respect to $\boldsymbol{\theta}$ and setting the derivative to zero, we have a system of linear equations such that

$$(\boldsymbol{H}_1 + u^2\boldsymbol{H}_2)\boldsymbol{\theta} = \boldsymbol{\xi} \tag{3.167}$$

Here

$$\boldsymbol{\xi} = (\xi_{jm_x+k}), \qquad 0 \le j < m_x, 0 \le k < m_y \tag{3.168}$$

$$\begin{aligned}\xi_{jm_x+k} &= \sum_i a_i^{(1)}\{\delta_i(j,k) - \delta_i(j-1,k) - \delta_i(j-1,k-1) + \delta_i(j,k-1)\}\\ &\quad + \sum_i a_i^{(2)}\{\delta_i(j,k) - \delta_i(j,k-1)\\ &\quad - \delta_i(j-1,k-1) + \delta_i(j-1,k)\}\end{aligned} \tag{3.169}$$

Matrices $\boldsymbol{H}_1$ and $\boldsymbol{H}_2$ are defined as follows. First, we write

$$\eta_{jk} = \begin{cases} \sum_i \delta_i(j,k), & \text{if } j \in [0, m_x) \text{ and } k \in [0, m_y) \\ 0, & \text{otherwise} \end{cases} \tag{3.170}$$

Provide the following $m_x \times m_y$ matrices $\boldsymbol{B}_{j0}$, $\boldsymbol{B}_{j1}$, $\boldsymbol{B}_{j2}$, $\boldsymbol{C}_0$, $\boldsymbol{C}_1$, $\boldsymbol{C}_2$:

$$\boldsymbol{B}_{j0} = \mathrm{diag}(\eta_{jk} + \eta_{(j-1)k} + \eta_{j(k-1)} + \eta_{(j-1)(k-1)}) \tag{3.171}$$

$$\boldsymbol{B}_{j1} = \begin{pmatrix} 0 & & & \boldsymbol{O} \\ \eta_{(j-1)0} & 0 & & \\ & \ddots & \ddots & \\ \boldsymbol{O} & & \eta_{(j-1)(m_y-2)} & 0 \end{pmatrix} \tag{3.172}$$

$$\boldsymbol{B}_{j2} = \begin{pmatrix} 0 & \eta_{j0} & & \boldsymbol{O} \\ & \ddots & \ddots & \\ & & 0 & \eta_{j(m_y-2)} \\ \boldsymbol{O} & & & 0 \end{pmatrix} \tag{3.173}$$

$$\boldsymbol{C}_0 = \begin{pmatrix} 2 & -2 & 1 & & & & \boldsymbol{O} \\ -1 & 4 & -3 & 1 & & & \\ 1 & -3 & 5 & -3 & 1 & & \\ & \ddots & \ddots & \ddots & \ddots & \ddots & \\ & & 1 & -3 & 5 & -3 & 1 \\ & & & 1 & -3 & 4 & -1 \\ \boldsymbol{O} & & & & 1 & -2 & 2 \end{pmatrix}, \tag{3.174}$$

$$\boldsymbol{C}_1 = \boldsymbol{C}_0 + 2\boldsymbol{I}, \qquad \boldsymbol{C}_2 = \boldsymbol{C}_0 + 3\boldsymbol{I} \tag{3.175}$$

Next, we define $m_x m_y \times m_x m_y$ matrices:

$$
H_1 = \begin{pmatrix}
B_{00} & -B_{02} & & & & O \\
-B_{11} & B_{10} & -B_{12} & & & \\
 & \ddots & \ddots & \ddots & & \\
 & & & -B_{(m_x-2)1} & B_{(m_x-2)0} & -B_{(m_x-2)2} \\
O & & & & -B_{(m_x-1)1} & B_{(m_x-1)0}
\end{pmatrix} \tag{3.176}
$$

$$
H_2 = \begin{pmatrix}
C_0 & -2I & I & & & & O \\
-I & C_1 & -3I & I & & & \\
I & -3I & C_2 & -3I & I & & \\
 & \ddots & \ddots & \ddots & \ddots & \ddots & \\
 & & I & -3I & C_2 & -3I & I \\
 & & & I & -3I & C_1 & -I \\
O & & & & I & -2I & C_0
\end{pmatrix} \tag{3.177}
$$

Note that the rank of the coefficient matrix in (3.167) is $m_x m_y - 1$. We impose the following condition to (3.167) in order to fix the average level of $h(x, y)$,

$$
(1 \ \ 1 \cdots 1)\boldsymbol{\theta} = 0 \tag{3.178}
$$

Given a value for the hyperparameter $u$, we can solve the system of linear equations, (3.167) and (3.178), under condition (3.166). Following Akaike (1980), we compute a set of estimates for the scalar potential $h$ for several values of $u$ and then select the one to minimize the *Akaike's Bayesian Information Criterion* (ABIC)

$$
\mathrm{ABIC} = -2\log\int \frac{1}{\sqrt{2\pi}} \exp\left(-\frac{1}{2\sigma^2}Q\right)\, d\theta \tag{3.179}
$$

### 3.5.2. Numerical Example

Given an asymmetric dissimilarity matrix $\boldsymbol{O}$, we performed decomposition (3.3), which is shown below.

$$\boldsymbol{O} = \boldsymbol{S} + \boldsymbol{T} = \begin{pmatrix} 0 & 1 & 2 & 1 & \sqrt{2} & \sqrt{5} & 2 & \sqrt{5} & \sqrt{2} \\ 1 & 0 & 1 & \sqrt{2} & 1 & \sqrt{2} & \sqrt{5} & 2 & \sqrt{5} \\ 2 & 1 & 0 & \sqrt{5} & \sqrt{2} & 1 & 2\sqrt{2} & \sqrt{5} & 2 \\ 1 & \sqrt{2} & \sqrt{5} & 0 & 1 & 2 & 1 & \sqrt{2} & \sqrt{5} \\ \sqrt{2} & 1 & \sqrt{2} & 1 & 0 & 1 & \sqrt{2} & 1 & \sqrt{2} \\ \sqrt{5} & \sqrt{2} & 1 & 2 & 1 & 0 & \sqrt{5} & \sqrt{2} & 1 \\ 2 & \sqrt{5} & 2\sqrt{2} & 1 & \sqrt{2} & \sqrt{5} & 0 & 1 & 2 \\ \sqrt{5} & 2 & \sqrt{5} & \sqrt{2} & 1 & \sqrt{2} & 1 & 0 & 1 \\ 2\sqrt{2} & \sqrt{5} & 2 & \sqrt{5} & \sqrt{2} & 1 & 2 & 1 & 0 \end{pmatrix}$$

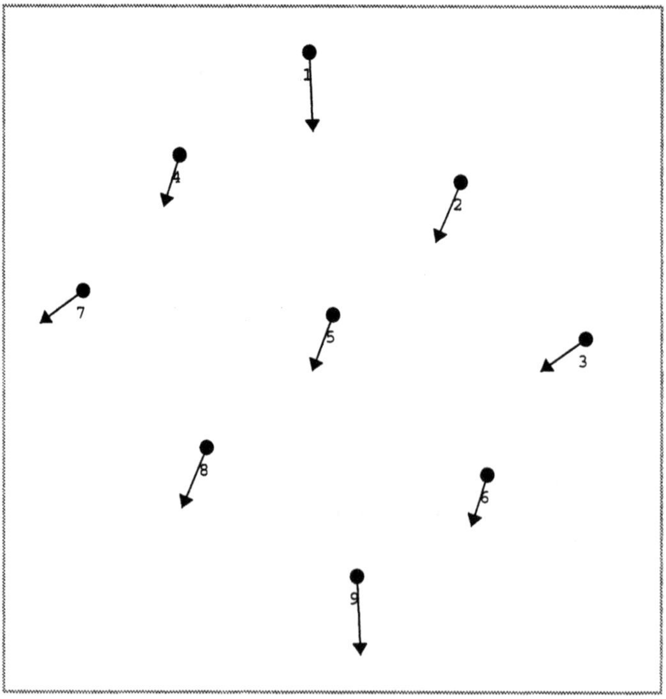


**FIGURE 3.15** Two-dimensional configuration of the nine objects

$$+\frac{1}{10}\begin{pmatrix} 0 & 1 & 1 & 1 & 1 & 1 & 1 & 1 & 1 \\ -1 & 0 & 1 & 1 & 1 & 1 & 1 & 1 & 1 \\ -1 & -1 & 0 & 1 & 1 & 1 & 1 & 1 & 1 \\ -1 & -1 & -1 & 0 & 1 & 1 & 1 & 1 & 1 \\ -1 & -1 & -1 & -1 & 0 & 1 & 1 & 1 & 1 \\ -1 & -1 & -1 & -1 & -1 & 0 & 1 & 1 & 1 \\ -1 & -1 & -1 & -1 & -1 & -1 & 0 & 1 & 1 \\ -1 & -1 & -1 & -1 & -1 & -1 & -1 & 0 & 1 \\ -1 & -1 & -1 & -1 & -1 & -1 & -1 & -1 & 0 \end{pmatrix} \tag{3.180}$$

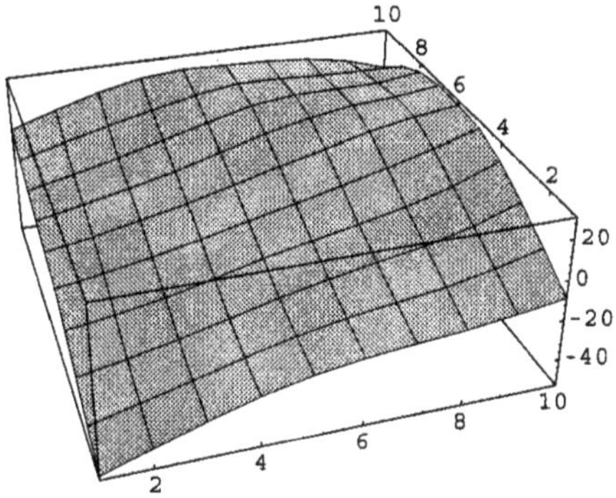


**FIGURE 3.16** Scalar potential of the configuration

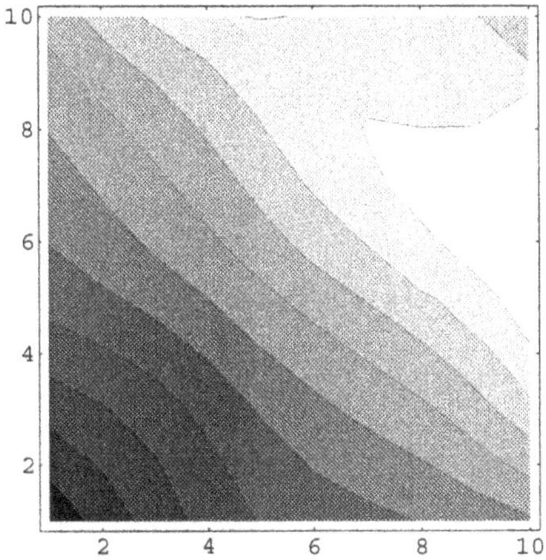


**FIGURE 3.17** Contour map of the scalar potential

**TABLE 3.10** Morse Code Confusion Data Collected by Rothkopf (1957) (Part 1)

| | A | B | C | D | E | F | G | H | I | J | K | L | M |
|---|---|---|---|---|---|---|---|---|---|---|---|---|---|
| A | 92 | 4 | 6 | 13 | 3 | 14 | 10 | 13 | 46 | 5 | 22 | 3 | 25 |
| B | 5 | 84 | 37 | 31 | 5 | 28 | 17 | 21 | 5 | 19 | 34 | 40 | 6 |
| C | 4 | 38 | 87 | 17 | 4 | 29 | 13 | 7 | 11 | 19 | 24 | 35 | 14 |
| D | 8 | 62 | 17 | 88 | 7 | 23 | 40 | 36 | 9 | 13 | 81 | 56 | 8 |
| E | 6 | 13 | 14 | 6 | 97 | 2 | 4 | 4 | 17 | 1 | 5 | 6 | 4 |
| F | 4 | 51 | 33 | 19 | 2 | 90 | 10 | 29 | 5 | 33 | 16 | 50 | 7 |
| G | 9 | 18 | 27 | 38 | 1 | 14 | 90 | 6 | 5 | 22 | 33 | 16 | 14 |
| H | 3 | 45 | 23 | 25 | 9 | 32 | 8 | 87 | 10 | 10 | 9 | 29 | 5 |
| I | 64 | 7 | 7 | 13 | 10 | 8 | 6 | 12 | 93 | 3 | 5 | 16 | 13 |
| J | 7 | 9 | 38 | 9 | 2 | 24 | 18 | 5 | 4 | 85 | 22 | 31 | 8 |
| K | 5 | 24 | 38 | 73 | 1 | 17 | 25 | 11 | 5 | 27 | 91 | 33 | 10 |
| L | 2 | 69 | 43 | 45 | 10 | 24 | 12 | 26 | 9 | 30 | 27 | 86 | 6 |
| M | 24 | 12 | 5 | 14 | 7 | 17 | 29 | 8 | 8 | 11 | 23 | 8 | 96 |
| N | 31 | 4 | 13 | 30 | 8 | 12 | 10 | 16 | 13 | 3 | 16 | 8 | 59 |
| O | 7 | 7 | 20 | 6 | 5 | 9 | 76 | 7 | 2 | 39 | 26 | 10 | 4 |
| P | 5 | 22 | 33 | 12 | 5 | 36 | 22 | 12 | 3 | 78 | 14 | 46 | 5 |
| Q | 8 | 20 | 38 | 11 | 4 | 15 | 10 | 5 | 2 | 27 | 23 | 26 | 7 |
| R | 13 | 14 | 16 | 23 | 5 | 34 | 26 | 15 | 7 | 12 | 21 | 37 | 14 |
| S | 17 | 24 | 5 | 30 | 11 | 26 | 5 | 59 | 16 | 3 | 13 | 10 | 5 |
| T | 13 | 10 | 1 | 5 | 46 | 3 | 6 | 6 | 14 | 6 | 14 | 7 | 6 |
| U | 14 | 29 | 12 | 32 | 4 | 32 | 11 | 34 | 21 | 7 | 44 | 32 | 11 |
| V | 5 | 17 | 24 | 16 | 9 | 29 | 6 | 39 | 5 | 11 | 26 | 43 | 4 |
| W | 9 | 21 | 30 | 22 | 9 | 36 | 25 | 15 | 4 | 25 | 29 | 18 | 15 |
| X | 7 | 64 | 45 | 19 | 3 | 28 | 11 | 6 | 1 | 35 | 50 | 42 | 10 |
| Y | 9 | 23 | 62 | 15 | 4 | 26 | 22 | 9 | 1 | 30 | 12 | 14 | 5 |
| Z | 3 | 46 | 45 | 18 | 2 | 22 | 17 | 10 | 7 | 23 | 21 | 51 | 11 |
| 1 | 2 | 5 | 10 | 3 | 3 | 5 | 13 | 4 | 2 | 29 | 5 | 14 | 9 |
| 2 | 7 | 14 | 22 | 5 | 4 | 20 | 13 | 3 | 25 | 26 | 9 | 14 | 2 |
| 3 | 3 | 8 | 21 | 5 | 4 | 32 | 6 | 12 | 2 | 23 | 6 | 13 | 5 |
| 4 | 6 | 19 | 19 | 12 | 6 | 25 | 14 | 16 | 7 | 21 | 13 | 19 | 3 |
| 5 | 8 | 45 | 15 | 14 | 2 | 45 | 4 | 67 | 7 | 14 | 4 | 41 | 2 |
| 6 | 7 | 80 | 30 | 17 | 4 | 23 | 4 | 14 | 2 | 11 | 11 | 27 | 6 |
| 7 | 6 | 33 | 22 | 14 | 5 | 25 | 6 | 4 | 6 | 24 | 13 | 32 | 7 |
| 8 | 3 | 23 | 40 | 6 | 3 | 15 | 15 | 6 | 2 | 33 | 10 | 14 | 3 |
| 9 | 3 | 14 | 23 | 3 | 1 | 6 | 14 | 5 | 2 | 30 | 6 | 7 | 16 |
| 0 | 9 | 3 | 11 | 2 | 5 | 7 | 14 | 4 | 5 | 30 | 8 | 3 | 2 |

**TABLE 3.11** Morse Code Confusion Data Collected by Rothkopf (1957) (Part 2)

| | N | O | P | Q | R | S | T | U | V | W | X | Y | Z |
|---|---|---|---|---|---|---|---|---|---|---|---|---|---|
| A | 34 | 6 | 6 | 9 | 35 | 23 | 6 | 37 | 13 | 17 | 12 | 7 | 3 |
| B | 10 | 12 | 22 | 25 | 16 | 18 | 2 | 18 | 34 | 8 | 84 | 30 | 42 |
| C | 3 | 9 | 51 | 34 | 24 | 14 | 6 | 6 | 11 | 14 | 32 | 82 | 38 |
| D | 7 | 9 | 27 | 9 | 45 | 29 | 6 | 17 | 20 | 27 | 40 | 15 | 33 |
| E | 4 | 5 | 1 | 5 | 10 | 7 | 67 | 3 | 3 | 2 | 5 | 6 | 5 |
| F | 6 | 10 | 42 | 12 | 35 | 14 | 2 | 21 | 27 | 25 | 19 | 27 | 13 |
| G | 13 | 82 | 52 | 23 | 21 | 5 | 3 | 15 | 14 | 32 | 21 | 23 | 39 |
| H | 8 | 8 | 14 | 8 | 17 | 37 | 4 | 36 | 59 | 9 | 33 | 14 | 11 |
| I | 30 | 7 | 3 | 5 | 19 | 35 | 16 | 10 | 5 | 8 | 2 | 5 | 7 |
| J | 3 | 21 | 63 | 47 | 11 | 2 | 7 | 9 | 9 | 9 | 22 | 32 | 28 |
| K | 12 | 31 | 14 | 31 | 22 | 2 | 2 | 23 | 17 | 33 | 63 | 16 | 18 |
| L | 2 | 9 | 37 | 36 | 28 | 12 | 5 | 16 | 19 | 20 | 31 | 25 | 59 |
| M | 62 | 11 | 10 | 15 | 20 | 7 | 9 | 13 | 4 | 21 | 9 | 18 | 8 |
| N | 93 | 5 | 9 | 5 | 28 | 12 | 10 | 16 | 4 | 12 | 4 | 6 | 11 |
| O | 8 | 86 | 37 | 35 | 10 | 3 | 4 | 11 | 14 | 25 | 35 | 27 | 27 |
| P | 6 | 21 | 83 | 43 | 23 | 9 | 4 | 12 | 19 | 19 | 19 | 41 | 30 |
| Q | 6 | 22 | 51 | 91 | 11 | 2 | 3 | 6 | 14 | 12 | 37 | 50 | 63 |
| R | 12 | 12 | 29 | 8 | 87 | 16 | 2 | 23 | 23 | 62 | 14 | 12 | 13 |
| S | 17 | 6 | 6 | 3 | 18 | 96 | 9 | 56 | 24 | 12 | 10 | 6 | 7 |
| T | 5 | 6 | 11 | 4 | 4 | 7 | 96 | 8 | 5 | 4 | 2 | 2 | 6 |
| U | 13 | 6 | 20 | 12 | 40 | 51 | 6 | 93 | 57 | 34 | 17 | 9 | 11 |
| V | 1 | 9 | 17 | 10 | 17 | 11 | 6 | 32 | 92 | 17 | 57 | 35 | 10 |
| W | 6 | 26 | 20 | 25 | 61 | 12 | 4 | 19 | 20 | 86 | 22 | 25 | 22 |
| X | 8 | 24 | 32 | 61 | 10 | 12 | 3 | 12 | 17 | 21 | 91 | 48 | 26 |
| Y | 6 | 14 | 30 | 52 | 5 | 7 | 4 | 6 | 13 | 21 | 44 | 86 | 23 |
| Z | 2 | 15 | 59 | 72 | 14 | 4 | 3 | 9 | 11 | 12 | 36 | 42 | 87 |
| 1 | 7 | 14 | 30 | 28 | 9 | 4 | 2 | 3 | 12 | 14 | 17 | 19 | 22 |
| 2 | 3 | 17 | 37 | 28 | 6 | 5 | 3 | 6 | 10 | 11 | 17 | 30 | 13 |
| 3 | 2 | 5 | 37 | 19 | 9 | 7 | 6 | 4 | 16 | 6 | 22 | 25 | 12 |
| 4 | 3 | 2 | 17 | 29 | 11 | 9 | 3 | 17 | 55 | 8 | 37 | 24 | 3 |
| 5 | 0 | 4 | 13 | 7 | 9 | 27 | 2 | 14 | 45 | 7 | 45 | 10 | 10 |
| 6 | 2 | 7 | 16 | 30 | 11 | 14 | 3 | 12 | 30 | 9 | 58 | 38 | 39 |
| 7 | 6 | 7 | 36 | 39 | 12 | 6 | 2 | 3 | 13 | 9 | 30 | 30 | 50 |
| 8 | 6 | 14 | 12 | 45 | 2 | 6 | 4 | 6 | 7 | 5 | 24 | 35 | 50 |
| 9 | 11 | 10 | 31 | 32 | 5 | 6 | 7 | 6 | 3 | 8 | 11 | 21 | 24 |
| 0 | 3 | 25 | 21 | 29 | 2 | 3 | 4 | 5 | 3 | 2 | 12 | 15 | 20 |

**TABLE 3.12** Morse Code Confusion Data Collected by Rothkopf (1957) (Part 3)

| | 1 | 2 | 3 | 4 | 5 | 6 | 7 | 8 | 9 | 0 |
|---|---|---|---|---|---|---|---|---|---|---|
| A | 2 | 7 | 5 | 5 | 8 | 6 | 5 | 6 | 2 | 3 |
| B | 12 | 17 | 14 | 40 | 32 | 74 | 43 | 17 | 4 | 4 |
| C | 13 | 15 | 31 | 14 | 10 | 30 | 28 | 24 | 18 | 12 |
| D | 3 | 9 | 6 | 11 | 9 | 19 | 8 | 10 | 5 | 6 |
| E | 4 | 3 | 5 | 3 | 5 | 2 | 4 | 2 | 3 | 3 |
| F | 8 | 16 | 47 | 25 | 26 | 24 | 21 | 5 | 5 | 5 |
| G | 15 | 14 | 5 | 10 | 4 | 10 | 17 | 23 | 20 | 11 |
| H | 3 | 9 | 15 | 43 | 70 | 35 | 17 | 4 | 3 | 3 |
| I | 2 | 5 | 8 | 9 | 6 | 8 | 5 | 2 | 4 | 5 |
| J | 67 | 66 | 33 | 15 | 7 | 11 | 28 | 29 | 26 | 23 |
| K | 5 | 9 | 17 | 8 | 8 | 18 | 14 | 13 | 5 | 6 |
| L | 12 | 13 | 17 | 15 | 26 | 29 | 36 | 16 | 7 | 3 |
| M | 5 | 7 | 6 | 6 | 5 | 7 | 11 | 7 | 10 | 4 |
| N | 5 | 2 | 3 | 4 | 4 | 6 | 2 | 2 | 10 | 2 |
| O | 19 | 17 | 7 | 7 | 6 | 18 | 14 | 11 | 20 | 12 |
| P | 34 | 44 | 24 | 11 | 15 | 17 | 24 | 23 | 25 | 13 |
| Q | 34 | 32 | 17 | 12 | 9 | 27 | 40 | 58 | 37 | 24 |
| R | 7 | 10 | 13 | 4 | 7 | 12 | 7 | 9 | 1 | 2 |
| S | 8 | 2 | 2 | 15 | 28 | 9 | 5 | 5 | 5 | 2 |
| T | 5 | 5 | 3 | 3 | 3 | 8 | 7 | 6 | 14 | 6 |
| U | 6 | 6 | 16 | 34 | 10 | 9 | 9 | 7 | 4 | 3 |
| V | 10 | 14 | 28 | 79 | 44 | 36 | 25 | 10 | 1 | 5 |
| W | 10 | 22 | 19 | 16 | 5 | 9 | 11 | 6 | 3 | 7 |
| X | 12 | 20 | 24 | 27 | 16 | 57 | 29 | 16 | 17 | 6 |
| Y | 26 | 44 | 40 | 15 | 11 | 26 | 22 | 33 | 23 | 16 |
| Z | 16 | 21 | 27 | 9 | 10 | 25 | 66 | 47 | 15 | 15 |
| 1 | 84 | 63 | 13 | 8 | 10 | 8 | 19 | 32 | 57 | 55 |
| 2 | 62 | 89 | 54 | 20 | 5 | 14 | 20 | 21 | 16 | 11 |
| 3 | 18 | 64 | 86 | 31 | 23 | 41 | 16 | 17 | 8 | 10 |
| 4 | 5 | 26 | 44 | 89 | 42 | 44 | 32 | 10 | 3 | 3 |
| 5 | 14 | 10 | 30 | 69 | 90 | 42 | 24 | 10 | 6 | 5 |
| 6 | 15 | 14 | 26 | 24 | 27 | 86 | 69 | 14 | 5 | 14 |
| 7 | 22 | 29 | 18 | 15 | 12 | 61 | 85 | 70 | 20 | 13 |
| 8 | 42 | 29 | 16 | 16 | 9 | 30 | 60 | 89 | 61 | 26 |
| 9 | 57 | 39 | 9 | 12 | 4 | 11 | 42 | 56 | 91 | 78 |
| 0 | 50 | 26 | 9 | 11 | 5 | 22 | 17 | 52 | 81 | 94 |

This pattern of the skew-symmetric part stands for a typical case of the jet-stream model in Gower (1977).

Applying our method described in the former section, we obtained a two-dimensional configuration (Fig. 3.15), where we used a procedure of MDS (Torgerson, 1952) for analyzing the symmetric part of the data. Dots and arrows represent the locations of the objects and their asymmetric vectors, respectively. Figures 3.16 and 3.17, respectively, give the estimated scalar potential of the vector field and its contour map.

Figures 3.16 and 3.17 show a linear tendency of asymmetry from the upper right to the lower left. It is emphasized that we have no prior information about the existence of the "linear drift of wind," unlike Gower (1977).

### 3.5.3. Application

We applied the method to the confusion data collected by Rothkopf (1957). The original data represent the confusion rates among the 36 Morse codes

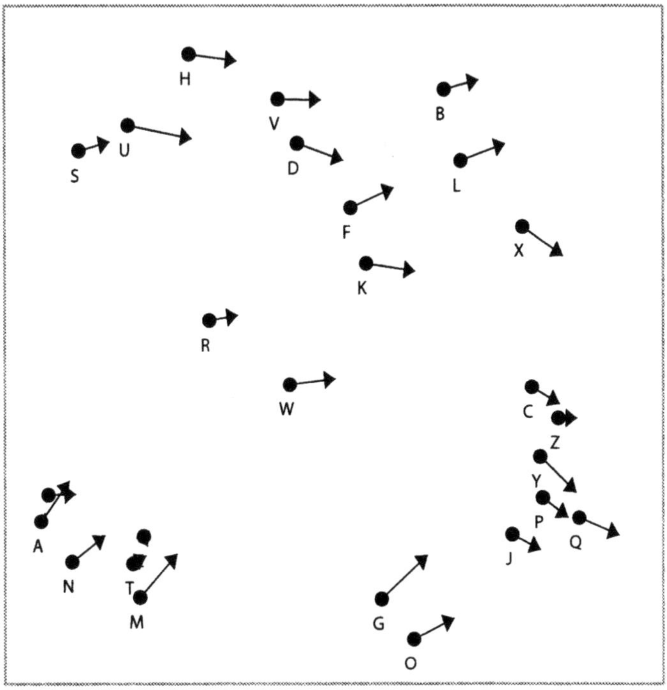


**FIGURE 3.18** Two-dimensional configuration of the confusion data

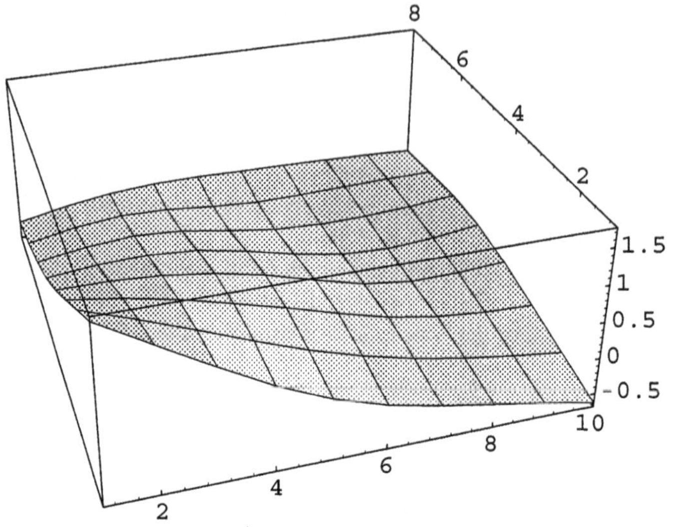


**FIGURE 3.19** Scalar potential of the configuration

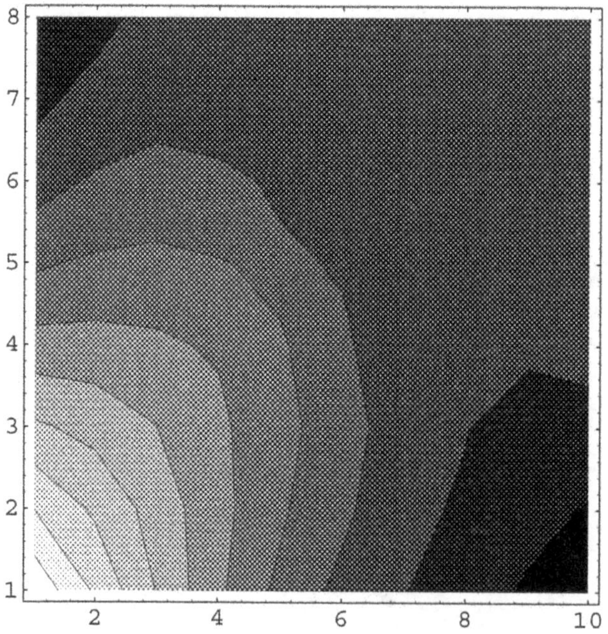


**FIGURE 3.20** Contour map of the scalar potential

(A to Z and 0 to 9) and have been analyzed by a number of researchers (e.g., Gower, 1977; Krumhansl, 1978; Saito and Takeda, 1990). For purposes of illustration, we used a portion of the data for the 26 alphabetic codes (A to Z), as shown in Tables 3.10 and 3.11.

After using Torgerson's procedure of MDS, we performed an analysis on the data and summarized the results in the three figures. Figure 3.18 shows the coordinates of the objects and the associated asymmetric vectors in a two-dimensional space. Figures 3.19 and 3.20 show the estimated scalar potential and the corresponding contour map, respectively.

Let us focus on the derived asymmetric structure. In view of the vector model, one may notice in Fig. 3.18 that there is a larger frequency of confusion from "M" to "W" than that from "W" to "M" without referring to the tables. In Figs. 3.19 and 3.20, we find the trend of asymmetry from the lower left side to the right side. Thus it is revealed that there is a stronger tendency of confusion from short codes (such as "A: ._," "N: _," and "M: _") to long codes (such as, "B: _... ," "H: ...." and "L: ._..") than the confusion in the opposite direction.

# 4

# Multidimensional Scaling of Asymmetric Data

## 4.1. OVERVIEW AND PRELIMINARIES

### 4.1.1. Asymmetry in the Subjects of Multidimensional Scaling

Multidimensional scaling (MDS) is a method of deriving a configuration of stimuli in a multidimensional space from a set of data, each entry of which represents a proximity measure between a pair of stimuli. Such a measure will include a variety of terms, as stated below. Many methods of symmetric MDS have been developed in terms of distance models or scalar product models, according to which proximity measure is assumed to be linearly or monotonically related to geometric distance (metric) or scalar product. As stated below at length, we are often faced with cases in which observed proximity may violate symmetry, one of the basic properties of distance or scalar product.

There are several approaches in MDS to cope with asymmetry. In the simplest way, we might ignore the asymmetry and fit a symmetric model to the lower triangular part and the upper triangular part of the data matrix separately. In the second treatment, the asymmetry is taken into account at data collection or at the preliminary analysis of the data and thereafter ignored.

For example in a psychological experiment, when spatial or temporal position of particular objects in the pair at (dis)similarity judgment is considered a source of bias to be eliminated, one will counterbalance the position by using an appropriate experimental design. Further, if one thinks in a preliminary analysis that the asymmetry occurs because of statistical fluctuation, one will construct a symmetric matrix from the original data matrix by averaging its $(j, k)$ and $(k, j)$ entries. Then one may apply symmetric MDS to the symmetrized data. However, when the asymmetry is prominent, these approaches may eliminate a significant asymmetric effect.

In the third approach, one will consider that such an effect may embody some important information of a systematic nature in particular situations, then the asymmetry is of interest in its own right. Among those, we set aside in this chapter the situation of PC judgment, which is assumed to be made on a unidimensional latent axis. Focusing on the asymmetry dealt with in MDS, we will distinguish, for descriptive purposes, two major types of data collection. In type (a), asymmetry is observed in cases of psychological judgments of proximity (similarity or dissimilarity, confusion, association, and so on). The judgments or any psychological tasks are assumed to be made with respect to more than one latent attribute of stimuli. In type (b), asymmetric proximity data are collected in a wide range of social phenomena at aggregated levels (social mobility, migration rates, brand switching and loyalty, pecking, journal citation, communication flows, and so on). The proximity is regarded as a composite measure in terms of more than one factor in connection with a particular phenomenon. Then, common to both types, it becomes a principal goal for data analysis to represent objects (or stimuli) in a latent space of multidimensionality.

Let us now consider such cases for which the data matrix is provided with a characteristic that row and column elements have different meanings or play different roles. For type (a), a typical example is confusion of one stimulus for another. This sort of data is collected in a stimulus identification experiment in which rows are associated with stimuli and columns with responses. For type (b), it is noted that almost all sorts of data matrices have the characteristic. Regarding social mobility data, the row element corresponds to father's occupation and column to son's occupation. In a flow matrix of foreign students by country of origin, the rows are associated with the host countries and columns with the countries of origin. Regarding the pecking frequency data, the individual specified by row pecks the individual specified by column.

Because psychology has been a traditional domain of MDS, for type (a) it has also been the subject of great interest for researchers to investigate the complex aspects of (dis)similarity judgment. It has been pointed out that similarity depends on attention shift (Shepard, 1964), context and frame of reference (Torgerson, 1965; Sjöberg, 1972). Micko (1970) proposed a kind of content model, which was called the halo-model. According to the model, a stimulus of concept is represented by a vector and its halo which takes the form of a multidimensional hypersphere. Gregson (1975) investigated the complexity of similarity by a set-theoretic approach and Rosch (1975) pointed out asymmetry in connection with cognitive tasks. Tversky (1977) examined and demonstrated the context effect in similarity judgment and questioned the metric and dimensional assumptions underlying the geometric representation of similarity data on both theoretical and empirical grounds. He proposed the feature-matching model based on a set-theoretic approach to similarity.

Krumhansl (1978) suggested that the distance model revised with an assumption might be compatible with those effects and some of the objections to the distance models raised by Tversky (1977). The assumption is that (dis)-similarity is a function of both interpoint distance and spatial density of other points in the surrounding region of the configuration. Then the assumption was incorporated into the distance-density model.

### 4.1.2. A Brief History

The studies mentioned above have brought about development of a diversity of scaling models and associated methods and procedures in MDS with asymmetry. Here we attempt to sketch the history briefly from our point of view.

First, we find that some models and procedures have been suggested with generalizations of distance models in a variety of ways. Regarding the method of nonmetric MDS developed mainly for analysis of symmetric proximities, Kruskal (1964) stated a treatment of asymmetric data. Young (1975) suggested a kind of weighted Euclidean distance model called ASYMSCAL, which accounted for asymmetry in dissimilarity by using weights to the related dimensions. On the other hand, analysis based on generalization of distance models has been proposed. Weeks and Bentler (1982) developed a scaling model in terms of a modified distance with a skew-symmetric form, while Okada and Imaizumi (1987) showed one in terms of distance with the same form for asymmetry. Imaizumi (1987) suggested another extension of distance model. Saito and Takeda (1990) proposed a scaling procedure based on a mixed model that may be viewed as a version of the distance-density model, and Saito (1991) developed procedures based on a model in terms of distance with additive terms. In a spetial context from those methods, Sato (1988) proposed an analysis of asymmetry in Minkowski space.

Secondly, let us consider development of models and procedures on extensions of scalar product models. Chino (1978) proposed another ASYMSCAL, in terms of vector representation of stimuli using two kinds of vector products. Chino (1990) extended it to a more general model, GIPSCAL, in dimensions higher than three. Kiers and Takane (1994) proposed a generalized GIPSCAL. As a different extension of vector models, Zielman and Heiser (1993) proposed analysis of asymmetry by slide-vector model, which might be viewed as a kind of generalization of distance models.

In a different generalization of scalar product models, Harshman (1978) presented a model called DEDICOM, which decomposes an asymmetric data matrix into the products of a weight matrix (rectangular) and asymmetric matrix of relations. For details, refer to Harshman et al. (1982). From this model (single domain DEDICOM), several models and associated algorithms have been developed (e.g., Harshman and Kiers, 1987; Kiers, 1989; Kiers et al., 1990). In the context of a vector space, Chino and Shiraiwa (1993) presented a comprehensive comparison of some models that may be regarded as a generalized family of scalar product models.

Thirdly, let us turn to unfolding-type models and procedures. Coombs (1964) proposed the unfolding model originally for analysis of preferences. Given a rectangular matrix $\boldsymbol{S} = (s_{ij})$ of two-mode data ($i$ and $j$ belong to different sets) in which observed $s_{ij}$ should be proportional to unknown distance $d_{ij}$, unfolding-type procedures provide row and column stimuli with a set of scale values, such as subject points and ideal points of objects. Setting aside the backgrounds of the model presentation and thinking of asymmetric data of type (b), one might perform data analysis using the unfolding model. Gower (1977) and Constantine and Gower (1978), for one-mode two-way data, suggested applying unfolding procedures to analysis of a square asymmetric matrix, so that row and column elements of the matrix are jointly represented in a space. It may be of interest to note that the slide-vector model is regarded a special case of the unfolding model as pointed out by Zielman and Heiser (1993).

Fourthly, we note that models incorporating asymmetric elements or asymmetric scaling procedures have been proposed in view of the complex aspects of similarity as mentioned above. DeSarbo et al. (1992) developed a procedure of TSCALE on the basis of the contrast model proposed by Tversky (1977). Other types of model and procedure were developed in connection with the distance-density model (Krumhansl, 1978), such as those by Saito and Takeda (1990), DeSarbo et al. (1990), and DeSarbo and Manrai (1992).

Finally, we consider the scale levels of data to which models and procedures are applied. Almost all procedures are formulated in metric approaches, that is, those which are concerned with the data measured on an interval or ratio scale level. To the authors' knowledge, Kruskal (1964) suggested a nonmetric treatment of asymmetric data for the first time. It leads to constructing asymmetric disparities that correspond to asymmetric dissimilarities and fitting the disparities to (symmetric) distances. Holman (1979) proposed a series of models for asymmetric proximities, each of which represents asymetric proximity as a monotone combination of a symmetric function and a bias function. For data analysis, he suggested a nonmetric procedure by using his algorithm for nonmetric MDS. Okada and Imaizumi (1987) suggested a nonmetric procedure based on their asymmetric model mentioned above.

Let us remark on studies of asymmetry in connection with square contingency tables. In dealing with those tables, the focus of attention is generally on analyzing subjects, such as main effect, independence, quasi-independence, rather than on analyzing subjects, such as quasi-asymmetry and association, related to asymmetry. For this reason, we will provide a very brief description on those studies in comparison with similarity and bias models.

### 4.1.3. Preliminaries

#### Metric Axioms

We consider a set of $n$ objects, $K = \{1, 2, \ldots, j, \ldots, n\}$. A real-valued function $d_{jk}$ defined for ordered pairs $(j, k)$ is a metric if it satisfies the following axioms.

*A1. Minimality.* For all ordered pairs $(j, k)$, it holds that

$$d_{jk} \geq d_{jj} = 0 \tag{4.1}$$

*A2. Symmetry.* For all ordered pairs $(j, k)$, the following equality holds,

$$d_{jk} = d_{kj} \tag{4.2}$$

*A3. Triangular Inequality.* For all ordered triples $(i, j, k)$, the following inequality holds,

$$d_{ij} + d_{jk} \geq d_{ik} \tag{4.3}$$

## Measurement Level

We would like to state the subject of asymmetry in MDS in connection with the measurement level of proximity. Many terms for proximity are related either to similarity or dissimilarity. The similarity $\psi_{jk}$ or dissimilarity $\omega_{jk}$ is a real valued function defined for an ordered pair $(j, k)$. Usually $\omega_{jk} \geq 0$, and self-dissimilarity $\omega_{jj}$ is sometimes measured to be zero. Usually $\psi_{jk} \geq 0$, and $\psi_{jk}$ is often measured so that the self-similarity $\psi_{jj}$ is maximum.

For a moment we take up the case of dissimilarity measure only. Suppose that we are given a set of measures of dissimilarity (distance-like), which we denote by an $n \times n$ matrix $\mathbf{\Omega} = (\omega_{jk})$ with $\omega_{jk} \geq 0$. We aim to obtain a configuration of objects in a multidimensional space from $\mathbf{\Omega}$.

Referring to Cox and Cox (1994), we consider four levels for $\omega_{jk}$, relaxing the metric requirement:

L1: $\omega_{jk}$ is a metric.
L2: $\omega_{jk}$ is symmetric and real-valued.
L3: $\omega_{jk}$ is real valued.
L4: $\omega_{jk}$ is completely or partially ordered.

At level L1, the purpose is exactly the problem of metric MDS, although Euclidean distance is not necessarily assumed for the metric. At level L2, the purpose is still handled by the approach of metric MDS. At level L3 for which the symmetry axiom is violated, the purpose becomes the principal subject of the present chapter. At level L4, losing the measurement (observation) of $\omega_{jk}$ on the ratio/interval scale leads to the so-called nommetric approach in MDS. We will mention this point briefly because the study of this has not been developed so much to date.

## Decomposition of Fitting a Model to Asymmetric Data

As stated in Sec. 3.1, the decomposition of an asymmetric data matrix by (3.3) leads to another decomposition of the matrix in terms of squared entries. These two kinds of decomposition serve to examine properties of models involved in asymmetric MDS. Consider a model representation $m_{ij}$ for asymmetry. It is decomposed as $m_{ij} = \gamma_{ij} + \tau_{ij}$, where $\gamma_{ij}$ is symmetric and $\tau_{ij}$ is skew-symmetric. Let $\boldsymbol{M} = (m_{ij})$, $\boldsymbol{\Gamma} = (\gamma_{ij})$, $\boldsymbol{T} = (\tau_{ij})$. Then, following a way similar to deriving (3.4) and (3.5), we see that

$$\mathrm{tr}((\boldsymbol{S} - \boldsymbol{\Gamma})(\boldsymbol{A} - \boldsymbol{T})) = \sum_{i=1}^{n} \sum_{j=1}^{n} (s_{ij} - \gamma_{ij})(a_{ij} - \tau_{ij}) = 0 \tag{4.4}$$

Thus we obtain

$$\sum_{i=1}^{n}\sum_{j=1}^{n}(o_{ij}-m_{ij})^2=\sum_{i=1}^{n}\sum_{j=1}^{n}(s_{ij}-\gamma_{ij})^2+\sum_{i=1}^{n}\sum_{j=1}^{n}(a_{ij}-\tau_{ij})^2 \tag{4.5}$$

which is alternatively expressed as

$$\|\boldsymbol{O}-\boldsymbol{M}\|^2=\|\boldsymbol{S}-\boldsymbol{\Gamma}\|^2+\|\boldsymbol{A}-\boldsymbol{T}\|^2 \tag{4.6}$$

This expression shows that a least squares problem for fitting the matrix $\boldsymbol{M}$ to the data $\boldsymbol{O}$ is separately treated as two problems of the least squares problem for fitting, one for the symmetric part and the other for the skew-symmetric part (Bove and Critchley, 1993).

## 4.2. GENERALIZATION OF SCALAR PRODUCT MODELS

In this section we are concerned with similarity matrix $\boldsymbol{\Psi}=(\psi_{jk})$ and its observation $\boldsymbol{O}=(o_{jk})$. We use $\boldsymbol{\Psi}$ for model representation, and $\boldsymbol{O}$ for model estimation.

### 4.2.1. DEDICOM

#### The Model

Harshman (1978) proposed the DEDICOM (DEcomposition into DIrectional COMponents) model. The error-free model is represented as

$$\boldsymbol{\Psi}=\boldsymbol{XRX}' \tag{4.7}$$

Here $\boldsymbol{\Psi}=(\psi_{jk})$ is an $n\times n$ asymmetric matrix of a similarity measure. For example, greater $\psi_{jk}$ indicates a closer relation between a pair of objects $j$ and $k$. $\boldsymbol{X}=(x_{jt})$ denotes an $n\times p$ matrix of weights for $n$ objects on $p$ latent dimensions (or components), $\boldsymbol{R}=(r_{ts})$ a $p\times p$ asymmetric matrix representing directional relations among the dimensions. The model is written elementwise as

$$\psi_{jk}=\sum_{s=1}^{p}\sum_{t=1}^{p}x_{jt}x_{ks}r_{ts} \tag{4.8}$$

The dimensions may be regarded as aspects to which objects are related, and then $\boldsymbol{X}$ indicates the weights of aspects for the objects. In what follows, we use the terms aspect, dimension, or component depending on the context. Given a data matrix $\boldsymbol{O}$, which is an observation matrix of $\boldsymbol{\Psi}$, one aims to estimate $\boldsymbol{X}$ and $\boldsymbol{R}$ by specifying $p$ through DEDICOM. The dimensionality $p$ may be provided on the basis of some external criterion.

## Indeterminacy

The model has a rotational indeterminacy for a given $p$ as follows. Define a $p \times p$ nonsingular matrix $\boldsymbol{T}$. Let $\boldsymbol{X}_* = \boldsymbol{XT}$ and $\boldsymbol{R}_* = \boldsymbol{T}^{-1}\boldsymbol{RT}'^{-1}$. Then

$$\boldsymbol{XRX}' = \boldsymbol{X}_*\boldsymbol{R}_*\boldsymbol{X}'_* \tag{4.9}$$

To deal with the indeterminacy, we have a choice of normalizing either $\boldsymbol{R}$ or $\boldsymbol{X}$. When $\boldsymbol{R}$ is normalized, one can, across the columns of $\boldsymbol{X}$, examine the relative values of the weights that an object loads on different dimensions. When $\boldsymbol{X}$ is normalized, we may impose either of the following conditions:

$$\sum_{j=1}^{n} x_{jt} = 1 \qquad (t = 1, 2, \ldots, p) \tag{4.10}$$

$$\sum_{j=1}^{n} x_{jt}^2 = 1 \qquad (t = 1, 2, \ldots, p) \tag{4.11}$$

$$\boldsymbol{X}'\boldsymbol{X} = \boldsymbol{I}_p \tag{4.12}$$

Those conditions would be imposed before otherwise after obtaining the solution in connection with estimation algorithms.

Imposing (4.10) may be useful when $x_{jt}$ are all positive. Summing (4.8) over all $(j, k)$ yields

$$\sum_{j=1}^{n}\sum_{k=1}^{n} \psi_{jk} = \sum_{s=1}^{p}\sum_{t=1}^{p} r_{ts}\left(\sum_{j=1}^{n} x_{jt}\right)\left(\sum_{k=1}^{n} x_{ks}\right) = \sum_{s=1}^{p}\sum_{t=1}^{p} r_{ts} \tag{4.13}$$

Thus $\boldsymbol{R}$ is regarded as a compressed version of $\boldsymbol{\Psi}$. To find the influence between aspects, we manipulate

$$\sum_{j=1}^{n}\sum_{k=1}^{n} x_{jt}x_{ks}r_{ts} = r_{ts}\left(\sum_{j=1}^{n} x_{jt}\right)\left(\sum_{k=1}^{n} x_{ks}\right) = r_{ts} \tag{4.14}$$

Then each $r_{ts}$ indicates the sum of influence from aspect $t$ to aspect $s$. When (4.12) is imposed, the problem to estimate $\boldsymbol{X}$ and $\boldsymbol{R}$ is conveniently dealt with using an alternating squares algorithm as stated below. Under (4.12), it follows from (4.7) that

$$\mathrm{tr}(\boldsymbol{\Psi}) = \mathrm{tr}(\boldsymbol{R}), \qquad \text{that is,} \qquad \sum_{j=1}^{n} \psi_{jj} = \sum_{t=1}^{p} r_{tt} \tag{4.15}$$

## Transformation of the Model

Applying decomposition (3.3) to $\boldsymbol{R}$ yields $\boldsymbol{R} = \boldsymbol{R}_s + \boldsymbol{R}_a$, where $\boldsymbol{R}_s$ is symmetric and $\boldsymbol{R}_a$ is skew-symmetric. Then we have

$$\boldsymbol{XRX}' = \boldsymbol{XR}_s\boldsymbol{X}' + \boldsymbol{XR}_a\boldsymbol{X}' \tag{4.16}$$

so that the model consists of an oblique factor model and a skew-symmetric model. The singular value decomposition of $\boldsymbol{R}_a$ in the form of (3.19) is represented as

$$\boldsymbol{R}_a = \boldsymbol{UMU}' \tag{4.17}$$

where diagonal entries of $\boldsymbol{M}$ are singular values of $\boldsymbol{R}_a$ and the last entry is zero when $p$ is odd. In view of the model specification, $p$ is much smaller than $n$, hence it is natural to assume only one singular value of zero. Let $q = [p/2]$. We can use the orthogonal matrix $\boldsymbol{U}$ as a rotation matrix. Let $\boldsymbol{X}_* = \boldsymbol{XU}$ and $\boldsymbol{R}_{s*} = \boldsymbol{U}'\boldsymbol{R}_s\boldsymbol{U}$. Substituting (4.17) into (4.16) gives the following result:

$$\begin{aligned}\boldsymbol{\Psi} &= \boldsymbol{XUU}'\boldsymbol{R}_s\boldsymbol{UU}'\boldsymbol{X}' + \boldsymbol{XUMU}'\boldsymbol{X}' \\ &= \boldsymbol{X}_*\boldsymbol{R}_{s*}\boldsymbol{X}_*' + \boldsymbol{X}_*\boldsymbol{MX}_*'\end{aligned} \tag{4.18}$$

In a way similar to the case of SSA, we can make $q$ plots of objects in two dimensions, using each sequential pair of columns in $\boldsymbol{X}$, $(\boldsymbol{x}_1^*, \boldsymbol{x}_2^*)$, $(\boldsymbol{x}_3^*, \boldsymbol{x}_4^*)$, and so on, which correspond to each singular value $\mu_t$. One would examine the part of an oblique factor model by referring to the two-dimensional configurations. When $p$ is odd, the last vector $\boldsymbol{x}_p^*$ contributes only to account for the symmetric part.

If we put strong conditions for structures of $\boldsymbol{R}_s$ or $\boldsymbol{R}_a$, the DEDICOM model becomes GIPSCAL or generalized GIPSCAL, as described later. Under a condition that $\boldsymbol{R}_s$ is positive definite, Kiers and Takane (1994) suggested that (4.18) becomes a simpler representation. To show that, we write $\boldsymbol{R}_s = \boldsymbol{TT}'$ under the condition where $\boldsymbol{T}$ is nonsingular. Then it follows from (4.16) that

$$\boldsymbol{\Psi} = \boldsymbol{XTT}'\boldsymbol{X} + \boldsymbol{XT} \cdot \boldsymbol{T}^{-1}\boldsymbol{R}_a(\boldsymbol{T}^{-1})' \cdot \boldsymbol{T}'\boldsymbol{X}' \tag{4.19}$$

Consider SVD of the skew-symmetric matrix such that

$$\boldsymbol{T}^{-1}\boldsymbol{R}_a(\boldsymbol{T}^{-1})' = \boldsymbol{VMV}' \tag{4.20}$$

where $\boldsymbol{V}$ is orthogonal. Define $\boldsymbol{Y} = \boldsymbol{XTV}$. Rewriting (4.19) yields

$$\boldsymbol{\Psi} = \boldsymbol{YY}' + \boldsymbol{YMY}' \tag{4.21}$$

This reduced form is called the generalized GIPSCAL.

## Estimation Procedures

For estimation of parameters, we set the error model of DEDICOM as

$$\boldsymbol{O} = \boldsymbol{XRX}' + \boldsymbol{E} \tag{4.22}$$

where $\boldsymbol{E} = (e_{jk})$ is a residual matrix. Following estimation procedures by Harshman (1978), several procedures were proposed. Harshman and Kiers (1987) provided a comparative study of them. Under constraint (4.12) with a specified $p$, we consider a least squares criterion to fit the model such as

$$\phi(\boldsymbol{X}, \boldsymbol{R}) = \mathrm{tr}(\boldsymbol{E}'\boldsymbol{E}) = \|\boldsymbol{O} - \boldsymbol{XRX}'\|^2 \tag{4.23}$$

For a fixed $\boldsymbol{X}$, the minimum of $\phi$ is attained by setting $\boldsymbol{R} = \boldsymbol{X}'\boldsymbol{OX}$. For a fixed $\boldsymbol{R}$, minimizing $\phi$ over $\boldsymbol{X}$ leads to maximizing

$$f(\boldsymbol{X}|\boldsymbol{R}) = \mathrm{tr}(\boldsymbol{X}'\boldsymbol{OXX}'\boldsymbol{O}'\boldsymbol{X}) \tag{4.24}$$

Using this framework, advanced procedures to fit the DEDICOM model have been developed by Kiers (1989) and Kiers et al. (1990).

Given estimates $\hat{\boldsymbol{X}}$ and $\hat{\boldsymbol{R}}$, we apply decompositions (4.6) and (4.16) for (4.23). Noting (4.4), we derive

$$\begin{aligned} \mathrm{tr}(\boldsymbol{E}'\boldsymbol{E}) &= \|\boldsymbol{S} + \boldsymbol{A} - \hat{\boldsymbol{X}}(\hat{\boldsymbol{R}}_s + \hat{\boldsymbol{R}}_a)\hat{\boldsymbol{X}}'\|^2 \\ &= \|\boldsymbol{S} - \hat{\boldsymbol{X}}\hat{\boldsymbol{R}}_s\hat{\boldsymbol{X}}'\|^2 + \|\boldsymbol{A} - \hat{\boldsymbol{X}}'\hat{\boldsymbol{R}}_a\hat{\boldsymbol{X}}'\|^2 \end{aligned} \tag{4.25}$$

Then we can examine the badness of fit separately for the symmetric and skew-symmetric parts. Given $\hat{\boldsymbol{R}}$, one can examine asymmetric relations among objects anyway, although its spatial representation is not provided by the original DEDICOM.

## Extension of the Model

As mentioned in the overview section, we may consider some sorts of asymmetric matrices, for which row elements and column elements have different meanings or play different roles. In view of such a matrix, an extended model in which the row space is allowed to be different from the column space was proposed in such a way as

$$\boldsymbol{\Psi} = \boldsymbol{XRY}' \tag{4.26}$$

(Harshman et al., 1982). This model is called dual domain DEDICOM, in contrast to which the primary model (4.7) is called single domain DEDICOM. Other extensions of the primary model have also been suggested (Harshman et al., 1982).

### 4.2.2. GIPSCAL

Chino (1978) proposed a model under the name of ASYMSCAL, independently of DEDICOM, to account for asymmetric data of similarity, giving two or three dimensions for analysis. The model is regarded as a variant of the DEDICOM model, replacing $\boldsymbol{R}$ in (4.7) by matrices of a very special structure such that $\boldsymbol{R} = a\boldsymbol{I}_p + b\boldsymbol{A}_*$, where $\boldsymbol{I}_p$ is an identity matrix of order $p$, and $\boldsymbol{A}_*$ is given by $\boldsymbol{A}_2$ for $p = 2$ and by $\boldsymbol{A}_3$ for $p = 3$:

$$\boldsymbol{A}_2 = \begin{pmatrix} 0 & 1 \\ -1 & 0 \end{pmatrix} \quad \text{and} \quad \boldsymbol{A}_3 = \begin{pmatrix} 0 & 1 & -1 \\ -1 & 0 & 1 \\ 1 & -1 & 0 \end{pmatrix} \tag{4.27}$$

Write the $j$th row of $\boldsymbol{X}$ as $\boldsymbol{x}'_j$. Chino developed an algorithm for fitting the model for off-diagonal entries of data. Then ASYMSCAL is expressed by

$$\psi_{jk} = \alpha \boldsymbol{x}'_j \boldsymbol{x}_k + \beta \boldsymbol{x}'_j \boldsymbol{A}_* \boldsymbol{x}_k + \gamma \qquad (j \neq k) \tag{4.28}$$

In the right-hand side, the first term indicates an inner product and the second term an outer product, which correspond to the symmetric and skew-symmetric parts of $\boldsymbol{O}$, respectively.

#### Representation and Estimation

For a general case of $p$, Chino (1980) and Gower (1984) suggested a skew-symmetric matrix $\boldsymbol{A}$ for $\boldsymbol{A}_*$ in (4.28) such that

$$\boldsymbol{A} = (a_{jk}), \qquad \text{where} \quad a_{jk} = -a_{kj} \tag{4.29}$$

$$a_{jk} = (-1)^{j+k-1}, \qquad \text{for} \quad j < k \tag{4.30}$$

For example, when $p = 5$, $\boldsymbol{A}$ takes the following pattern of a cyclic matrix:

$$\boldsymbol{A} = \begin{pmatrix} 0 & 1 & -1 & 1 & -1 \\ -1 & 0 & 1 & -1 & 1 \\ 1 & -1 & 0 & 1 & -1 \\ -1 & 1 & -1 & 0 & 1 \\ 1 & -1 & -1 & 1 & 0 \end{pmatrix} \tag{4.31}$$

Setting $\boldsymbol{A}$ instead of $\boldsymbol{A}_*$ in (4.28) gives a variant of DEDICOM as

$$\boldsymbol{\Psi} = \alpha \boldsymbol{X}\boldsymbol{X}' + \beta \boldsymbol{X}\boldsymbol{A}\boldsymbol{X}' + \gamma \mathbf{1}\mathbf{1}' \tag{4.32}$$

Independently, Chino (1990) proposed this model under the name GIPSCAL (Generalized Inner Product SCALing) for a general case of $p$. The name comes from setting $\boldsymbol{Y} = \boldsymbol{R}\boldsymbol{X}$ where $\boldsymbol{R} = \alpha \boldsymbol{I}_p + \beta \boldsymbol{A}$ and rewriting the model in the form of an inner product as

$$\boldsymbol{\Psi} = \boldsymbol{X}\boldsymbol{Y}' + \gamma\mathbf{1}\mathbf{1}' \tag{4.33}$$

As Kiers and Takane (1994) pointed out, the model has indeterminancy of scale units for $\alpha$, $\beta$, and $\boldsymbol{X}$ as shown by

$$\boldsymbol{\Psi} = \frac{\alpha}{c}\sqrt{c}\boldsymbol{X}(\sqrt{c}\boldsymbol{X})' + \frac{\beta}{c}\sqrt{c}\boldsymbol{X}\boldsymbol{A}(\sqrt{c}\boldsymbol{X})' + \gamma\mathbf{1}\mathbf{1}' \tag{4.34}$$

where $c$ is a positive scalar. Chino provided an alternating least-squares procedure, giving estimates for $\boldsymbol{X}$ and parameters $\alpha$, $\beta$, and $\gamma$. Although not explicitly stated in the algorithm, those estimates seem to be determined up to the multiplicative constant.

Given estimates $\hat{\boldsymbol{X}} = (\hat{\boldsymbol{x}}_j)$, $\hat{\alpha}$, $\hat{\beta}$, and $\hat{\gamma}$ by scaling, write the row vector of $\boldsymbol{X}$ as $\boldsymbol{x}_j'$. The terms $\hat{\alpha}\hat{\boldsymbol{x}}_j'\hat{\boldsymbol{x}}_k + \hat{\gamma}$ and $\hat{\boldsymbol{x}}_j'\hat{\boldsymbol{A}}\hat{\boldsymbol{x}}_k$ account for the symmetric component and the skew-symmetric component in data matrix $\boldsymbol{O}$, respectively. For utilization of the scaling result, Chino suggested making a two-dimensional plot of $\boldsymbol{X}$ for each pair of dimensions, hence a requirement of $p(p-1)/2$ plots in total. On each plane, one would interpret the projection of $\hat{\boldsymbol{x}}_j'\hat{\boldsymbol{x}}_k$ on the basis of an inner product and that of $\hat{\boldsymbol{x}}_j'\hat{\boldsymbol{A}}\hat{\boldsymbol{x}}_k$ in terms of a triangle. This sort of inconvenience for interpretation is common to the case of using (3.81) or (3.98) in E–G procedure.

### 4.2.3. Generalized GIPSCAL

Kiers and Takane (1994) proposed an extension of GIPSCAL. To cope with the indeterminancy of the GIPSCAL model, we set $\tilde{\boldsymbol{X}} = \sqrt{\alpha}\boldsymbol{X}$ and $\tilde{\boldsymbol{A}} = \boldsymbol{A}/\beta$ and rewrite (4.32) in such a way that

$$\boldsymbol{\Psi} = \tilde{\boldsymbol{X}}\tilde{\boldsymbol{X}}' + \tilde{\boldsymbol{X}}\tilde{\boldsymbol{A}}\boldsymbol{X}' + \gamma\mathbf{1}\mathbf{1}' \tag{4.35}$$

Applying SVD of $\tilde{\boldsymbol{A}} = \boldsymbol{U}\boldsymbol{M}\boldsymbol{U}'$ as in (4.17) and (4.18), we find that

$$\boldsymbol{\Psi} = \boldsymbol{Y}\boldsymbol{Y}' + \boldsymbol{Y}\boldsymbol{M}\boldsymbol{Y}' + \gamma\mathbf{1}\mathbf{1}' \tag{4.36}$$

Here $\boldsymbol{Y} = \tilde{\boldsymbol{X}}\boldsymbol{U}$ and $\boldsymbol{U}$ is an orthogonal matrix consisting of the singular vectors. $\boldsymbol{M}$ is a block-diagonal matrix in the form of (3.17) fixed with singular values of $\tilde{\boldsymbol{A}}$, which are proportional to those of $\boldsymbol{A}$ with $\beta$. For example, for matrix $\boldsymbol{A}$ in (4.31), the singular values are fixed as follows:

$$\mu_1 = 2(\sin\theta + 2\sin 2\theta) \tag{4.37}$$

$$\mu_2 = 2(\sin\theta - 2\sin 2\theta) \tag{4.38}$$

$$\mu_3 = 0, \qquad \mu_1^2 + \mu_2^2 = 10 \tag{4.39}$$

where $\theta = 2\pi/5$.

Now we may regard (4.36) as a parameterized form of GIPSCAL. For this generalization, we will treat the diagonal elements as free parameters. Thus the generalized GIPSCAL has been formulated in the form of (4.36) with parameters $\boldsymbol{Y}$, $\boldsymbol{M}$, and $\gamma$. Note that the number of free parameters in $\boldsymbol{M}$ is $q = [p/2]$.

Kiers and Takane showed some uniqueness of the representation of the model under the following conditions: (i) $\boldsymbol{Y}$ has full rank and does not contain $\mathbf{1}$ in its column space, (ii) $\boldsymbol{M}$ has distinct singular values. Given a solution to satisfy the conditions, $\boldsymbol{M}$ and $\gamma$ are determined uniquely, and also $\boldsymbol{Y}$ is up to rotations for each sequential pair of columns of $\boldsymbol{Y}$, that is, for $(\boldsymbol{y}_1, \boldsymbol{y}_2)$, $(\boldsymbol{y}_3, \boldsymbol{y}_4)$, and so on.

Regarding a relation between DEDICOM solution and generalized GIPSCAL solution, Kiers and Takane presented the following theorem. Note that the theorem is not stated explicitly in terms of $\boldsymbol{O}$. We already used the result partly to describe (4.19) and (4.21).

*Theorem 4.1.* Given a DEDICOM solution with full rank matrices $\boldsymbol{X}$ and $\boldsymbol{R}$, its representation $\boldsymbol{XRX}'$ is transformed in the form of generalized GIPSCAL with $\gamma = 0$, if and only if the symmetric part of $\boldsymbol{R}$ is positive definite.

Let us consider the feature of the model. Denote the pair of vectors in $\boldsymbol{Y}$ associated with $\mu_h$ by an $n \times 2$ matrix $\boldsymbol{Y}_h$. We focus on the structure part of (4.36) in terms of $\boldsymbol{Y}_h$, such that

$$\boldsymbol{\Psi}_h = \boldsymbol{Y}_h(\boldsymbol{I} + \boldsymbol{M}_h)\boldsymbol{Y}_h' \qquad (h = 1, 2, \ldots, q) \tag{4.40}$$

For a graphical display in two dimensions, we define a scalar $\beta_h$, a $2 \times 2$ matrix $\boldsymbol{T}_h$, and $n \times 2$ matrices $\boldsymbol{U}_h$ and $\boldsymbol{V}_h$ as follows:

$$\beta_h = (1 + \mu_h^2)^{\frac{1}{2}} \qquad \text{and} \qquad \boldsymbol{T}_h = \beta_h^{-1}(\boldsymbol{I}_2 - \boldsymbol{M}_h) \tag{4.41}$$

$$\boldsymbol{U}_h = \sqrt{\beta_h}\boldsymbol{Y}_h \qquad \text{and} \qquad \boldsymbol{V}_h = \sqrt{\beta_h}\boldsymbol{Y}_h\boldsymbol{T}_h \tag{4.42}$$

Rewriting (4.40) with these quantities yields

$$\boldsymbol{\Psi}_h = \beta_h\boldsymbol{Y}_h\boldsymbol{T}_h'\boldsymbol{Y}_h' = \boldsymbol{U}_h\boldsymbol{V}_h' \tag{4.43}$$

This expression serves to show asymmetric data on the basis of scalar products, providing an illustration by means of the biplot technique. Write $\boldsymbol{\Psi}_h = (\psi_{jk}(h))$. Let the $j$th row of $\boldsymbol{U}_h$ be $\boldsymbol{u}_j(h)'$ and that of $\boldsymbol{V}_h$ be $\boldsymbol{v}_j(h)'$. Then

$$\psi_{jk}(h) = \boldsymbol{u}_j(h)'\boldsymbol{v}_k(h) \qquad \text{and} \qquad \psi_{kj}(h) = \boldsymbol{u}_k(h)'\boldsymbol{v}_j(h) \tag{4.44}$$

Note that $T_h$ is an orthogonal matrix to give a clockwise rotation with angle $\phi_h = \arctan(\mu_h)$. Since $V_h = U_h T_h$ in (4.42), we find that the configuration of the row points is the same as that of the column points up to the rotation over the angle $\phi_h$. In this respect the biplot differs from ordinary biplot and also from the graphical representation of (4.33).

Let us summarize characteristic points of the generalized GIPSCAL in comparison with GIPSCAL. It is based on a less restricted model than GIPSCAL. Under some conditions, it has unique representation of $X$, $M$, and $\gamma$ in contrast to the indeterminacy in the case of GIPSCAL. For a graphical representation of scaling result, it is provided with some advantageous points to GIPSCAL. First, one needs only $q$ plots in two dimensions for interpretation of the derived configuration. Secondly, weighting those plots is possible in view of the singular values $\mu_t$. Thirdly, for the two-dimensional configuration illustrated by means of the biplot, one can examine the inner products visually, finding the symmetry by the sum of the two products in (4.44) and the skew-symmetry by the difference of the two.

## Estimation Algorithms

Assume an asymmetric data matrix $O$ that is an observation of $\Psi$ with error, that is, $O = \Psi + E$. We consider an error model

$$O = YY' + YMY' + r\mathbf{11}' + E \tag{4.45}$$

We aim to estimate $Y$ and $\mu_1, \mu_2, \ldots, \mu_q$ and $\gamma$. This estimation problem leads to minimizing a loss function such as

$$\begin{aligned} f(A, \mu_1, \mu_2, \ldots, \mu_q, \gamma) &= \mathrm{tr}(E'E) \\ &= \|O - YY' - YMY' - \gamma\mathbf{11}'\|^2 \end{aligned} \tag{4.46}$$

A procedure based on an alternating least-squares algorithm was suggested. Let us consider decomposition (3.3) and (4.6). Given estimates $\hat{Y}$, $\hat{M}$, and $\hat{\gamma}$, we can examine the badness of fit separately:

$$\mathrm{tr}(E'E) = \|S - \hat{Y}\hat{Y}' - \hat{r}\mathbf{11}'\|^2 + \|A - \hat{Y}\hat{M}\hat{Y}'\|^2 \tag{4.47}$$

Another version to deal with missing data was developed. Write the $j$th row of $Y$ as $y_j$. Then the loss function is given as

$$g(A, \mu_1, \mu_2, \ldots, \mu_q, \gamma) = \sum_{j=1}^{n}\sum_{k=1}^{n} \varepsilon_{jk}(o_{jk} - y_j'y_k - y_j'My_k - \gamma) \tag{4.48}$$

where $\varepsilon_{jk}$ denotes an indicator (or design) matrix.

## Relationship of Models in Connection with DEDICOM

Kiers and Takane (1994) discussed the relationship of the E–G procedure to DEDICOM and generalized GIPSCAL. To follow their argument, we supplement two points: (a) utilization of the E–G procedure with some assumption and (b) conditions on some related vectors.

First, let us regard E–G as a procedure for approximating the asymmetric data matrix $\boldsymbol{O}$ in some dimensions, unlike the canonical analysis in terms of all the dimensions. Now refer to the formulas related to the present discussion: eigenvalue problem (3.82) of a symmetric matrix $\boldsymbol{E}$, definition (3.83) and spectral decomposition (3.84). Although there may be negative eigenvalues, we assume that the largest $2q$ eigenvalues are all positive, and we will take those eigenvectors associated with them. In terms of those eigenvectors and eigenvalues, we consider the low rank matrix approximation of $\boldsymbol{O}$.

Next, using component vectors $\boldsymbol{x}_t$, $\boldsymbol{y}_t$ of eigenvectors $z_{1t}$, $z_{2t}$ of (3.83) associated with $\lambda_t$, we define $\boldsymbol{W} = (\boldsymbol{w}_1, \boldsymbol{w}_2, \ldots, \boldsymbol{w}_{2q})$ where

$$\boldsymbol{w}_{2i-1} = \sqrt{\lambda_i}\boldsymbol{y}_i \qquad \text{and} \qquad \boldsymbol{w}_{2i} = \sqrt{\lambda_i}\boldsymbol{x}_i \qquad (i = 1, 2, \ldots, q) \tag{4.49}$$

Then approximations of (3.79) and (3.80) are stated as

$$\boldsymbol{S} = \boldsymbol{W}\boldsymbol{W}' + \boldsymbol{E}_s \tag{4.50}$$

$$\boldsymbol{A} = \boldsymbol{W}\boldsymbol{J}\boldsymbol{W}' + \boldsymbol{E}_a \tag{4.51}$$

where $\boldsymbol{J}$ is the $2q \times 2q$ matrix in the form of (3.17), and $\boldsymbol{E}_s$ and $\boldsymbol{E}_a$ are residual matrices. Consider a least squares criterion to (4.50) and (4.51) for the asymmetric matrix $\boldsymbol{O}$, such that

$$h(\boldsymbol{W}) = \|\boldsymbol{S} - \boldsymbol{W}\boldsymbol{W}'\|^2 + \|\boldsymbol{A} - \boldsymbol{W}\boldsymbol{J}\boldsymbol{W}'\|^2 \tag{4.52}$$

Applying relation (4.6) leads to

$$h(\boldsymbol{W}) = \|\boldsymbol{S} + \boldsymbol{A} - (\boldsymbol{W}\boldsymbol{W}' + \boldsymbol{W}\boldsymbol{J}\boldsymbol{W}')\|^2 \tag{4.53}$$

$$= \|\boldsymbol{O} - \boldsymbol{W}(\boldsymbol{I} + \boldsymbol{J})\boldsymbol{W}'\|^2 \tag{4.54}$$

Setting $\boldsymbol{R} = \boldsymbol{I} + \boldsymbol{J}$, we find that (4.54) is equivalent to the criterion for fitting DEDICOM. In order that the least squares solution $\{\boldsymbol{w}_i\}$ of this DEDICOM version should coincide with the solution of the original E–G procedure, it is required that vectors $\{\boldsymbol{w}_i\}$ satisfy the following conditions:

$$\boldsymbol{w}_{2k-1}\boldsymbol{w}_{2k-1} + \boldsymbol{w}_{2k}\boldsymbol{w}_{2k} = \lambda_k \qquad (k = 1, 2, \ldots, q) \tag{4.55}$$

$$\boldsymbol{w}_{2i-1}\boldsymbol{w}_{2j-1} + \boldsymbol{w}_{2i}\boldsymbol{w}_{2j} = 0 \qquad (i \neq j; \quad i, j = 1, 2, \ldots, q) \tag{4.56}$$

$$\boldsymbol{w}_{2i-1}\boldsymbol{w}_{2j} - \boldsymbol{w}_{2i}\boldsymbol{w}_{2j-1} = 0 \qquad (i \neq j; \quad i, j = 1, 2, \ldots, q) \tag{4.57}$$

Thus the data analysis in $q$ dimensions by utilizing the E–G procedure under the assumption mentioned is equivalent to fitting the DEDICOM model under constraints (4.55) to (4.57). From the form of (4.54), such an analysis is also equivalent to a constraint variant of the generalized GIPSCAL with $c = 0$. Here the constraint means to specify $\mu_1, \mu_2, \ldots, \mu_q$ to unity in (4.40) and impose (4.55) to (4.57).

## 4.2.4. Scalar Product Model in a Vector Space

From the argument in Sec. 3.3.3, it is clear that we can construct a metric space regarding the complex data matrix $\boldsymbol{H}$, which includes the symmetric and the skew-symmetric information of $\boldsymbol{O}$. On the basis of this result, Chino and Shiraiwa (1993) suggested the possibility of asymmetric MDS and proposed a generalization of the theorem of Young and Householder (1938). However, we find it difficult to follow their arguments, including some unclear points. To begin, we give some supplements to clarify their arguments. Then we set out alternative treatments of the generalization of the Young and Householders' theorem, in a good correspondence to the role of the theorem in symmetric MDS (Saito, 2003a). Finally, we examine the advantageous points in a substantial viewpoint of data analysis.

### Diagonal Entry, Similarity, and Inner Product

Although not explicitly done by Chino and Shiraiwa, it is clearly pointed out, from the discussion in Sec. 3.3.4, that all the diagonal entries of the asymmetric data matrix should be positive in order that the definitions of the norm and the metric are valid. In this connection we explained, around (3.123), the direction of the measure which asymmetric $o_{jk}$ values indicate. It is required for the conformity of a possible model that the proximity measure should be similarity, not dissimilarity (or psychological distance). The reason can be understood by referring to the relation such as

$$o_{jj} + o_{kk} - o_{jk} - o_{kj} = q_{jk} = \sum_{t=1}^{r} \lambda_t d_{jk}^2(t) = D_{jk}^2 \tag{4.58}$$

The reason is further confirmed by noting in (3.116) that the interpoint distance $D_{jk}$ is inversely related to the observation $o_{jk}$. The expression (3.115) looks like a form of distance model extended in the vector space. We can take a different view that (3.115) has followed from the correspondence of $h_{jk}$ in terms of transformed observations to the inner product defined by (3.104) such that

$$\tfrac{1}{2}(o_{jk} + o_{kj}) + i\tfrac{1}{2}(o_{jk} - o_{kj}) = h_{jk} = \boldsymbol{w}_j \boldsymbol{\Lambda} \boldsymbol{w}_k^* \tag{4.59}$$

Thus, we can say that (3.115) is an inner product model for representation of asymmetric similarity in a vector space. In this connection, it is remembered that vector models based on a scalar product were suggested for analysis of similarity in the case of symmetric MDS (e.g., Ekman, 1963). In comparison, symmetric dissimilarity $o_{jk}$ corresponds to Euclidean distance in the traditional MDS procedure (Torgerson, 1952).

## Extension of Young and Householder's Theorem

Now we are in a position to summarize the arguments so far given in the form of a theorem.

*Theorem 4.2.* Given an $n \times n$ asymmetric data matrix $\boldsymbol{O} = (o_{jk})$ where $o_{jk}$ indicates the similarity between a pair of objects $(j, k)$ and all $o_{jj}$ are positive, define a Hermitian matrix $\boldsymbol{H}$ by (3.76). If $\boldsymbol{H}$ is of rank $r$ and positive semidefinite, then vectors $\{\boldsymbol{w}_1, \boldsymbol{w}_2, \ldots, \boldsymbol{w}_n\}$ defined under (3.73) span the vector space in $r$ dimensions with the inner product, which is defined by (3.104). Hence the $n$ objects are represented by $\{\boldsymbol{w}_1, \boldsymbol{w}_2, \ldots, \boldsymbol{w}_n\}$. The interpoint distance is given by (3.106), which is expressed in terms of observations by (3.117).

This is a generalization of the statement of sufficiency in the Young and Householders' theorem, which implies that the measure of dissimilarity (psychological distance) should be Euclidean distance. Given a set of asymmetric similarity data, one can check the sufficiency condition. Conversely, let us consider a matrix $\boldsymbol{H}$ that is given in terms of quantities such as (3.106) and (3.110), which are defined on the vector space with the norm. It is shown that $\boldsymbol{H}$ becomes positive semidefinite. To state that without respect to observations, we like to use similar but different notation.

*Theorem 4.3.* Assume a vector space spanned by complex (row) vectors $\boldsymbol{v}_1, \boldsymbol{v}_2, \ldots, \boldsymbol{v}_n$ with an inner product defined by $(\boldsymbol{v}_j, \boldsymbol{v}_k) = \boldsymbol{v}_j \boldsymbol{\Lambda} \boldsymbol{v}_k^*$ where $\boldsymbol{\Lambda}$ is a diagonal matrix with non-negative entries. Define the norm by $\|\boldsymbol{v}_j\| = (\boldsymbol{v}_j \boldsymbol{\Lambda} \boldsymbol{v}_j^*)^{1/2}$. Suppose that the following quantities $D_{jk}$, $\bar{D}_{jk}$, $D_{0j}$

are provided:

$$D_{jk} = \|v_j - v_k\|, \qquad \bar{D}_{jk} = \|v_j - iv_k\|, \qquad D_{0j} = \|v_j\| \tag{4.60}$$

Define an $n \times n$ matrix $\boldsymbol{H} = (h_{jk})$ as

$$h_{jk} = \tfrac{1}{2}(D_{0j}^2 + D_{0k}^2 - D_{jk}^2) + \tfrac{1}{2}i(D_{0j}^2 + D_{0k}^2 - \bar{D}_{jk}^2) \tag{4.61}$$

Then $\boldsymbol{H}$ is a Hermitian matrix and positive semidefinite and the dimension of the space is equal to rank ($\boldsymbol{H}$).

We provide a proof of Theorem 4.3. From definitions (4.60) and (4.61), it follows that

$$h_{jk} = (v_j, v_k) = v_j \boldsymbol{\Lambda} v_k^* \tag{4.62}$$

Then $h_{jk} = h_{kj}^*$, hence $\boldsymbol{H}$ is a Hermitian matrix. Since $h_{jj} = v_j \boldsymbol{\Lambda} v_j^* \geq 0$ ($j = 1, 2, \ldots, n$), $\boldsymbol{H}$ is positive semidefinite; accordingly the space is of rank ($\boldsymbol{H}$). This concludes the proof.

**Remark.** The expression (3.115), which is an alternative expression of (3.98), indicates a geometrical relationship clearly. It shows that $o_{jk}$ is the sum of the symmetric component (consisting of the first three terms) and the skew-symmetric component (the fourth term). Naturally, (3.98) provides the same implication.

Expression (3.115) and Theorems 4.2 and 4.3 are of interest from a theoretical point of view. The theorems are meaningful in discussing the property of the complex vector space associated with $\boldsymbol{H}$. However, it is almost impossible for us to comprehend the spatial configuration of objects in the space by (3.115). Then, in a practical working, one has to interpret the configuration by using the plot of objects in two dimensions for every $z_t$ ($t = 1, 2, \ldots, r$). This process is quite the same as that one does in using the E–G procedure. Hence from a substantive perspective, the scalar product model in the vector space serves theoretical interests rather than practical purposes. Finally, it should be noted that as far as our theoretical statements are concerned, they have been provided without referring to a finite Hilbert space.

### 4.2.5. Example

#### Example 1

Here is given an example of a DEDICOM application, which is extracted from the full description in Harshman et al. (1982). The study used car switching data of 16 types of cars in 1979. Table 4.2 shows the data taken from Harshman et al. (1982), regarding 16 car types in Table 4.1. The analysis by the

**Table 4.1** Sixteen Car Types

| Abbreviation | | Car type |
|---|---|---|
| 1 | SUBD | Subcompact/domestic |
| 2 | SUBC | Subcompact/captive imports |
| 3 | SUBI | Subcompact/imports |
| 4 | SMAD | Small specialty/domestic |
| 5 | SMAC | Small specialty/captive imports |
| 6 | SMAI | Small specialty/imports |
| 7 | COML | Low price compact |
| 8 | COMM | Medium price compact |
| 9 | COMI | Import compact |
| 10 | MIDD | Midsize domestic |
| 11 | MIDI | Midsize import |
| 12 | MIDS | Midsize specialty |
| 13 | STDL | Low price standard |
| 14 | STDM | Medium price standard |
| 15 | LUXD | Luxury domestic |
| 16 | LUXI | Luxury imports |

model of DEDICOM (4.7) was performed under the specified dimensionality $p\ (= 1, 2, \ldots, 5)$, utilizing a closed-form algorithm that gives an approximation to a least squares solution. For comparison, the data were also analyzed by a symmetric DEDICOM (factor analytic model).

Figure 4.1 shows the dependence of fit values on dimension $p$ given by the two models. As is seen, the asymmetric model provides higher degrees of fit than the symmetric model, suggesting an elbow point at $p = 2$. For the difference between the fit values of the two models, a statistical test of significance was performed, using a randomization test. It concluded that there are systematic asymmetries in the data. In Fig 4.1, the bars indicate the 95% intervals under the null hypothesis of 110 systematic asymmetries.

The four-dimensional configuration is illustrated in Figs. 4.2 and 4.3. According to the authors' view, the derived dimensions are all interpretable, the first dimension labeled "plain large-midsize," the second "speciality," the third "fancy large," and the fourth "small."

Through the analysis, the $\boldsymbol{R}$ matrix was estimated. The largest asymmetry was found between $\hat{r}_{14} = 226,400$ and $\hat{r}_{41} = 83,069$. It indicates the direction of the asymmetry, that the first aspect evokes the fourth aspect far greater than the fourth does the first in the car brand switching. In passing, for an interesting application of DEDICOM, refer to Dawson and Harshman (1986).

**Table 4.2** Car Switching Data Among 16 Car Types

| From type | | To type 1 | 2 | 3 | 4 | 5 | 6 | 7 | 8 |
|---|---|---|---|---|---|---|---|---|---|
| 1 | SUBD | 23272 | 1487 | 10501 | 18994 | 49 | 2319 | 12349 | 4061 |
| 2 | SUBC | 3254 | 1114 | 3014 | 2656 | 23 | 551 | 959 | 894 |
| 3 | SUBI | 11344 | 1214 | 25986 | 9803 | 47 | 5400 | 3262 | 1353 |
| 4 | SMAD | 11740 | 1192 | 11149 | 38434 | 69 | 4880 | 6047 | 2335 |
| 5 | SMAC | 47 | 6 | 0 | 117 | 4 | 0 | 0 | 49 |
| 6 | SMAI | 1772 | 217 | 3622 | 3453 | 16 | 5249 | 1113 | 313 |
| 7 | COML | 18441 | 1866 | 12154 | 15237 | 65 | 1626 | 27137 | 6182 |
| 8 | COMM | 10359 | 693 | 5841 | 6368 | 40 | 610 | 6223 | 7469 |
| 9 | COMI | 2613 | 481 | 6981 | 1853 | 10 | 1023 | 1305 | 632 |
| 10 | MIDD | 33012 | 2323 | 22029 | 29623 | 110 | 4193 | 20997 | 12155 |
| 11 | MIDI | 1293 | 114 | 2844 | 1242 | 5 | 772 | 1507 | 452 |
| 12 | MIDS | 12981 | 981 | 8271 | 18908 | 97 | 3444 | 3693 | 1748 |
| 13 | STDL | 27816 | 1890 | 12980 | 15993 | 34 | 1323 | 18928 | 5836 |
| 14 | STDM | 17293 | 1291 | 11243 | 11457 | 41 | 1862 | 7731 | 6178 |
| 15 | LUXD | 3733 | 430 | 4647 | 5913 | 6 | 622 | 1652 | 1044 |
| 16 | LUXI | 105 | 40 | 997 | 603 | 0 | 341 | 75 | 558 |

| From type | | To type 9 | 10 | 11 | 12 | 13 | 14 | 15 | 16 |
|---|---|---|---|---|---|---|---|---|---|
| 1 | SUBD | 545 | 12622 | 481 | 16329 | 4253 | 2370 | 949 | 127 |
| 2 | SUBC | 223 | 1672 | 223 | 2012 | 926 | 540 | 246 | 37 |
| 3 | SUBI | 2257 | 5195 | 1307 | 8347 | 2308 | 1611 | 1071 | 288 |
| 4 | SMAD | 931 | 8503 | 1177 | 23898 | 3238 | 4422 | 4114 | 410 |
| 5 | SMAC | 0 | 110 | 0 | 10 | 0 | 0 | 0 | 0 |
| 6 | SMAI | 738 | 1631 | 1070 | 4937 | 338 | 901 | 1310 | 459 |
| 7 | COML | 835 | 20909 | 566 | 15342 | 9728 | 3610 | 910 | 170 |
| 8 | COMM | 564 | 9620 | 435 | 9731 | 3601 | 5498 | 764 | 85 |
| 9 | COMI | 1536 | 2738 | 1005 | 990 | 454 | 991 | 543 | 127 |
| 10 | MIDD | 2533 | 53002 | 2140 | 61350 | 28006 | 33913 | 9808 | 706 |
| 11 | MIDI | 565 | 3820 | 3059 | 2357 | 589 | 1052 | 871 | 595 |
| 12 | MIDS | 935 | 11551 | 1314 | 56025 | 10959 | 18688 | 12541 | 578 |
| 13 | STDL | 1182 | 28324 | 938 | 37380 | 67964 | 28881 | 6585 | 300 |
| 14 | STDM | 1288 | 20942 | 1048 | 30189 | 15318 | 81808 | 21974 | 548 |
| 15 | LUXD | 476 | 3068 | 829 | 8571 | 2964 | 9187 | 63509 | 1585 |
| 16 | LUXI | 176 | 151 | 589 | 758 | 158 | 756 | 1234 | 3124 |

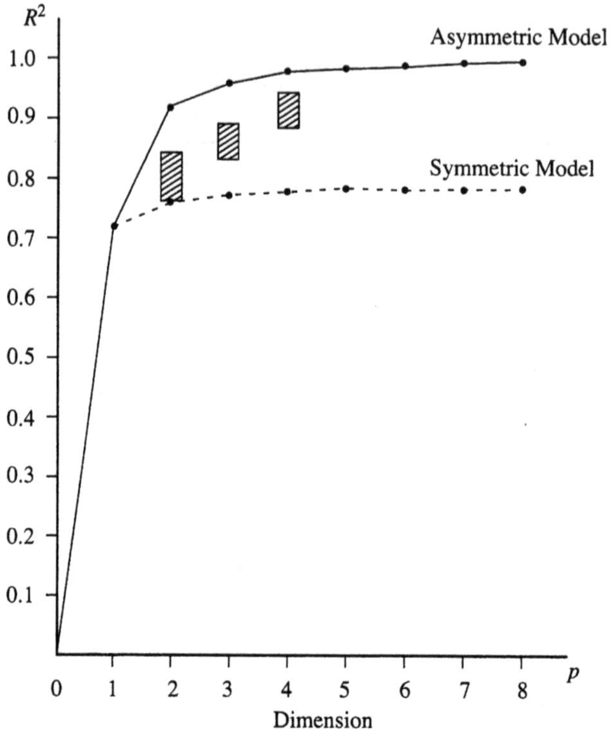


**FIGURE 4.1** Fit for dimensionality (DEDICOM)

## Example 2

Here is given an example extracted from Kiers and Takane (1994). It presents a numerical demonstration of Theorem 4.1, not a result of the execution of the generalized GIPSCAL algorithms. Applying the DEDICOM algorithm (Kiers et al., 1990) to the car switching data (Harshman et al., 1982), they obtained solutions in two, three, and four dimensions. Those solutions accounted for 77.2, 86.4, and 92.0% of the total sum of squares of the data, respectively. We take up the four-dimensional solution only. The estimated $\boldsymbol{R}$ was positive definite, indicating that DEDICOM is equivalent to generalized GIPSCAL for the data.

Then the solution was transformed by Theorem 4.1 into the generalized GIPSCAL form (4.36), with $\hat{\mu}_1 = 0.4479$ and $\hat{\mu}_2 = 0.0904$. These values indicate that $\hat{\boldsymbol{Y}}_1$ accounts for the data to a much greater extent than $\hat{\boldsymbol{Y}}_2$. From $\hat{\boldsymbol{Y}}_1$, $\hat{\boldsymbol{U}}_1$ and $\hat{\boldsymbol{V}}_1$ were computed. Figure 4.4 plots the two-dimensional configuration of

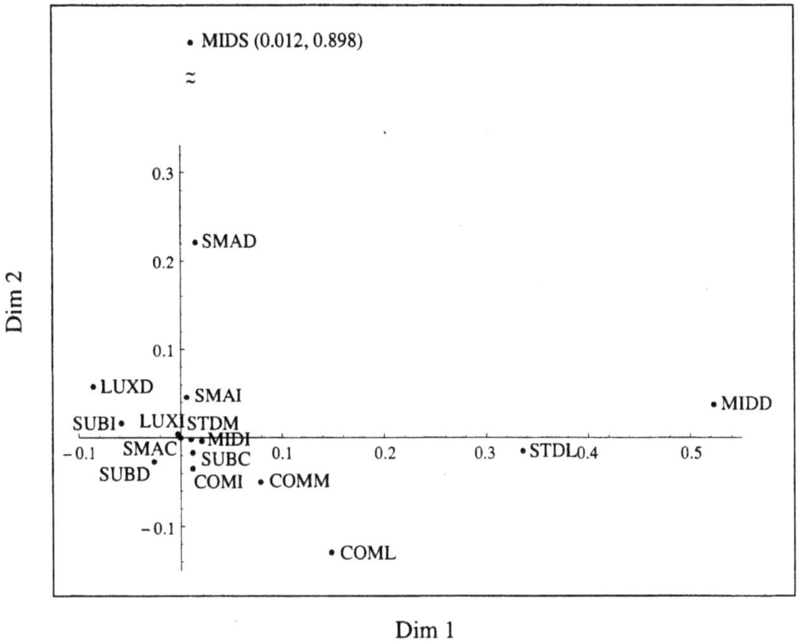


**FIGURE 4.2** Two-dimensional configuration of 16 car types in Dim 1 vs. Dim 2 (DEDICOM). (Note: MIDS is out of the range)

$\hat{U}_1$ where capital letter labels refer to rows of the data. In the figure, the configuration of $\hat{V}_1$ is also superimposed, where small letter labels refer to columns of the data, which is a rotated matrix of $\hat{U}_1$ with angle $\phi_1 = 24.1°$.

The largest switching frequencies are found for the pairs (stdl, MIDD), (stdl, STDL), (mids, MIDS) and so on, which involve the car types with the largest market shares: MIDD, MIDS, and STDL. The largest asymmetric relations are found for pairs (STDL, MIDS), (STDL, MIDD), (STDL, SMAD), (MIDD, MIDS). The authors claim that the indication of (STDL, MIDD) was a result of modeling error. Seven car types that were too close to the origin are omitted for illustration.

## 4.3. SIMILARITY AND BIAS MODEL

Let $\boldsymbol{\Psi} = (\psi_{jk})$ be a similarity matrix where $\psi_{jk}$ represents the similarity for a pair of objects $(j, k)$. Holman (1979) considered a model for which a transitive

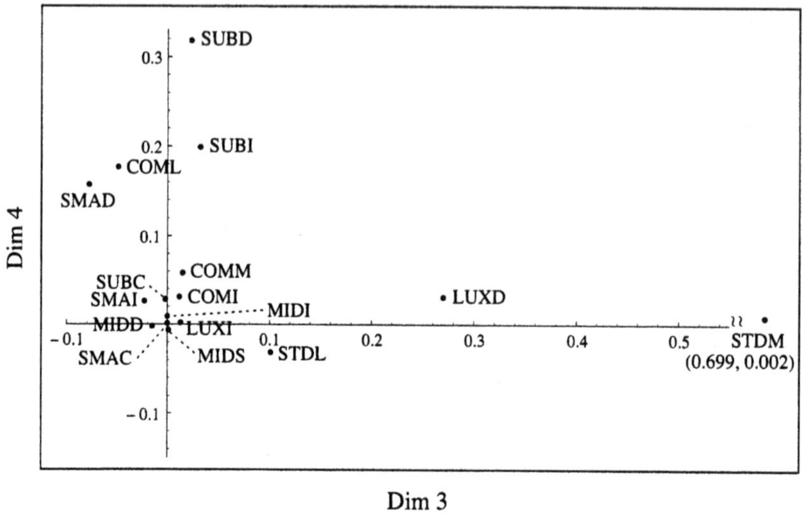


**FIGURE 4.3** Two-dimensional configuration of 16 car types in Dim 3 vs. Dim 4 (DEDICOM). (Note: STDM is out of range)

relation holds as follows:

$$\text{If } \psi_{ij} > \psi_{ji} \quad \text{and} \quad \psi_{jk} > \psi_{kj}, \quad \text{then} \quad \psi_{ik} > \psi_{ki} \tag{4.63}$$

for all triples $(i, j, k)$ (see also Nosofsky, 1991). This kind of model was called the weak bias model. Expressing $\psi_{ij}$ as the sum of symmetric $\gamma_{ij}$ and skew-symmetric $\tau_{ij}$, that is, $\psi_{ij} = \gamma_{ij} + \tau_{ij}$, we restate (4.63) as transitivity in terms of skew-symmetry in such a way that

$$\text{If } \tau_{ij} > \tau_{ji} \quad \text{and} \quad \tau_{jk} > \tau_{kj}, \quad \text{then} \quad \tau_{ik} > \tau_{ki} \tag{4.64}$$

Let $\alpha_{ij} = \tau_{ij} - \tau_{ji}$, which is skew-symmetric. Then we have transitivity in terms of dominance,

$$\text{If } \alpha_{ij} > 0 \quad \text{and} \quad \alpha_{jk} > 0, \quad \text{then} \quad \alpha_{ik} > 0 \tag{4.65}$$

As a special case of the weak bias model, Holman presented the similarity and bias model, which represents the similarity as

$$\psi_{ij} = F(\gamma_{ij} + r_i + c_j) \tag{4.66}$$

Here $F$ is a general monotonic function, $\gamma_{ij}$ symmetric similarity, $r_i$ a bias function on the $i$th row and $c_j$ a bias function on the $j$th column.

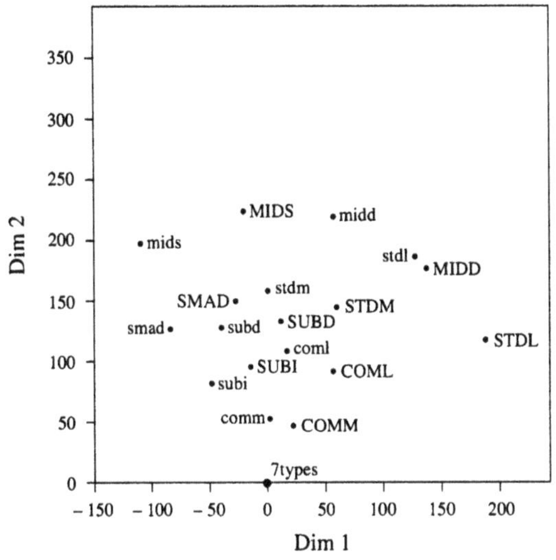


**FIGURE 4.4** Transformation of DEDICOM solution to generalized GIPSCAL form. (Note: seven car types located on the origin)

Let $\lambda_i = r_i - c_i$ and $p_{ij} = c_i + c_j + \gamma_{ij}$. Then (4.66) is rewritten as

$$\psi_{ij} = F(\lambda_i + p_{ij}) \tag{4.67}$$

It shows that the row and column bias model (4.66) leads to a row bias model with a symmetric function $p_{ij}$. Let $\mu_j = c_j - r_j$ and $q_{ij} = r_i + r_j + \gamma_{ij}$. Then (4.66) is rewritten as

$$\psi_{ij} = F(\mu_j + q_{ij}) \tag{4.68}$$

showing a column bias model with a symmetric function $q_{ij}$.

Zielman and Heiser (1996) presented a comprehensive review on models for asymmetric proximities on the basis of the similarity and bias model. In the review, they classified a variety of models into three classes: (i) distance models, (ii) similarity choice models, and (iii) feature models. As a matter of interest, they restated the bias components as the sum of symmetric and skew-symmetric components as follows. Define

$$u_i = \tfrac{1}{2}(\gamma_i + c_i) \qquad \text{and} \qquad v_i = \tfrac{1}{2}(\gamma_i - c_i) \tag{4.69}$$

Then we see that

$$\gamma_i + c_j = u_i + u_j + v_i - v_j \tag{4.70}$$

For a moment, let us set $F$ as the identity function and let $w_{ij} = v_i - v_j$. Then model (4.66) becomes

$$\psi_{ij} = \gamma_{ij} + u_i + u_j + w_{ij} \tag{4.71}$$

Hence, for the skew-symmetric part of the model, we have an expression of transitivity (or additivity) as

$$w_{ij} + w_{jk} = w_{ik} \tag{4.72}$$

We remark on a similar relation that was derived in connection with SSA in Sec. 3.2. Given observations of $\psi_{ij}$, which is denoted by $o_{ij}$, we consider decomposition (3.3) as $o_{ij} = s_{ij} + a_{ij}$. Theorem 3.2 indicates the unidimensional structure inherent in the skew-symmetric data $a_{ij}$, if the following additivity holds as

$$a_{ij} + a_{jk} = a_{ik} \tag{4.73}$$

In this regards, it may be of interest to refer to the treatment in Sec. 2.3, where the skew-symmetric component is represented as the sum of linear and non-linear components in (2.11).

## Comparison with Analysis of Contingency Table

If we set an exponential function for $F$ in (4.66) we have a multiplicative version, such that

$$\psi_{ij} = K \exp(\gamma_{ij}) \exp(r_i) \exp(c_j) \tag{4.74}$$

and its logarithmic form as

$$\log \psi_{ij} = \gamma_{ij} + r_i + c_j + \kappa \tag{4.75}$$

where $\kappa = \log K$. At this point, let us think of a *square* contingency table where $\pi_{ij}$ may indicate the mobility from $i$ to $j$ in terms of frequency or represents the probability of an observation being in the cell $(i, j)$. For analyzing the contingency table, we remember a log-linear model such as

$$\pi_{ij} = KA_iB_jC_{ij} \tag{4.76}$$

Comparing (4.66), (4.74), (4.75), and (4.76), we find the common form of model expressions. However, we note the difference in the focus of analysis in those approaches. In the similarity and bias model, the principal interest will often be on the symmetric component $\gamma_{ij}$, while in the analysis of the square contingency table, the interaction $C_{ij}$ will be of less interest than the row and column effects.

Let us review research in this connection. Levin and Brown (1979) suggested two least squares procedures on a quasi-symmetry model. Caussinus and de Falguerolles (1987) suggested analyzing quasi-independence and quasi-symmetry in both multiplicative and additive models. On closely related subjects, Bove (1992) studied the relation between asymmetric MDS and correspondence analysis, followed by Bove and Critchley (1993). Van der Heijden and Mooijaart (1995) proposed log-bilinear models with focus on the asymmetry. Turning to social mobility, residential mobility or spatial interaction models, Goldthorpe (1980), Willekens (1983), and Scholten (1984) developed analysis related to asymmetry in the respective subject of research.

## 4.4. GENERALIZATION OF DISTANCE MODELS

In this section we are concerned with an $n \times n$ dissimilarity matrix $\boldsymbol{\Omega} = (\omega_{jk})$ and its observation matrix $\boldsymbol{O} = (o_{jk})$. We use $\boldsymbol{\Omega}$ for representation of models and $\boldsymbol{O}$ for estimation of model parameters. To begin, we classify asymmetric data matrices and consider some conditions associated with those matrices (Saito, 1991, 1993a; Saito and Takeda, 1990).

### 4.4.1. Types of Asymmetric Data Matrices

Noting diagonal entries, we like to classify asymmetric matrices $\boldsymbol{\Omega} = (\omega_{jk})$ into three types.

T1: Diagonal entries are not all zero.
T2: Diagonal entries are not all zero and not all identical.
T3: Diagonal entries are all zero.

Apparently, type T2 is a special case of type T1, but we like to differentiate them for conveniences of model description. Matrices of type T3 arise from some cases in which the observation of a phenomenon defines zero values for diagonals (for example, pecking frequency matrices), or from other cases in which diagonal entries are defined but missing diagonal entries are simply filled with zero values before data analysis.

In what follows, we describe two models based on generalization of distance, Model I for matrices of type T1 and Model II for matrices of type T2, and the slide-vector model for matrices of type T1. For models to accommodate matrices of type T3, we will provide some descriptions on the related subjects.

## 4.4.2. Some Conditions of Asymmetric Data

For a matrix type T1 or T2, we define the following quantities:

$$\omega_{jk}^{-} = \tfrac{1}{2}(\omega_{jk} + \omega_{kj} - \omega_{jj} - \omega_{kk}) \tag{4.77}$$

$$\omega_{jk}^{+} = \tfrac{1}{2}(\omega_{jk} - \omega_{kj} + \omega_{jj} + \omega_{kk}) \tag{4.78}$$

$$\tilde{\omega}_{jk} = \tfrac{1}{2}(\omega_{jk} - \omega_{kj} + \omega_{jj} - \omega_{kk}) \tag{4.79}$$

$$\omega_{jk}^{*} = \omega_{jk} - \omega_{\cdot k} - \omega_{j\cdot} + \omega_{\cdot\cdot} \tag{4.80}$$

Here $\omega_{j\cdot}$, $\omega_{\cdot k}$, and $\omega_{\cdot\cdot}$ are given as:

$$\omega_{j\cdot} = \frac{1}{n}\sum_{k=1}^{n} \omega_{jk}, \qquad \omega_{\cdot k} = \frac{1}{n}\sum_{j=1}^{n} \omega_{jk}, \qquad \omega_{\cdot\cdot} = \frac{1}{n^2}\sum_{j=1}^{n}\sum_{k=1}^{n} \omega_{jk} \tag{4.81}$$

In a similar way, $\omega_{j\cdot}^{+}$, $\omega_{\cdot k}^{+}$, $\omega_{\cdot\cdot}^{+}$ are given as:

$$\omega_{j\cdot}^{+} = \frac{1}{n}\sum_{k=1}^{n} w_{jk}^{+}, \qquad \omega_{\cdot k}^{+} = \frac{1}{n}\sum_{j=1}^{n} \omega_{jk}^{+}, \qquad \omega_{\cdot\cdot}^{+} = \frac{1}{n^2}\sum_{j=1}^{n}\sum_{k=1}^{n} \omega_{jk}^{+} \tag{4.82}$$

From (4.78) and (4.79) we have

$$\tilde{\omega}_{jk} = \omega_{jk}^{+} - \omega_{kk}^{+} \tag{4.83}$$

$$\omega_{j\cdot}^{+} + \omega_{\cdot j}^{+} = \omega_{jj}^{+} + \omega_{\cdot\cdot}^{+} \tag{4.84}$$

The mean over diagonal elements is written as

$$\omega_{d} = \frac{1}{n}\sum_{j=1}^{n} \omega_{jj} = \frac{1}{n}\sum_{j=1}^{n} \omega_{jj}^{+} = \omega_{\cdot\cdot}^{+} \tag{4.85}$$

For convenience of exposition, we state conditions for all ordered triples $(i, j, k)$ or ordered pairs $(j, k)$ as follows.

### Generalized Triangular Inequality

For all ordered triples $(i, j, k)$, the following inequality holds:

$$\omega_{ij} + \omega_{jk} - \omega_{ik} \geq \omega_{jj} \tag{4.86}$$

## Additivity

For all ordered triples $(i, j, k)$, the following equality holds:

$$\tilde{\omega}_{ij} + \tilde{\omega}_{jk} = \tilde{\omega}_{ik} \tag{4.87}$$

It is alternatively expressed by either of the following:

$$\omega_{ij} + \omega_{jk} + \omega_{ki} = \omega_{ji} + \omega_{ik} + \omega_{kj} \tag{4.88}$$

$$\omega_{ij}^{+} + \omega_{jk}^{+} + \omega_{ki}^{+} = \omega_{ji}^{+} + \omega_{ik}^{+} + \omega_{kj}^{+} \tag{4.89}$$

Hence (4.87), (4.88), and (4.89) are mutually equivalent. Equation (4.89) implies that

$$\omega_{jk}^{+} = \omega_{j\cdot}^{+} + \omega_{\cdot j}^{+} - \omega_{jj}^{+} \tag{4.90}$$

From (4.87) it follows that

$$\tilde{\omega}_{jk} = \tilde{\omega}_{j\cdot} - \tilde{\omega}_{k\cdot} = \tilde{\omega}_{\cdot k} - \tilde{\omega}_{\cdot j} \tag{4.91}$$

Thus, when the additivity holds, $\tilde{\mathbf{\Omega}} = (\tilde{\omega}_{jk})$ is a matrix of rank 2.

## Centered Symmetry Condition

For all ordered pairs $(j, k)$, the following equality holds:

$$\omega_{jk}^{*} = \omega_{kj}^{*} \tag{4.92}$$

## Constant Ratio Condition

For all ordered pairs $(j, k)$ such that $\omega_{jj} \neq \omega_{kk}$, the following equality holds:

$$\frac{\omega_{jk} - \omega_{kj}}{\omega_{jj} - \omega_{kk}} = c \text{ (const.)} \tag{4.93}$$

The generalized triangular inequality, the additivity, and the centered symmetry condition are applicable to matrices of type T1 or T2. The constant ratio condition is meaningful for data matrices of type T2. Before generalization of distance models, we describe a theorem presented by Saito (1991), which plays an important role in the subsequent models. It states the metric property involved in an asymmetric matrix.

*Theorem 4.4.* Given an asymmetric matrix $\mathbf{\Omega}$ of type T1, the additivity holds with $\mathbf{\Omega}$ if and only if the centered symmetry holds with $\mathbf{\Omega}$. If the generalized triangular inequality holds with $\mathbf{\Omega}$, $\omega_{jk}^{-}$, of (4.77) satisfies the metric axioms.

The first statement is derived easily by a simple manipulation (see Saito, 1991). Let us prove the latter statement of the theorem. First, it is seen that the symmetry obviously holds, because

$$\omega_{\overline{jk}} = \omega_{\overline{kj}} \qquad \text{and} \qquad \omega_{\overline{jj}} = 0 \tag{4.94}$$

Assuming the generalized triangular inequality, we have two inequalities for $(i, j, k)$ and $(i, k, j)$. Combining them gives

$$\omega_{jk} + \omega_{kj} \geq \omega_{jj} + \omega_{kk} \tag{4.95}$$

which leads to the non-negativity as $\omega_{\overline{jk}} \geq 0$. Since

$$\begin{aligned} 2(\omega_{\overline{ij}} + \omega_{\overline{jk}} - \omega_{\overline{ik}}) &= (\omega_{ij} + \omega_{jk} - \omega_{ik} - \omega_{jj}) \\ &\quad + (\omega_{kj} + \omega_{ji} - \omega_{ki} - \omega_{jj}) \end{aligned} \tag{4.96}$$

the two terms in parentheses are non-negative by the assumption, hence

$$\omega_{\overline{ij}} + \omega_{\overline{jk}} - \omega_{\overline{ik}} \geq 0 \tag{4.97}$$

From the results above, it follows that $\omega_{\overline{jk}}$ satisfies the metric axioms of a metric. This concludes the proof.

### 4.4.3. Generalized Distance Model I

Supposing an asymmetric matrix of dissimilarity $\mathbf{\Omega} = (\omega_{jk})$ of type T1 where $\omega_{jk} \geq 0$, we represent a model in terms of parameters such as

$$\omega_{jk} = f_{jk} + \theta_j + \phi_k + \gamma \qquad (j, k = 1, 2, \ldots, n) \tag{4.98}$$

where $f_{jk}$ is symmetric, that is, $f_{jk} = f_{kj}$, and $f_{jj} = 0$. We think of two reasons for setting this model. First, when we regard the symmetric component as revealing major information of the data, we may approximate the residual $r_{jk}(= \omega_{jk} - f_{jk})$ by an additive structure such that

$$r_{jk} = \theta_j + \phi_k + \gamma \tag{4.99}$$

among possible others. Next, we consider asymmetry that might be observed in a situation where row and column elements in a proximity matrix play different roles. For such cases model (4.98) seems to serve to explore the structure of asymmetry.

Substituting (4.98) into (4.77) gives $\omega_{\overline{jk}} = f_{jk}$. Assuming that the generalized triangular inequality holds with $\mathbf{\Omega}$, we find, by Theorem 4.4, that $\omega_{\overline{jk}}$ is a metric. Then we denote $f_{jk}$ by $d_{jk}$ to mean distance, and represent the model as

$$\omega_{jk} = d_{jk} + \theta_j + \phi_k + \gamma \qquad (j, k = 1, 2, \ldots, n) \tag{4.100}$$

On the standpoint of MDS, it is of main interest to obtain a configuration of objects in a multidimensional space, which is expressed by an $n \times r$ matrix $\boldsymbol{X} = (x_{jt})$. For this purpose, let us set the Minkowsky $p$ metric as for $d_{jk}$ and restate the model as

$$\omega_{jk} = d_{jk}(\boldsymbol{X}) + \theta_j + \phi_k + \gamma \qquad (j, k = 1, 2, \ldots, n) \tag{4.101}$$

where

$$d_{jk}(\boldsymbol{X}) = \left( \sum_{t=1}^{r} |x_{jt} - x_{kt}|^p \right)^{1/p} \tag{4.102}$$

Saito (1991) proposed this model, which we call the *Generalized distance model I* (GDM-I) in what follows. Model parameters are $d_{jk}$, $\theta_j$, $\phi_k$, $\gamma$ in (4.100), but $\boldsymbol{X}$, $\theta_j$, $\phi_k$, $\gamma$ for specified $p$ and $r$ in (4.101).

Let us consider (4.98) by referring to Zielman and Heiser (1996). Define

$$\begin{aligned} b_j &= \tfrac{1}{2}(\theta_j + \phi_j) \\ c_j &= \frac{1}{2}(\theta_j - \phi_j) \\ g_{jk} &= f_{jk} + b_j + b_k + \gamma \end{aligned} \tag{4.103}$$

Then the model is expressed by the sum of symmetric $g_{jk}$ and skew-symmetric components $c_j - c_k$ in such a way that

$$\omega_{jk} = g_{jk} + c_j - c_k \qquad (j, k = 1, 2, \ldots, n) \tag{4.104}$$

For a moment, assume a situation in which $o_{jk}$, observations of $\omega_{jk}$, are given. Consider the decomposition (3.3), such as $o_{jk} = s_{jk} + a_{jk}$, and apply the relation of (4.5). When $f_{jk}$ is provided with an explicit expression, we can deal with two problems of least squares fitting separately, one for fitting $g_{jk}$ to $s_{jk}$ and the other for fitting $c_j - c_k$ to $a_{jk}$. However, in the first problem, setting a distance function for $g_{jk}$ would not attain a good fit for a case in which $\omega_{\bar{jk}}$ is a metric, which means that $f_{jk}$ is a metric. The requirement for $g_{jk}$ to be a metric leads to the contradiction that all the $\omega_{jj}$ should be zero. In such a case, it is not reasonable to treat the first LS problem on $\{s_{jk}\}$ by setting $g_{jk}$ to a distance function.

## Properties of GDM-I

There is indeterminacy of translation for the parameters of the additive part in model (4.100) or (4.101), as shown by

$$\begin{aligned}\omega_{jk}^{+} &= (\theta_j + c\gamma) + (\phi_k - c\gamma) + \gamma \\ &= (\theta_j - a) + (\phi_k - b) + (\gamma + a + b)\end{aligned} \tag{4.105}$$

where $a$, $b$, $c$ are unknown constants. Then we may put $\gamma = 0$ simply. However, we will retain the form with $\gamma$ for the model.

When model (4.100) holds, it is decomposed as

$$\omega_{jk}^{-} = d_{jk} \tag{4.106}$$

$$\omega_{jk}^{+} = \theta_j + \phi_k + \gamma \tag{4.107}$$

The model implies that following relation:

$$\tilde{\omega}_{jk} = \theta_j - \theta_k \tag{4.108}$$

Comparing it with (4.91), we see that $\theta_j = \tilde{\omega}_{j\cdot} + \alpha$, where $\alpha$ is a constant.

*Theorem 4.5.* Necessity and sufficiency of the model. If model (4.100) holds with $\boldsymbol{\Omega} = (\omega_{jk})$, then both the generalized triangular inequality and the additivity hold. Conversely, if the two conditions hold with $\boldsymbol{\Omega}$, then a representation of $\omega_{jk}$ is provided with model (4.100).

When the two conditions hold, the additive part of the model is represented in several ways, as stated below.

## Algebraic Solutions for GDM-I

Suppose that we are given dissimilarities $\boldsymbol{\Omega} = (\omega_{jk})$ that satisfy the generalized triangular inequality (4.86) and the additivity (4.87). We like to obtain a solution of parameters for model (4.101) by setting $p = 2$ for (4.102). The model decomposition of (4.106) to (4.108), indicates that we can treat $\boldsymbol{X}$ and the other parameters independently. First, for a set of distances $\boldsymbol{\Omega}^{-} = (\omega_{jk}^{-})$, we aim to obtain an $n \times r$ matrix $\boldsymbol{X} = (x_{jt})$ to represent object coordinates in an $r$-dimensional Euclidean space. Then it leads to the problem of metric multidimensional scaling (Torgerson, 1952). It is easily handled by using a well-known theorem due to Young and Householder (1938). We can determine $\boldsymbol{X}$ up to rotational and translational indeterminacy. The dimensionality $r$ is determined through the analysis. If we set a weighted Euclidean distance for $d_{jk}$, the problem can be treated algebraically under some constraints, using the procedure by Schönemann (1972).

Next we aim to obtain $\theta_j$, $\phi_k$, and $\gamma$ to meet (4.107) and (4.108). To cope with the indeterminacy of the additive parameters, we may put two constraints appropriately. Here are given three solutions. Related subjects were studied by Bove and Critchley (1993) in connection with a square contingency table.

*Case 1.* Under constraints that

$$\sum_{j=1}^{n} \theta_j = \sum_{j=1}^{n} \phi_j = 0 \tag{4.109}$$

the solution is given as

$$\theta_j = \omega_{j\cdot}^{+} - \omega_{\cdot\cdot}^{+}, \qquad \phi_j = \omega_{\cdot j}^{+} - \omega_{\cdot\cdot}^{+}, \qquad \gamma = \omega_{\cdot\cdot}^{+}$$
$$(j = 1, 2, \ldots, n) \tag{4.110}$$

*Case 2.* Under constraints that

$$\sum_{j=1}^{n} \theta_j = \sum_{j=1}^{n} \phi_j \qquad \text{and} \qquad \gamma = 0 \tag{4.111}$$

the solution is given as

$$\theta_j = \omega_{j\cdot} - \omega_{\cdot j} + \omega_{jj}, \qquad \phi_j = \omega_{\cdot j} - \omega_{j\cdot} + \omega_{jj} \tag{4.112}$$

*Case 3.* Under constraints that

$$\sum_{j=1}^{m} \theta_j = 0 \qquad \text{and} \qquad \gamma = 0 \tag{4.113}$$

the solution is given as

$$\theta_j = \tfrac{1}{2}(\omega_{j\cdot} - \omega_{\cdot j} + \omega_{jj} - \omega_d),$$
$$\phi_j = \tfrac{1}{2}(\omega_{\cdot j} - \omega_{j\cdot} + \omega_{jj} + \omega_d) \tag{4.114}$$

Across the three solutions, we find that when the degree of asymmetry $\kappa_1$ is small and the diagonal entries are large, $\theta$ values become nearly equal to $\phi$ values.

## Estimation Procedures

Given an asymmetric data matrix $\boldsymbol{O}$, which is an observation of $\boldsymbol{\Omega}$ with error $\boldsymbol{E} = (e_{jk})$, that is, $\boldsymbol{O} = \boldsymbol{\Omega} + \boldsymbol{E}$, we set the error model

$$\begin{aligned} o_{jk} &= \omega_{jk} + e_{jk} \\ &= d_{jk}(\boldsymbol{X}) + \theta_j + \phi_k + \gamma + e_{jk} \qquad (j, k = 1, 2, \ldots, n) \end{aligned} \tag{4.115}$$

where

$$d_{jk}(\boldsymbol{X}) = \left( \sum_{t=1}^{r} |x_{it} - x_{jt}|^p \right)^{1/p} \tag{4.116}$$

From $\boldsymbol{O}$ we construct $\boldsymbol{O}^{-} = (o_{jk}^{-})$, of which definition conforms to that of $\boldsymbol{\Omega}^{-}$. If $\boldsymbol{O}^{-}$ meets the metric conditions approximately, it may be allowed to set $d_{jk}(\boldsymbol{X})$ in the model. For estimation of parameters, two procedures are described briefly.

### (I) Two-Stage Estimation Algorithm

Setting Euclidean distance for $d_{jk}(\boldsymbol{X})$, two-stage algorithm is suggested for estimation. Applying the procedure of metric MDS of Torgerson (1952) to $\boldsymbol{O}^{-}$, we obtain a least squares estimate of $\boldsymbol{X}$ and determine the dimensionality $r$ to attain the best fit. Write the estimate as $\hat{\boldsymbol{X}}$, from which estimates of distance $\hat{d}_{jk}$ are derived. Let $a_{jk} = o_{jk} - \hat{d}_{jk}$. A solution to minimize

$$Q = \sum_{j=1}^{n} \sum_{k=1}^{n} (a_{jk}(\hat{\boldsymbol{X}}) - \theta_j - \phi_k - \gamma)^2 \tag{4.117}$$

is given in view of the indeterminacy in such a way that

$$\begin{aligned} &\hat{\gamma} = a_{..}, \qquad \hat{\theta}_j = a_{j.} - a_{..}, \qquad \hat{\phi}_j = a_{.j} - a_{..}, \\ &(j = 1, \ldots, n) \end{aligned} \tag{4.118}$$

Here $a_{j.}$, $a_{.j}$, and $a_{..}$ refer to the row mean, the column mean, and the total mean on $\boldsymbol{A} = (a_{jk})$, respectively.

## (II) Iterative Algorithm for Nonlinear Minimization

We aim to obtain estimates of parameters by fitting the model to the data, minimizing a least squares criterion defined by

$$L_1(\boldsymbol{X}, \boldsymbol{\theta}, \boldsymbol{\phi}, \gamma) = \sum_{j=1}^{n}\sum_{k=1}^{n}(o_{jk} - (d_{jk}(\boldsymbol{X}) + \theta_j + \phi_k + \gamma))^2 \tag{4.119}$$

under some constraints to the additive parameters. Those constraints will be such as (4.109), (4.111), or (4.113). If Euclidean distance is set for $d_{jk}(\boldsymbol{X})$, we may regard (4.118) as conditional estimates of $\boldsymbol{\theta}$, $\boldsymbol{\phi}$, and $\boldsymbol{\gamma}$ on an arbitrary $\boldsymbol{X}$ (not a null matrix). Using them, we consider a loss function $L_2$ in terms of $\boldsymbol{X}$ only, so that we minimize

$$L_2(\boldsymbol{X}) = \sum_{j=1}^{n}\sum_{k=1}^{n}(o_{jk} - (d_{jk}(\boldsymbol{X}) + \theta_j(\boldsymbol{X}) + \phi_k(\boldsymbol{X}) + \gamma(\boldsymbol{X})))^2 \tag{4.120}$$

Hence estimates of parameters are obtained by solving the unconstrained minimization of $L_2(\boldsymbol{X})$ for a prescribed dimension $r$.

### 4.4.4. Generalized Distance Model II

Considering an asymmetric matrix of dissimilarity $\boldsymbol{\Omega} = (\omega_{jk})$ of type T2 where $\omega_{jk} \geq 0$, we represent a variant of model I such as

$$\omega_{jk} = d_{jk}(\boldsymbol{X}) + a\theta_j + b\theta_k + \gamma \qquad (j, k = 1, 2, \ldots, n) \tag{4.121}$$

where $d_{jk}$ is a metric. For MDS, we set the Minkowsky distance of (4.102) for it. Saito and Takeda (1990) proposed this model, which we call the *Generalized distance model II* (GDM-II) in what follows.

The assumption of type T2 is equivalent to putting that $a + b \neq 0$ and not all $\theta_j$ are equal. The asymmetry of $\boldsymbol{\Omega}$ implies that $a \neq b$ and not all $\theta_j$ are equal. Thus, we can define a constant $c$ by

$$c = \frac{a - b}{a + b} \tag{4.122}$$

From (4.121), it follows that the constant ratio condition holds with $c$ given above, in such a way as

$$\frac{\omega_{jk} - \omega_{kj}}{\omega_{jj} - \omega_{kk}} = c \tag{4.123}$$

Noting that

$$\omega_{jj} = (a+b)\theta_j + \gamma \tag{4.124}$$

we find that (4.121) is equivalent to the following expression:

$$\omega_{jk} = d_{jk}(\boldsymbol{X}) + \tfrac{1}{2}(1+c)\omega_{jj} + \tfrac{1}{2}(1-c)\omega_{kk} \tag{4.125}$$

Hence the model accounts for the asymmetry by a single parameter $c$.

## Property of GDM-II

When model (4.121) holds, it is decomposed as

$$\omega_{jk}^{-} = d_{jk} \tag{4.126}$$

$$\omega_{jk}^{+} = a\theta_j + b\theta_k + \gamma \tag{4.127}$$

The model implies the following relations:

$$\omega_{jk}^{+} = \tfrac{1}{2}(1+c)\omega_{jj} + \tfrac{1}{2}(1-c)\omega_{kk} \tag{4.128}$$

$$\tilde{\omega}_{jk} = a(\theta_j - \theta_k) \tag{4.129}$$

*Theorem 4.6.* Necessity and sufficiency of the model. If model (4.121) holds with $\boldsymbol{\Omega} = (\omega_{jk})$, then the generalized triangular inequality, the additivity, and the constant ratio condition hold. Conversely, if the three conditions hold with $\boldsymbol{\Omega}$, then a representation of $\omega_{jk}$ is provided with model (4.121).

From the constant ratio condition it follows that

$$\frac{\omega_{j\cdot} - \omega_{\cdot j}}{\omega_{jj} - \omega_{d}} = c \qquad (j = 1, 2, \ldots, n) \tag{4.130}$$

To identify the parameters, we have to specify some constraints. The additive part of the model is rewritten as:

$$a\theta_j + b\theta_k + \gamma = a\theta'_j + b\theta'_k = \phi_j + \lambda\phi_k \tag{4.131}$$

where

$$\theta'_j = \theta_j + \frac{\gamma}{a+b}, \qquad \phi_j = a\theta'_j, \qquad \text{and} \qquad \lambda = \frac{b}{a} \neq 1 \tag{4.132}$$

Thus there is indeterminacy of translation and multiplication for the additive part. Using $\lambda$, (4.122) is rewritten as

$$c = \frac{1-\lambda}{1+\lambda} \tag{4.133}$$

## Algebraic Solutions for GDM-II

Suppose that we are given dissimilarities $\boldsymbol{\Omega} = (\omega_{jk})$ that satisfy the generalized triangular inequality (4.86), the additivity (4.87), and the constant ratio condition (4.93). We like to obtain a solution of parameters by setting Euclidean distance $d_{jk}(\boldsymbol{X})$ for model (4.121). According to the model decomposition of (4.126) to (4.127), we can treat $\boldsymbol{X}$ and the other parameters independently. As is the case for Model I, we can obtain $\boldsymbol{X}$ with the indeterminacy of rotation and translation by metric MDS. Now consider obtaining $\theta_j$ and $\gamma$ to meet (4.127) and (4.128). To cope with the indeterminacy of the additive parameters, we may put two constraints appropriately. Here are given two solutions. First, it should be noted that $c$ is given by (4.130). Let

$$\sigma_\omega = \left(\sum_{j=1}^{n} (\omega_{jj} - \omega_d)^2\right)^{1/2} \quad \text{and} \quad \omega_{dr} = \left(\sum_{j=1}^{n} \omega_{jj}^2\right)^{1/2} \tag{4.134}$$

*Case 1.* Under constraints that

$$\sum_{j=1}^{n} \theta_j^2 = 1 \quad \text{and} \quad \sum_{j=1}^{n} \theta_j = 0 \tag{4.135}$$

a solution is given as

$$\theta_j = \pm \frac{1}{\sigma_\omega}(\omega_{jj} - \omega_d) \quad (j = 1, 2, \ldots, n), \quad \gamma = \omega_d \tag{4.136}$$

$$a = \pm \frac{1}{2}(1 + c)\sigma_\omega \quad \text{and} \quad b = \pm \frac{1}{2}(1 - c)\sigma_\omega \tag{4.137}$$

*Case 2.* Under constraints that

$$\sum_{j=1}^{n} \theta_j^2 = 1 \quad \text{and} \quad \gamma = 0 \tag{4.138}$$

a solution is given as

$$\theta_j = \pm \frac{\omega_{jj}}{\omega_{dr}} \quad (j = 1, 2, \ldots, n) \tag{4.139}$$

$$a = \pm \tfrac{1}{2}(1 + c)\omega_{dr} \quad \text{and} \quad b = \pm \tfrac{1}{2}(1 - c)\omega_{dr} \tag{4.140}$$

## Estimation Procedures

Given an asymmetric data matrix $\boldsymbol{O}$ that is an observation of $\boldsymbol{\Omega}$, we set the error model $o_{jk} = \omega_{jk} + e_{jk}$ where $\omega_{jk}$ is specified by model (4.121). In view of the indeterminacy of the additive part, we may set

$$o_{jk} = d_{jk}(\boldsymbol{X} + \theta_j + \lambda\theta_k + \gamma + e_{jk} \tag{4.141}$$

$$= d_{jk}(\boldsymbol{X}) + \phi_j + \lambda\phi_k + e_{jk} \qquad (j, k = 1, 2, \ldots, n) \tag{4.142}$$

where

$$\phi_j = \theta_j + \frac{\gamma}{1+\lambda} \qquad (j = 1, 2, \ldots, n) \tag{4.143}$$

and

$$d_{jk}(\boldsymbol{X}) = \left( \sum_{t=1}^{r} |x_{it} - x_{jt}|^p \right)^{1/p} \tag{4.144}$$

If $\boldsymbol{O}^-$ meets the metric axioms approximately, one can regard $\boldsymbol{O}^-$ as a metric. Then it may be allowed to set $d_{jk}(\boldsymbol{X})$ in the model.

When Euclidean distance is specified for $d_{jk}(\boldsymbol{X})$, we can estimate $c$ (or equivalently $\lambda$) and then $\boldsymbol{X}$. First, regarding (4.130) as an approximate relation, we have a least squares estimate $\hat{c}$, from which follows an estimate $\hat{\lambda}$ by (4.133). Next, let us assume that (4.126) holds approximately with $o_{jk}^-$ and also (4.127) does with $o_{jk}^+$. Then we have

$$o_{jk}^- = d_{jk}(\boldsymbol{X}) \tag{4.145}$$

$$o_{jk} = d_{jk}(\boldsymbol{X}) + \tfrac{1}{2}(1+c)o_{jj} + \tfrac{1}{2}(1-c)o_{kk} \tag{4.146}$$

$$o_{jk}^+ = \phi_j + \lambda\phi_k \tag{4.147}$$

As a simple but rough way to obtain $\boldsymbol{X}$, we might apply the procedure of metric MDS to (4.145). Another way is to compute $d_{jk}(\boldsymbol{X})$ by substituting $\hat{c}$ in (4.146) and applying the procedure to the derived distances in order to obtain $\boldsymbol{X}$. When $d_{jk}(\boldsymbol{X})$ with $p \neq 2$ is specified, it is required to estimate $\boldsymbol{X}$ and $c$ by a nonlinear minimization $L(\boldsymbol{X}, c)$ based on (4.146).

Given $\hat{\boldsymbol{X}}$ and $\hat{c}$, $\hat{\lambda}$ is computed by (4.133). Finally, substituting $\hat{\lambda}$ in (4.147), we derive least squares estimates for $\phi_j$.

**Remark.** Let us mention some treatments of variants of GDM-I and GDM-II with Euclidean distance specified in the distance function. Denote a centering matrix by $\boldsymbol{H} = \boldsymbol{I}_n - \boldsymbol{1}\boldsymbol{1}'/n$. For either model, it holds that

$$\boldsymbol{H\Omega H} = \boldsymbol{HDH} \tag{4.148}$$

It is not convenient to work on the matrix of doubly centered distances. Now consider variants of the two models such that

$$\omega_{jk} = d_{jk}^2(\boldsymbol{X}) + \theta_j + \phi_k + \gamma \tag{4.149}$$

$$\omega_{jk} = d_{jk}^2(\boldsymbol{X}) + a\theta_j + b\theta_k + \gamma \tag{4.150}$$

For either model, it holds that

$$\boldsymbol{H\Omega H} = \boldsymbol{HD}^{(2)}\boldsymbol{H} \qquad \text{where} \qquad \boldsymbol{D}^{(2)} = (d_{jk}^2) \tag{4.151}$$

Then it is possible to apply the metric MDS procedure directly to the doubly centered matrix of squared Euclidean distances. According to this treatment, we can derive $\boldsymbol{X}$ from $\boldsymbol{\Omega}$ or $\hat{\boldsymbol{X}}$ from $\boldsymbol{O}$.

### 4.4.5. Distance-Density Model and Other Models

Krumhansl (1978) proposed the distance-density model, which is based on a hypothesis that (dis)similarity is a function of both interpoint distance and the spatial density of other points in the surrounding region of the configuration. For the sake of exposition, we will use stimuli and the term dissimilarity, which we write as $\omega_{jk}$ here. Regarding asymmetric dissimilarity $\omega_{jk}$ of type T2, Krumhansl originally presented a nonmetric model, but here we may state a metric version as

$$\omega_{jk} = d_{jk}(\boldsymbol{X}) + \alpha\rho_j(\boldsymbol{X}) + \beta\rho_k(\boldsymbol{X}) + \gamma \tag{4.152}$$

where $d_{jk}(\boldsymbol{X})$ is distance in terms of coordinates $\boldsymbol{X}$, density $\rho_j(\boldsymbol{X})$, $\alpha$ and $\beta$ positive parameters. Two definitions of the density were suggested. First, it is defined as

$$\rho_j(\boldsymbol{X}) = \sum_{k \neq j}^{n} g(d_{jk}(\boldsymbol{X})) \tag{4.153}$$

where $g(\cdot)$ is a decreasing function of distance, such as $g(d) = d^{-h}$ or $g(d) = \exp(-hd)$. Another expression is

$$\rho_j(\boldsymbol{X}) = \sum_{k=1}^{n} \varepsilon_{jk} \tag{4.154}$$

where $\varepsilon_{jk} = 1$ if $d_{jk}(\boldsymbol{X}) \leq \rho$ and $\varepsilon_{jk} = 0$ otherwise. That is, $\rho_j(\boldsymbol{X})$ indicates the number of stimulus points included within a fixed radius $\rho$.

According to the model, the self-dissimilarity is given as

$$\omega_{jj} = (\alpha + \beta)\rho_j + \gamma \tag{4.155}$$

Since both $\alpha$ and $\beta$ are positive, we see that $\omega_{jj} > \omega_{kk}$ if and only if $\rho_j > \rho_k$. Note that

$$\omega_{jk} - \omega_{kj} = (\alpha - \beta)(\rho_j - \rho_k) \tag{4.156}$$

When $\alpha > \beta$, then $\omega_{jk} > \omega_{kj}$ if and only if $\omega_{jj} > \omega_{kk}$. In other words, if object $j$ is more dissimilar to $k$ than $k$ to $j$, then self-dissimilarity of $j$ is larger than that of $k$, and also the converse is true. When $\alpha < \beta$, then $\omega_{jk} < \omega_{kj}$ if and only if $\omega_{jj} > \omega_{kk}$. Hence diagonal elements are related to the direction of the asymmetry.

## Property of the Model

Given an asymmetric dissimilarity matrix $\mathbf{\Omega} = (\omega_{jk})$ of type T2, let us consider the property of model (4.152). Besides the concept of density, the dissimilarity is defined by a form of the sum of a distance and two terms dependimig on stimuli and an additive constant. Then it should be noted that the argument from (4.121) to (4.130) still holds for model (4.152), by replacing $\theta_j$ by $\rho_j(\boldsymbol{X})$. For clarity of the exposition, we restate a theorem for the present model and its decomposition.

It is necessary and sufficient for model (4.152) to hold with $\mathbf{\Omega} = (\omega_{jk})$ of type T2, that the generalized triangular inequality, the additivity, and the constant ratio condition hold.

Model (4.152) is decomposed into two parts:

$$\omega_{jk}^- = d_{jk}(\boldsymbol{X}) \tag{4.157}$$

$$\omega_{jk}^+ = \alpha\rho_j(\boldsymbol{X}) + \beta\rho_k(\boldsymbol{X}) + \gamma \tag{4.158}$$

## Algebraic Treatment

Given dissimilarities $\mathbf{\Omega} = (\omega_{jk})$ that satisfy the generalized triangular inequality (4.86), the additivity (4.87), and the constant ratio condition (4.93), we like to obtain $\boldsymbol{X}$ and parameters or $\rho_j(\boldsymbol{X})$. It is not tractable to deal with (4.152) as a set of nonlinear equations in terms of $\boldsymbol{X}$, $\alpha$, $\beta$, and $\gamma$, for which the function forms of $d_{jk}(\boldsymbol{X})$ and $\rho_j(\boldsymbol{X})$ are specified. Now let us specify Euclidean distance for $d_{jk}(\boldsymbol{X})$. In the same way as in the case of Model II, we can determine dimensionality $r$ and an $n \times r$ matrix $\boldsymbol{X}$ from (4.157), using the procedure of Torgerson. Given $\boldsymbol{X}$ and a nonlinear form of $\rho_j(\boldsymbol{X})$, it would be difficult to solve $\alpha$, $\beta$, and $\gamma$. Now we regard (4.158) as an equation to obtain parameters and function values of $\rho_j(\boldsymbol{X})$. First, we can determine $c$ defined by (4.122), using (4.123) or (4.130). Note that $c < 1$ due to the positivity of $\alpha$ and $\beta$.

The density values should not have translational indeterminacy so we set $\gamma = 0$. To cope with the product term such as $\alpha\rho_j(\boldsymbol{X})$, two treatments are conceivable among others.

*Case 1.* Set either $\alpha$ or $\beta$ to be unity. Simply, imposing $\alpha = 1$, we have

$$\beta = \frac{1-c}{1+c} \tag{4.159}$$

Thus we obtain from (4.158) with $\gamma = 0$,

$$\rho_j(\boldsymbol{X}) = \frac{1}{1+\beta}\omega_{jj} = \frac{1}{2}(1+c)\omega_{jj} \qquad (j = 1, 2, \ldots, n) \tag{4.160}$$

*Case 2.* To determine the relative values of density, it is imposed that

$$\sum_{j=1}^{n} \rho_j(\boldsymbol{X})^2 = 1 \tag{4.161}$$

Retain definition (4.134). Then the solution is given as

$$\rho_j(\boldsymbol{X}) = \frac{\omega_{jj}}{\omega_{dr}} \qquad (j = 1, 2, \ldots, n) \tag{4.162}$$

$$\alpha = \tfrac{1}{2}(1+c)\omega_{dr} \qquad \text{and} \qquad \beta = \tfrac{1}{2}(1-c)\omega_{dr} \tag{4.163}$$

For both cases the product terms are determined as

$$\alpha\rho_j(\boldsymbol{X}) = \frac{1}{2}(1+c)\omega_{jj} \qquad \text{and} \qquad \beta\rho_j(\boldsymbol{X}) = \frac{1}{2}(1-c)\omega_{jj} \tag{4.164}$$

Only one parameter $c$ accounts for the asymmetry. It would be useful to examine the relationship of $\boldsymbol{X}$ and the function values $\rho_j(\boldsymbol{X})$ $(j = 1, 2, \ldots, n)$, for detection of the form of density (4.153).

In passing, the distance-density hypothesis may be incorporated in clustering methods. DeSarbo et al. (1990) proposed a tree-fitting algorithm based on the hypothesis, and showed an example on the confusion data regarding ten stimuli, extracted from that collected by Rothkopf (1957). It will be meaningful to examine the concept of density on data with more stimuli.

## Weeks and Bentler's Model

Consider an asymmetric dissimilarity matrix $\boldsymbol{\Omega} = (\omega_{jk})$ for which all $\omega_{jk} \geq 0$ and all $\omega_{jj} = c$. For this sort of matrix, Weeks and Bentler (1982) proposed a model such that

$$\omega_{jk} = \beta d_{jk} + \eta_j - \eta_k + \gamma \qquad (j, k = 1, 2, \ldots, n) \tag{4.165}$$

where $\beta > 0$ and $d_{jk}$ is distance in a $p$-dimensional space as defined below, and $p$ is a prescribed dimension. In a similar way to the case of Model I, the following statement is established. It is necessary and sufficient for model (4.165) to hold that both the generalized triangular inequality and the additivity hold with $\boldsymbol{\Omega}$.

A supplement should be given on the model. For the case of a similarity measure $\omega_{jk}$, setting $\beta = -1$ was suggested. Also, an alternative expression with replacing $d_{jk}$ by $d_{jk}^2$ for (4.165) was suggested in view of some appropriate cases. These two points will be mentioned in the example, but the present description is confined to (4.165).

Let $\boldsymbol{x}_j'$ be the $j$th row of $\boldsymbol{X}$. In the formulation by Weeks and Bentler, the distance is specified by

$$d_{jk}(\boldsymbol{X}, \boldsymbol{W}) = ((\boldsymbol{x}_j - \boldsymbol{x}_k)'\boldsymbol{W}(\boldsymbol{x}_j - \boldsymbol{x}_k))^{1/2} \tag{4.166}$$

where $\boldsymbol{W}$ is a symmetric and positive definite matrix of order $p \times p$. Given $o_{jk}$, which are observations of $\omega_{jk}$, one aims to estimate parameters $\eta_j$, $\gamma$, $\boldsymbol{X}$, and $\boldsymbol{W}$ under some constraints, where $\beta$ is usually prespecified. Some are imposed to deal with the indeterminacy of parameters, others are set for allowing arbitrary constraints on parameters for data analysis. They proposed a procedure based on nonlinear least squares under those constraints. However, because of the simple description of the algorithm, it is not clear how one can treat the positivity of $\beta$, and the requirement of $\boldsymbol{W}$ to be positive definite.

Consider decomposition (3.3) $o_{jk} = s_{jk} + a_{jk}$. Let $\boldsymbol{\eta} = (\eta_1, \eta_2, \ldots, \eta_n)$. Using (4.6), the estimation problem is decomposed into two least squares problems:

$$Q_1(\boldsymbol{X}, \boldsymbol{W}, \gamma) = \sum_{j=1}^{n}\sum_{k=1}^{n}(s_{jk} - \beta d_{jk}(\boldsymbol{X}, \boldsymbol{W}) + \gamma)^2 \tag{4.167}$$

$$Q_2(\boldsymbol{\eta}) = \sum_{j=1}^{n}\sum_{k=1}^{n}(a_{jk} - \eta_j + \eta_k)^2 \tag{4.168}$$

Setting the Euclidean distance for $d_{jk}$ and $\beta = 1$, the estimation problem of $\boldsymbol{X}$ is reduced to the well-known problem of metric MDS with additive constant. The second problem (3.3) leads to approximating $\boldsymbol{A}$ by a skew-symmetric matrix of rank 2.

### 4.4.6. Related Subjects

Of principal concern in this section have been the asymmetric matrices of type T1 or T2 and also the distance-based view of the models. Here we describe the case of asymmetric matrices of type T3, and for that case, we present a mixed view of distance and content models.

#### Asymmetric Matrix with Null Diagonals

In connection with the asymmetric matrices of type T3, we reconsider the models and scaling procedures so far mentioned. At first, setting aside asymmetry, it is interesting to note a theorem due to Gower and Legendre (1986).

*Theorem 4.7.* Given a dissimilarity matrix $\boldsymbol{\Omega} = (\omega_{jk})$ with $\omega_{jj} = 0$ for all $j$. If the following inequality holds for all triples $(i, j, k)$,

$$\omega_{ik} + \omega_{ik} \geq \omega_{jk} \tag{4.169}$$

then $\omega_{jk} \geq 0$ and $\omega_{jk} = \omega_{kj}$; hence $\omega_{jk}$ is a metric.

It should be noticed that (4.169) is not the triangular inequality that is one of the metric axioms. Next we start with an asymmetric dissimilarity matrix $\boldsymbol{\Omega}$ with all diagonal entries being zero. On this matrix, the definitions from (4.77) to (4.79) become simple:

$$\omega_{jk}^{-} = \tfrac{1}{2}(\omega_{jk} + \omega_{kj}) \tag{4.170}$$

$$\omega_{jk}^{+} = \tfrac{1}{2}(\omega_{jk} - \omega_{kj}) = \tilde{\omega}_{jk} \tag{4.171}$$

and $\omega_{jk}^{*}$ is of the same form as (4.80). The generalized triangular inequality is reduced to the triangular inequality. Through a simple manipulation on the matrix $\boldsymbol{\Omega}$, we find that Theorem 4.4 still holds. Rewrite the result as a theorem for comparison with Theorem 4.5.

*Theorem 4.8.* Given an asymmetric dissimilarity matrix $\boldsymbol{\Omega} = (\omega_{jk})$ with $\omega_{jj} = 0$ for all $j$. The additivity holds if and oniy if the centered symmetry condition holds. If the triangular inequality holds with $\boldsymbol{\Omega}$, $\omega_{jk}^{-}$ satisfies the metric axioms.

Regarding GDM-I, it is seen that $\omega_{jj} = \theta_j + \phi_j + \gamma = 0$, thus the model is rewritten as

$$\omega_{jk} = d_{jk} + \theta_j - \theta_k \tag{4.172}$$

Regarding GDM-II, noting that $\omega_{jj} = a\theta_j + b\theta_j + \gamma = 0$ and replacing $a\theta_j$ by $\theta_j$, we have the same form as (4.172). Its decomposition is stated as

$$\omega_{jk}^{-} = d_{jk}(\boldsymbol{X}) \qquad \text{and} \qquad \omega_{jk}^{+} = \theta_j - \theta_k = \tilde{\omega}_{jk} \tag{4.173}$$

For (4.172), a statement similar to Theorem 4.5 is given as follows. If the model holds, the generalized triangular inequality and the additivity hold. If the two conditions hold with $\boldsymbol{\Omega} = (\omega_{jk})$, an expression of $\omega_{jk}$ is provided with (4.172).

## Mixed Model for Asymmetry

To account for dissimilarity judgment among a set of stimuli, Saito (1983, 1986) proposed a procedure of MDS based on a model, such that

$$\omega_{jk} = d_{jk}(\boldsymbol{X}) - r_j - r_k \tag{4.174}$$

where $d_{jk}$ is Euclidean distance and $r_j$ is scalar. For the model to hold, it is required that

$$r_j \geq \frac{1}{2} \max_{i,k} (\omega_{ik} - \omega_{ij} - \omega_{jk}) \tag{4.175}$$

In a case for which diagonal entries $\omega_{jj}$ are all zero, the model is defined only for off-diagonal entries. For the model, suggested are three interpretations of $r_j$ with or without restrictions on $r_j$. According to the first, $r_j$ is a free parameter and it indicates a psychological effect associated with stimulus $j$. Second, in connection with the distance-density hypothesis, one may interpret $-r_j$ as indicating the density if estimates are positive (requiring that all $r_j < 0$), although $r_j$ is not specified explicitly in terms of a function of $\boldsymbol{X}$ but a function value in this case.

Third, for some phenomena, the free parameter $r_j$ would represent a measure of the region (the radius of a supersphere) that stimulus $j$ occupies in the multidimensional psychological space. Then (4.174) is regarded as a hypothetical model, mixing distance and content models. The sign of $r_j$ would reflect the reference type of a stimulus. We might think of three cases of psychological distances, depending on the combination of the signs of $r_j$ and $r_k$ (Fig. 4.5): (a) symmetric model with nearest distance ($r_j > 0$, $r_k > 0$), (b) symmetric model with furthest distance ($r_j < 0$, $r_k < 0$),

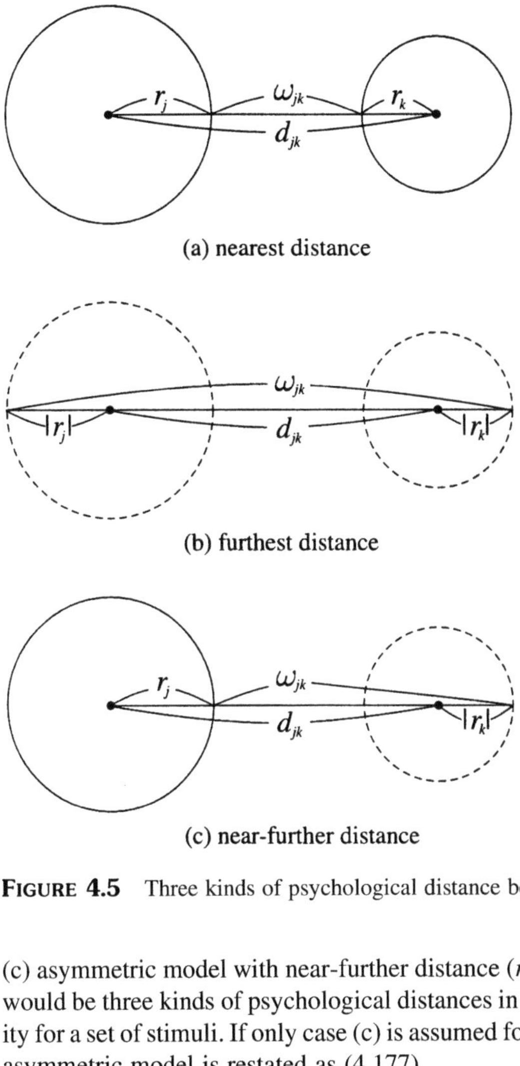


**FIGURE 4.5** Three kinds of psychological distance between stimulus regions

(c) asymmetric model with near-further distance ($r_j < 0$, $r_k > 0$). Hence there would be three kinds of psychological distances in the judgment of dissimilarity for a set of stimuli. If only case (c) is assumed for all ordered pairs $(j, k)$, the asymmetric model is restated as (4.177).

Winsberg and Carroll (1989) developed a quasi-nonmetric method of MDS based on a model closely related to (4.174), such that

$$\omega_{jk} = (d_{jk}(\boldsymbol{X})^2 + s_j + s_k)^{1/2} \qquad (j \neq k) \tag{4.176}$$

In their interpretation $s_j$ represents specificity of object $j$.

Okada and Imaizumi (1987) proposed a nonmetric procedure on the basis of a variant of Weeks and Bentler's model (off-diagonal elements only, and $\gamma = 0$ for (4.165)), such that

$$\begin{aligned} \omega_{jk} &= d_{jk}(X) + r_j - r_k \qquad (j \neq k) \\ r_j &\geq 0 \qquad (j = 1, 2, \ldots, n) \end{aligned} \tag{4.177}$$

where $d_{jk}(X)$ is distance in a multidimensional space. To cope with the translational indeterminacy of $r_j$, it is imposed that $\min_j (r_j) = 0$. For the display of the derived configuration, the authors adopted representing stimuli according to the mixed model above, thus $r_j$ shows the radius of a supersphere (object) $j$.

However, a question may be posed for visually inspecting that sort of geometrical display for case (c) and (4.177). Because of the indeterminacy of the origin of $r_j$ values, a unidimensional display of estimates of $r_j$, such as Fig 4.5, is better and less misleading than the display using spheres. One would identify easily the difference such as $r_j + \alpha - (r_k + \alpha)$ in Fig. 4.5, but from the sphere representation above one would tend to read the difference in proportion to the size of area such as $(r_j + \alpha)^2 - (r_k + \alpha)^2$ rather than the unidimensional difference.

## 4.4.7. Slide-Vector Model

Zielman and Heiser (1993) presented slide-vector model to account for asymmetric dissimilarity data of type T1. It represents asymmetry by a uniform shift (or translation) of the difference vector between the two points in a multidimensional space. The error-free model is stated as

$$\omega_{jk}^2(X, z) = \sum_{t=1}^{r} (x_{jt} - x_{kt} + z_t)^2 \tag{4.178}$$

The dissimilarity is a function of $X$ and $z$, where $X = (x_{jt})$ is an $n \times r$ matrix of coordinates, and $z = (z_t)$ is the slide-vector. Denote $\omega_{jk}(X, z)$ by $\omega_{jk}$ simply. According to the model, the self-dissimilarity is not zero but a constant $l_z$,

$$\omega_{jj} = \left( \sum_{t=1}^{r} z_t^2 \right)^{1/2} = l_z \qquad (j = 1, 2, \ldots, n) \tag{4.179}$$

If points $j$ and $k$ coincide, their dissimilarity is equal to $l_z$. Let $\boldsymbol{Y} = (y_{jt})$ where $y_{jt} = x_{jt} - z_t$. Then (4.178) becomes

$$\omega_{jk} = \left( \sum_{t=1}^{r} (x_{jt} - y_{kt})^2 \right)^{1/2} \tag{4.180}$$

Thus, as far as the model form is concerned, the slide-vector model is regarded as a restricted case of the unfolding model in that the configuration of column points ($\boldsymbol{Y}$) is a translation of the configuration of row points ($\boldsymbol{X}$), such that $\boldsymbol{Y} = \boldsymbol{X} - \mathbf{1}\boldsymbol{z}'$. The model involves a larger degree of asymmetry the more dissimilar are the two objects.

For a graphical representation of the model, we refer to Fig 4.6. Two objects are illustrated as vectors, $\boldsymbol{a}$ and $\boldsymbol{b}$. The discrepancy of the termini of the vectors is drawn by a dashed segment. The difference vectors, $\boldsymbol{a} - \boldsymbol{b}$ and $\boldsymbol{b} - \boldsymbol{a}$, are drawn by bold arrows in the figure. Write their identical length by $l(\boldsymbol{a} - \boldsymbol{b}) = l(\boldsymbol{b} - \boldsymbol{a})$. Adding the slide-vector $z$ to these difference vectors yields $\boldsymbol{a} - \boldsymbol{b} + z$ and $\boldsymbol{b} - \boldsymbol{a} + z$, respectively. The length of each vector indicates the asymmetric dissimilarity, such that

$$\omega_{ab} = l(\boldsymbol{a} - \boldsymbol{b} + z) \qquad \text{and} \qquad \omega_{ba} = l(\boldsymbol{b} - \boldsymbol{a} + z) \tag{4.181}$$

To know the properties of the model in more detail, we expand (4.178)

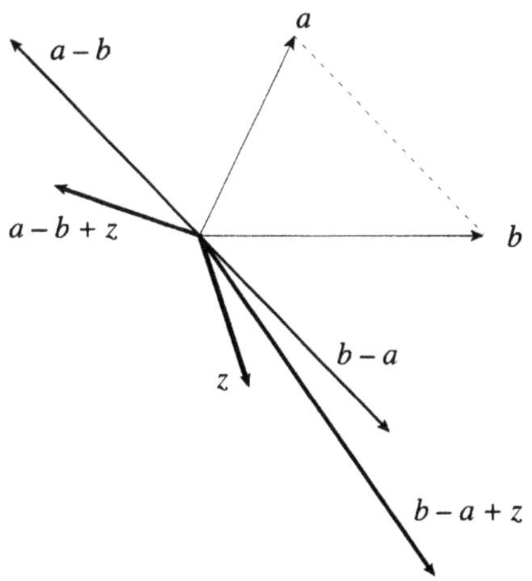


**FIGURE 4.6** Illustration of the slide-vector model

as

$$\omega_{jk}^2 = \sum_{t=1}^{r}(x_{jt} - x_{kt})^2 + \sum_{t=1}^{r} z_t^2 + 2\sum_{t=1}^{r} z_t(x_{jt} - x_{kt}) \tag{4.182}$$

On the right-hand side, the first term shows the squared Euclidean distance and the second the squared norm of $l_z$, both of which contribute to the symmetric component of the dissimilarity. In contrast, the third term accounts for the skew-symmetric component in such a way that

$$\boldsymbol{z}'\boldsymbol{x}_{(j)} - \boldsymbol{z}'\boldsymbol{x}_{(k)} = \sum_{t=1}^{r} z_t(x_{jt} - x_{kt}) = -\sum_{t=1}^{r} z_t(x_{kt} - x_{jt}) \tag{4.183}$$

where $\boldsymbol{x}_{(j)} = (x_{j1}, x_{j2}, \ldots, x_{jr})'$.

From this relation and Fig. 4.6, we find that the relative dominance of one object to the other in the skew-symmetry is shown by the projection of the object to the slide-vector. In other words, objects positioned in directions similar to the slide-vector dominate the others in the asymmetrical relationship.

The slide-vector plays a role as reference point in the space of psychological judgment. In this viewpoint, it may be of interest to compare the slide-vector model with the psychological scaling model in Sec 2.3. Comparing model expressions, we see that model (4.180) is a multidimensional version of (2.50). As shown below, the model can be regarded as a variant of Weeks and Bentler's (1982) model. We like to clarify the properties of the model, considering indeterminacy of the model parameters, the structure of the symmetric part, and that of the skew-symmetric part on the basis of Saito (2003b).

## Indeterminacy of the Parameters

Define a symmetric matrix $\boldsymbol{S} = (s_{jk})$ and a skew-symmetric matrix $\boldsymbol{A} = (a_{jk})$ as:

$$s_{jk} = \tfrac{1}{2}(\omega_{jk}^2 + \omega_{kj}^2) \tag{4.184}$$

$$a_{jk} = \tfrac{1}{2}(\omega_{jk}^2 - \omega_{kj}^2) \tag{4.185}$$

Then we have a decomposition of the squared dissimilarity as

$$\omega_{jk}^2 = s_{jk} + a_{jk} \tag{4.186}$$

to which corresponds the model decomposition as

$$s_{jk} = d_{jk}^2 + l_z^2 \tag{4.187}$$

$$a_{jk} = 2z'(\boldsymbol{x}_{(j)} - \boldsymbol{x}_{(k)}) \tag{4.188}$$

where

$$d_{jk}^2 = \sum_{t=1}^{r} (x_{jt} - x_{kt})^2 \tag{4.189}$$

Let us focus on the indeterminacy of parameters. Let $\boldsymbol{T}$ be an orthogonal matrix and $\boldsymbol{\xi}$ be a translation vector. Consider transformations such as $\tilde{\boldsymbol{X}} = \boldsymbol{XT} + \boldsymbol{\xi}$ and $\tilde{z} = \boldsymbol{T}'z$. Noting that

$$d_{jk}(\tilde{\boldsymbol{X}}) = d_{jk}(\boldsymbol{X}) \qquad \text{and} \qquad z'z = \tilde{z}'\tilde{z} = l_z^2 \tag{4.190}$$

We find that expressions of $s_{jk}$ and $a_{jk}$ are invariant under the rotations and translations. Thus, any constraints or assumptions are required to cope with the indeterminancy.

## Structure of the Skew-Symmetric Part

From (4.185) and (4.188) it follows that

$$\omega_{jk}^2 - \omega_{kj}^2 = 4z'(\boldsymbol{x}_{(j)} - \boldsymbol{x}_{(k)}) \tag{4.191}$$

Summing equations of this type over pairs $(i, j)$, $(j, k)$, and $(k, i)$ gives

$$\omega_{ij}^2 + \omega_{jk}^2 + \omega_{ki}^2 = \omega_{ik}^2 + \omega_{kj}^2 + \omega_{ji}^2 \tag{4.192}$$

Thus a relation of additivity holds with $\{\omega_{ij}^2\}$. Alternatively it is expressed as

$$a_{ik} = a_{ij} + a_{jk} \tag{4.193}$$

Let us define $\boldsymbol{\tau} = \boldsymbol{X}z$, an element of which is

$$\tau_j = z'\boldsymbol{x}_{(j)} = \sum_{t=1}^{r} z_t x_{jt} \qquad (j = 1, 2, \ldots, n) \tag{4.194}$$

Thus $\tau_j$ shows a kind of similarity of object $j$ to the reference point specified by the slide-vector $z$, being proportional to the projection of vector $\boldsymbol{x}_{(j)}$ on vector $z$. From (4.188), $\boldsymbol{A}$ is stated in such a way that

$$\boldsymbol{A} = 2(\boldsymbol{X}z\mathbf{1}' - \mathbf{1}z'\boldsymbol{X}') = 2(\boldsymbol{\tau}\mathbf{1}' - \mathbf{1}\boldsymbol{\tau}') \tag{4.195}$$

showing that $\boldsymbol{A}$ is of rank 2 and $a_{jk} = 2(\tau_j - \tau_k)$. Combining this with (4.186) and (4.187) yields

$$\omega_{jk}^2 = d_{jk}^2 + l_z^2 + 2(\tau_j - \tau_k) \tag{4.196}$$

The form is seemingly the same as (4.165) and its variant (Weeks and Bentler 1982, p. 202). However, expression (4.196) should be given with (4.194), in other words, $\boldsymbol{\tau}$ is not independent from $\boldsymbol{X}$ and $z$. Thus the slide-vector model may be viewed as a variant of (4.165) with the constraint, but it is distinguished from the model by Weeks and Bentler.

By some manipulation on $\boldsymbol{A}$, we derive the following equation:

$$\boldsymbol{AA1} = \lambda \boldsymbol{1} \tag{4.197}$$

Here $\lambda$, $\theta_1$, and $\theta_2$ are given by

$$\lambda = 4(\theta_1^2 - n\theta_2), \qquad \theta_1 = \boldsymbol{1}'\boldsymbol{\tau}, \qquad \theta_2 = \boldsymbol{\tau}'\boldsymbol{\tau} \tag{4.198}$$

The SVD of $\boldsymbol{A}$ gives only a pair of singular vectors. According to Theorem 3.1, it is found that condition (4.197) is necessary and sufficient in order that a line structure exists on the plane spanned by the singular vectors of $\boldsymbol{A}$.

## Structure of the Symmetric Part

Summing equations of type (4.182) over pairs $(j, k)$ and $(k, j)$ gives

$$\tfrac{1}{2}(\omega_{jk}^2 + \omega_{kj}^2 - \omega_{jj}^2 - \omega_{kk}^2) = d_{jk}^2 \tag{4.199}$$

Define a transformation of $\omega_{jk}^2$ by

$$q_{jk} = \tfrac{1}{2}(\omega_{jk}^2 + \omega_{kj}^2 - \omega_{jj}^2 - \omega_{kk}^2) \tag{4.200}$$

From the model structure for $\boldsymbol{\Omega}$, $q_{jk}$ should be non-negative, which leads to

$$\omega_{jk}^2 + \omega_{kj}^2 \geq \omega_{jj}^2 + \omega_{kk}^2 \qquad \text{for all pairs } (j, k) \tag{4.201}$$

When $q_{jk}$ is non-negative, define

$$p_{jk} = \sqrt{q_{jk}} \tag{4.202}$$

Write matrices in terms of dissimilarities as $\boldsymbol{P} = (p_{jk})$ and $\boldsymbol{Q} = (q_{jk})$. By definition, $p_{jk} = p_{kj}$ and $p_{jj} = 0$. It is necessary for $p_{jk}$ to be a metric that the triangular imnequality is satisfied. By the way, it is of interest to note that (4.200) is of the same form as (4.77) and (4.201) of the same form as (4.95). Accordingly, when (4.201) holds, $q_{jk}$ becomes a metric by Theorem 4.4.

When the model holds, $p_{jk}$ should satisfy not only the metric axioms but also the requirement of Euclidean distance. Denote a centering matrix by $\boldsymbol{H}$ and define $\boldsymbol{B}$ in terms of $\{\omega_{jk}\}$ as

$$\boldsymbol{B} = \boldsymbol{HQH} \tag{4.203}$$

Using a theorem due to Young and Householder (1983), we find that $\boldsymbol{B}$ should be positive semidefinite with rank $r$ in order that $p_{jk}$ is Euclidean distance in $r$ dimensions. Now we are in a position to summarize necessary conditions.

*Theorem 4.9.* Necessary conditions. Given a dissimilarity matrix $\boldsymbol{\Omega} = (\omega_{jk})$ of type T1, necessary conditions for the slide-vector model to hold in $r$-dimensional Euclidean space are stated as follows.

Condition 1: additivity. The dissimilarity matrix satisfies the additivity, that is, $\boldsymbol{\Omega}$ satisfies (4.192) or $\boldsymbol{A}$ does (4.193).

Condition 2: line pattern. Matrix $\boldsymbol{A}$ is of rank 2. The two-dimensional plot of objects in terms of the singular vectors of $\boldsymbol{A}$ reveals a line pattern on the plane. The spacing of points on the line corresponds to the projection of object points on the slide-vector.

Condition 3: distance properties. The $\{q_{jk}\}$ are non-negative, equivalently, $\boldsymbol{\Omega}$ satisfy (4.201). Matrix $\boldsymbol{B}$ is positive semidefinite and of rank $r$.

Under some assumptions, it is sufficient for the model that both condition 1 and condition 3 hold with $\boldsymbol{\Omega}$ (Saito, 2003b). Accordingly, a necessary and sufficient condition consists of these two conditions.

### Estimation Procedure

Given observations $o_{jk}$, we set the error model $o_{jk} = \omega_{jk} + e_{jk}$. For model (4.178), we aim to estimate a coordinate matrix $\boldsymbol{X}$ and a slide-vector $\boldsymbol{z}$ by optimizing a criterion. For this purpose, Zielman and Heiser suggested a procedure by using a restricted unfolding algorithm. It consists of two steps, one to improve the location of points and the other to solve a metric projection problem. It is provided with an additional step for scaling of the dissimilarity.

## 4.4.8. Example

### Example 1

We present an example taken from Saito (1993a). Analysis based on GDM-I was applied to the soft drinks brand switching data of Table 3.3. We treat the entries divided by 1000, and the resultant data matrix is denoted by $\boldsymbol{P} = (p_{jk})$.

**TABLE 4.3** Journals in Citation Data

| Abbreviations | | Journal |
|---|---|---|
| 1 | AJP | American Journal of Psychology |
| 2 | JABN | Journal of Abnormal Psychology |
| 3 | JPSP | Journal of Personality and Social Psychology |
| 4 | JAPPL | Journal of Applied Psychology |
| 5 | JCPP | Journal of Comparative and Physiological Psychology |
| 6 | JCCP | Journal of Consulting and Clinical Psychology |
| 7 | JEDP | Journal of Educational Psychology |
| 8 | JEP | Journal of Experimental Psychology |
| 9 | PKA | Psychometrika |
| 10 | PB | Psychological Bulletin |
| 11 | PR | Psychological Review |
| 12 | MBR | Multivariate Behavioral Research |

Then the sum becomes unity for each row. Prior to the analysis, data $p_{jk}$ were converted to dissimilarities $o_{jk} = 1 - p_{jk}$. Further we transformed $o_{jk}$ into $o_{jk}^+$ using (4.78), and into $\tilde{o}_{jk}$ using (4.79). The necessary conditions for the model, (4.86) and (4.87), were checked on $o_{jk}$ and $\tilde{o}_{jk}$. Since it is not meaningful to check strictly the conditions on real data involving some errors, we checked whether the following inequalities are satisfied or not for triples:

$$o_{ij} + o_{jk} - o_{ik} - o_{jj} > -\varepsilon_1 \tag{4.204}$$

$$|\tilde{o}_{ij} + \tilde{o}_{jk} - \tilde{o}_{ik}| < \varepsilon_2 \tag{4.205}$$

Here $\varepsilon_1 = 0.01\sigma_1$ and $\varepsilon_2 = \sigma_2$, and $\sigma_1$ and $\sigma_2$ are the standard deviations of $o_{jk}$ and of $\tilde{o}_{jk}$, respectively. Condition (4.204) was satisfied to the degree of 92.6% and condition (4.205) to the degree of 85.7% for all the meaningful triples.

Applying the LS algorithm to the data $o_{jk}$, we obtained a high degree of fit of the model for a two-dimensional solution. Figure 4.7 illustrates the derived configuration. We see four clusters of brands, such as lemon-lime (Sprite, 7-Up), diet (Diet Pepsi, Tab, Like, Fresca), and two isolated brands, Pepsi and Coke. It is interesting to find that this result is common to that given by the clustering algorithm of DeSarbo and DeSoete (1984, Fig. H).

Regarding the derived $\theta$-scale and $\phi$-scale, Cor$(\theta, \phi)$ was near unity. The reason is understood, referring to the degree of asymmetry $\kappa_1 = 0.072$ (see Sec. 3.3.3) and seeing that the diagonal entries are generally large in comparison with off-diagonal entries. This suggests that asymmetric $\boldsymbol{O}^+$ is accounted for by the scale of $\theta$. The tendency was observed on the data. The scale values are proportional to the diagonal entries in the original data

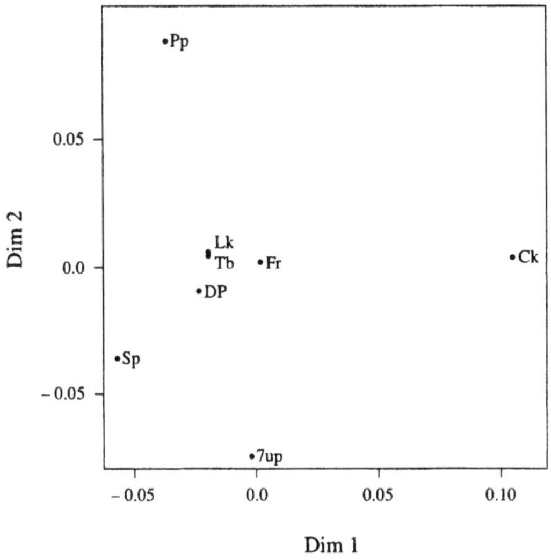


**FIGURE 4.7** Two-dimensional configuration of eight soft drinks (GDM-I)

matrix. Figure 4.8 shows a plot of $\theta$ vs. the diagonal entries in $\boldsymbol{O}_{jk}$, indicating that $\theta$ (or $\phi$)-scale represents the brand loyalty.

## Example 2

An example is taken from Saito and Takeda (1990). Analysis based on GDM-II was applied to the confusion data on 36 Morse codes (26 alphabets and 10 numerals) collected by Rothkopf (1957). The data are provided in Tables 3.10 to 3.12. Krumhansl (1978) referred to the two-dimensional configuration derived from the data obtained by Shepard (1963), as an example to support the distance-density hypothesis. Saito (1983, 1986) analyzed the data by using the symmetric model (4.174) with constraints that all $r_j < 0$, in connection with the hypothesis.

Prior to the analysis by using GDM-II, necessary conditions for the model were examined on the data. It was found, in the same way as in the case of Example 1, that condition (4.86) was satisfied to the degree of 99.5% and condition (4.87) to the degree of 60.5% for all the meaningful triples. To check condition (4.93), we computed $c_{jk}$, which is the ratio of the left-hand side of (4.93) for pair $(j, k)$ such that $o_{jj} \neq o_{kk}$. For 54.1% of those pairs, the $c_{jk}$ values scattered in a narrow interval. As is known, it is often difficult to verify exactly whether a certain equality condition holds or

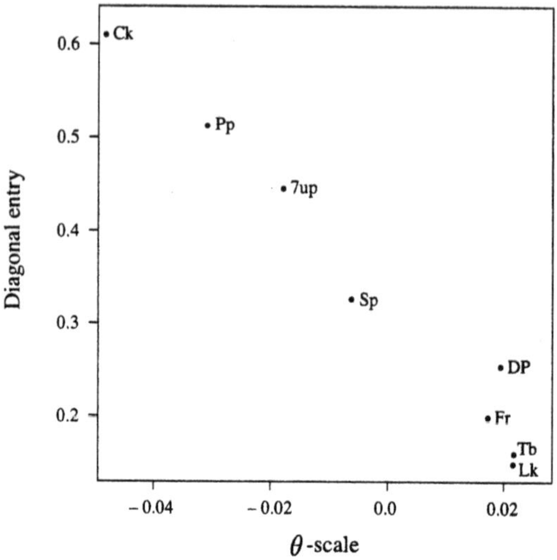


**FIGURE 4.8** Correspondence of diagonal entry $\theta$-scale

not on some data that might involve a fair degree of noise. From this consideration, the examination result allows assuming GDM-II for data analysis.

Setting Euclidean distance for asymmetric model (4.121), the analysis was performed in one to five dimensions. Figure 4.9 shows the fit for each dimensionality, revealing an elbow point at $r = 2$. Figure 4.10 gives a plot of 36 stimuli in two dimensions, illustrating estimates $\hat{\theta}$ by vertical segments on each point. Comparing the configuration with that of Shepard (1963) in view of the rotational indeterminacy, we have the same interpretation as Shepard's, that the first dimension indicates the number of components in the signal and the second dimension the relative frequency of dots vs. dashes.

Inspecting the segment of $\hat{\theta}$, we find roughly higher segments in relatively dense domains and lower segments in less dense domains in the configuration. Thus we may put the distance-density hypothesis for the data, although in a rough sense. Comparing the display of $\hat{\theta}$ in Fig. 4.10 with the corresponding one of Saito (1986, p. 57), we find that the hypothesis holds better with the present result.

In order to assess the appropriateness of the hypothesis for the configuration in a more objective way, we assumed the density form for a given $h$ as

$$\delta_j(h) = \sum_{k \neq j}^{n} \exp(-d_{jk}(\boldsymbol{X})h) \qquad (j = 1, 2, \ldots, n) \tag{4.206}$$

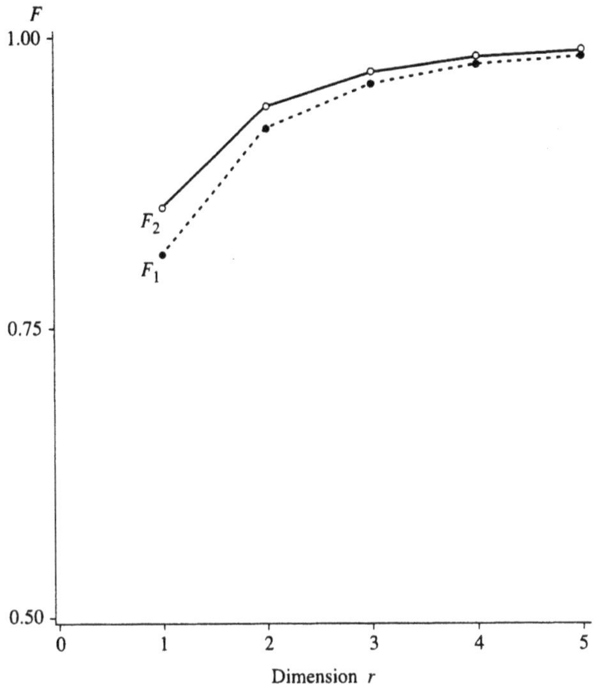


**FIGURE 4.9** Fit for dimensionality (GDM-II)

and then computed values of correlation coefficient $\boldsymbol{R}(h)$ between $\theta_j$ and $\delta_j(h)$ where $h$ changed from 0.01 to 50. The correlation showed the largest value of 0.791 at $h = 14.1$. Figure 4.11 illustrates the change of $\boldsymbol{R}(h)$. As long as the density is given by the form of (4.206), we consider that the distance-density hypothesis holds with the data.

Turn to the estimate of $c$. We obtained $\lambda$ as 0.4931, which gives $c = 0.3395$ by (4.133). Note that $|c|$ shows departure from symmetry ($\lambda = 1$) for which $c = 0$. Thus the present $c$ indicates some degree of asymmetry.

## Example 3

An example is taken from Weeks and Bentler (1982). The data are the frequency of citations of 12 psychological journals to each other in 1979 (Tables 4.3 and 4.4). Prior to the analysis, data were converted to proportions of citations. Since the data are similarity-like, the analysis was performed by

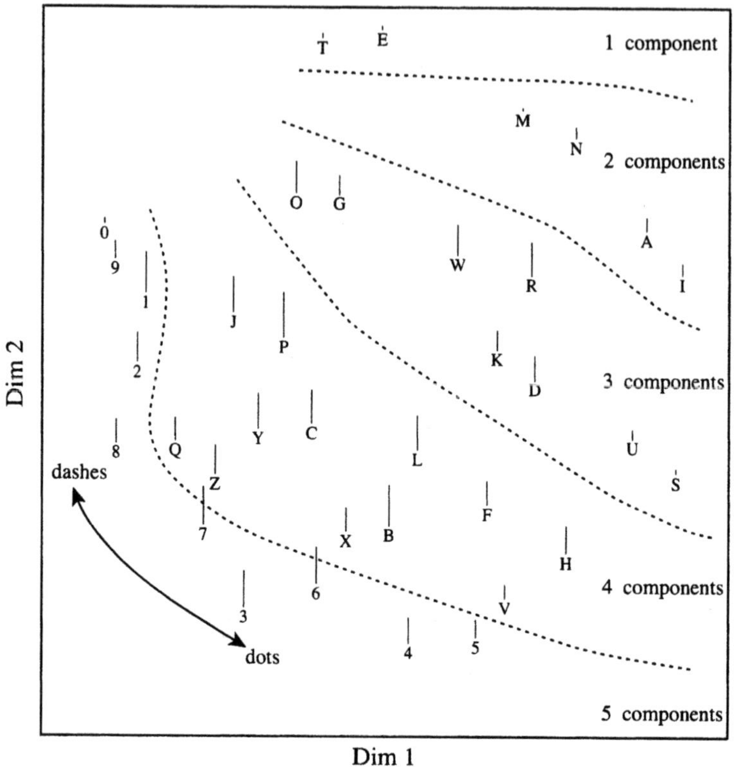


**FIGURE 4.10** Two-dimensional configuration of Morse codes (GDM-II)

setting $\beta = -1$ and squared distance with $\boldsymbol{W} = \boldsymbol{I}_p$ in the model. To cope with indeterminacy of the parameters, one element of $\boldsymbol{\eta}$ was set to zero, and an upper triangle of $\boldsymbol{X}$ set to zero to fix the origin and orientation.

Specifying dimensionality $p$ from one to three, a two-dimensional solution was adopted as the best solution in view of the fit and the interpretability. Making use of the advantage of the restricted MDS, they performed an analysis of the 1979 citation data by fixing the 10 journal locations to those of the solution derived from the 1960 citation data (Coombs, 1964, Table 22.2). Figure 4.12 shows the derived configuration, where dots indicate locations of the 10 journals and triangles indicate the locations of the four journals appearing only in the 1979 data. There is a hard–soft or clinical–experimental dimension roughly parallel to the line between JCPP and JCP. A dimension orthogonal to that separates PKA and MBR from the rest.

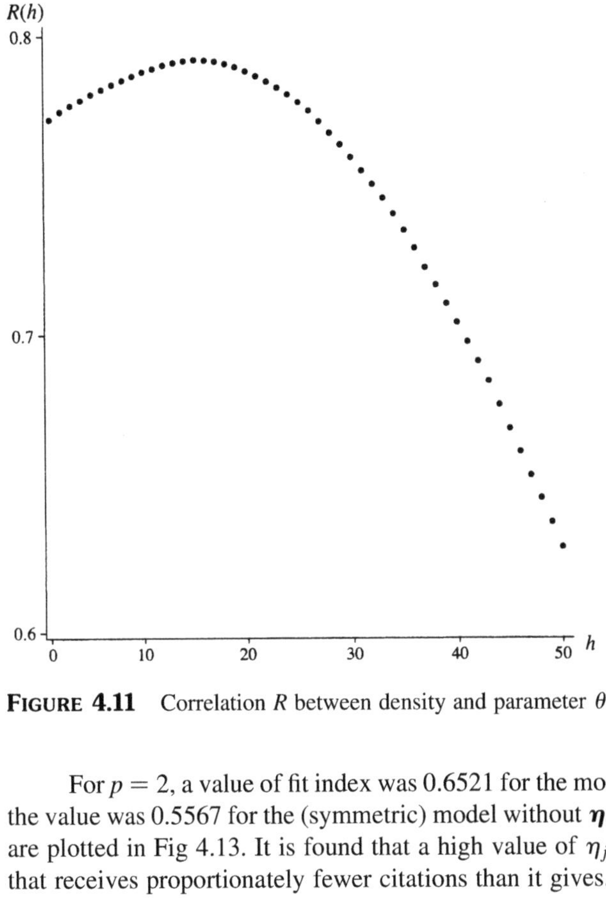


**FIGURE 4.11** Correlation $R$ between density and parameter $\theta$

For $p = 2$, a value of fit index was 0.6521 for the model with $\boldsymbol{\eta}$, whereas the value was 0.5567 for the (symmetric) model without $\boldsymbol{\eta}$. The estimates of $\boldsymbol{\eta}$ are plotted in Fig 4.13. It is found that a high value of $\eta_j$ indicates a journal that receives proportionately fewer citations than it gives.

## Example 4

Here an example is taken from Zielman and Heiser (1993), applying the slide-vector model to the car switching data (Tables 4.1 and 4.2). The slide-vector indicates the direction in a multidimensional space of car types, where the most popular type will be revealed. Prior to the analysis by the model, data transformation was performed to the raw data matrix by several steps, so as to give an appropriate measure of dissimilarity. The transformed data matrix was analyzed in two dimensions. Figure 4.14 shows the derived configuration of types. The first three letters of the labels indicate the size of the cars, and the last letter the distinctive feature for a size category. The first dimension

**TABLE 4.4** 1979 Citation Data

| | Cited journal | | | | | | | | | | | |
|---|---|---|---|---|---|---|---|---|---|---|---|---|
| Citing journal | 1 | 2 | 3 | 4 | 5 | 6 | 7 | 8 | 9 | 10 | 11 | 12 |
| 1 AJP | 31 | 10 | 10 | 1 | 36 | 4 | 1 | 119 | 2 | 14 | 36 | 0 |
| 2 JABN | 7 | 235 | 55 | 0 | 13 | 4 | 65 | 25 | 3 | 50 | 31 | 0 |
| 3 JPSP | 16 | 54 | 969 | 28 | 15 | 21 | 89 | 62 | 16 | 149 | 141 | 16 |
| 4 JAPPL | 3 | 2 | 30 | 310 | 0 | 8 | 5 | 7 | 6 | 71 | 14 | 0 |
| 5 JCPP | 4 | 0 | 2 | 0 | 386 | 0 | 2 | 13 | 1 | 22 | 35 | 1 |
| 6 JCCP | 1 | 7 | 61 | 10 | 2 | 100 | 6 | 5 | 4 | 18 | 9 | 2 |
| 7 JEDP | 0 | 105 | 55 | 7 | 3 | 10 | 331 | 3 | 19 | 89 | 22 | 8 |
| 8 JEP | 9 | 20 | 16 | 0 | 32 | 6 | 1 | 120 | 2 | 18 | 46 | 0 |
| 9 PKA | 2 | 0 | 0 | 0 | 0 | 6 | 0 | 6 | 152 | 31 | 7 | 10 |
| 10 PB | 23 | 46 | 124 | 117 | 138 | 7 | 86 | 84 | 62 | 186 | 90 | 7 |
| 11 PR | 9 | 2 | 21 | 6 | 3 | 0 | 0 | 51 | 30 | 32 | 104 | 2 |
| 12 MBR | 0 | 7 | 14 | 4 | 0 | 0 | 24 | 3 | 95 | 46 | 2 | 56 |

indicates an important domestic distinction and the second dimension discriminates the small cars from the large cars. It is found that the imported cars compete more with each other than with the domestic cars. The slide-vector is drawn as a bold arrow. The long vector indicates the "switch from" vector. We see the trend to buy a luxury imported car instead of a luxury domestic car, and the trend that small cars are winning in the market.

## 4.5. FEATURE-MATCHING MODEL AND TSCALE

### 4.5.1. Contrast Model

Tversky developed a feature-matching model by a set-theoretic approach to proximity. For exposition we consider a pair of ordered stimuli $(i, j)$, which are associated with feature sets $I$ and $J$, respectively. Let $\psi_{ij}$ be a measure of similarity, which is stated as a function $g$ of three arguments as follows:

$$\psi_{ij} = g(I - J, J - I, I \cap J) \tag{4.207}$$

Here $I \cap J$ denotes the set of common features. $I - J$ denotes the set of features distinctive to $i$, that is, features associated with $i$ but not with $j$. $J - I$ denotes the set of features distinctive to $j$.

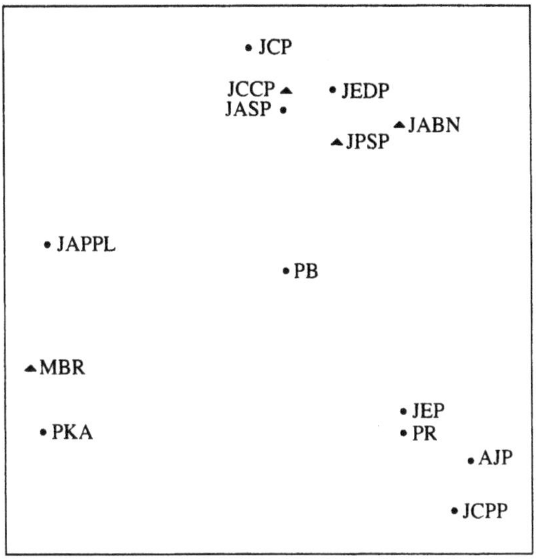


**FIGURE 4.12** Two-dimensional configuration of 12 journals (Weeks & Bentler's model)

The additive case of $g$ is called the contrast model, and the dissimilarity of $(i, j)$ is represented as

$$\omega_{ij} = \alpha f(I - J) + \beta f(J - I) - \theta f(I \cap J) \qquad (i \neq j) \tag{4.208}$$

The function $f(\cdot)$ indicates the number of features of a set. Parameters $\theta$, $\alpha$, $\beta$ are assumed to be positive. If $\alpha \neq \beta$, the model represents asymmetry of $\omega_{ij}$.

Using the relation

$$f(I - J) = f(I) - f(I \cap J) \tag{4.209}$$

we have an alternative expression such that

$$\omega_{ij} = \alpha f(I) + \beta f(J) - (\theta + \alpha + \beta) f(I \cap J) \tag{4.210}$$

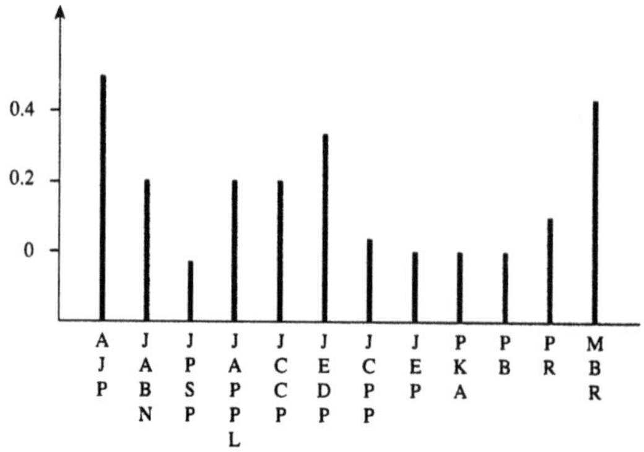


**FIGURE 4.13** Representation of unidimensional scale $\eta$

When we think of a sort of decomposition (3.3) to the model except for the case of $i = j$, we obtain the symmetric and skew-symmetric components as

$$\gamma_{ij} = \frac{1}{2}(\alpha + \beta)(f(I) + f(J)) - (\theta + \alpha + \beta)f(I \cap J) \tag{4.211}$$

$$\tau_{ij} = \frac{1}{2}(\alpha - \beta)(f(I) - f(J)) \tag{4.212}$$

Then, the skew-symmetry means the difference in the number of features.

## 4.5.2. Formulation of TSCALE

DeSarbo et al. (1992) proposed an MDS procedure called TSCALE, which is based on Tversky's contrast model. It starts with the presumption of a dimensional structure $\boldsymbol{X} = (x_{jt})$ underlying the proximity judgment, where $x_{jt}$ denotes the $t$th coordinate of object $j$. Remembering that the feature-matching model was proposed with questions on the metric and dimensional assumptions of MDS, it is an interesting idea to incorporate the model to develop a method of MDS. In the formulation, they introduced another form of dissimilarity in terms of a ratio of distinctive features to the total ones,

$$\omega_{ij} = \frac{\alpha f(I - J) + \beta f(J - I)}{\alpha f(I - J) + \beta f(J - I) + \theta f(I \cap J)} \tag{4.213}$$

Setting $\theta = \alpha = \beta = 1$, it is closely related to the similarity model in Gregson (1975). As the authors pointed out, the linear model (4.208) has some advantages over the ratio model (4.213) for a wide range of applied studies. In

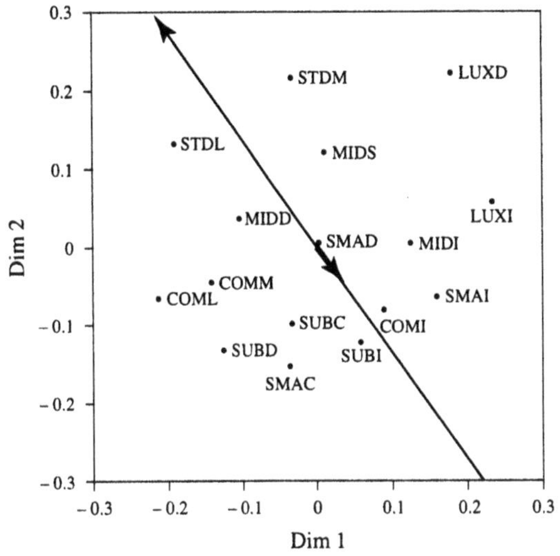


**FIGURE 4.14** Two-dimensional configuration of 16 car-types (slide-vector model)

TSCALE, one assumes that the latent dimensional structure has a corresponding representation such that

$$f(I \cap J) = \sum_{t=1}^{r} \theta \min(x_{it}, x_{jt}) \tag{4.214}$$

$$f(I - J) = \sum_{t=1}^{r} \alpha(x_{it} - x_{jt})_{+} \tag{4.215}$$

$$f(J - I) = \sum_{t=1}^{r} \beta(x_{jt} - x_{it})_{+} \tag{4.216}$$

where $(a - b)_{+} = \max(a - b, 0)$. Then, the common features of two stimuli are represented as the minimum or intersection along dimensions, while the distinctive features are represented as differences.

Now let us consider a two-mode three-way dissimilarity $\omega_{ijk}$ that denotes $\omega_{ij}$ for the $k$th condition (or subject) $(i, j = 1, 2, \ldots, n;\ k = 1, 2, \ldots, m)$. Using (4.214) to (4.216), we extend Tversky's model to a latent dimensional structure as

$$\omega_{ijk} = \sum_{t=1}^{r} \alpha_k (x_{it} - x_{jt})_+ + \sum_{t=1}^{r} \beta_k (x_{jt} - x_{it})_+ - \sum_{t=1}^{r} \theta_k \min(x_{it}, x_{jt}) \tag{4.217}$$

for the linear model (4.208), and

$$\omega_{ijk} = \frac{\sum_{t=1}^{r} \alpha_k (x_{it} - x_{jt})_+ + \sum_{t=1}^{r} \beta_k (x_{jt} - x_{it})_+}{\sum_{t=1}^{r} \alpha_k (x_{it} - x_{jt})_+ + \sum_{t=1}^{r} \beta_k (x_{jt} - x_{it})_+ + \sum_{t=1}^{r} \theta_k \min(x_{it}, x_{jt})} \tag{4.218}$$

for the ratio model (4.213). It is assumed that $x_{jt}, \alpha_k, \beta_k, \theta_k \geq 0$ for both (4.217) and (4.218), and so $0 \leq \omega_{ijk} \leq 1$ for (4.218). Note that there is a multiplicative indeterminacy for $\boldsymbol{X}$, $\boldsymbol{\alpha}$, $\boldsymbol{\beta}$, and $\boldsymbol{\theta}$. More specifically, $\omega_{ijk}$ is invariant to the scale unit of $\boldsymbol{X}$ for (4.218).

### 4.5.3. Scaling Procedure

Given $\boldsymbol{O} = (o_{ijk})$, which are values of $\omega_{ijk}$ $(i \neq j)$ observed on a ratio scale. For a specified $r$, we aim to find a matrix $\boldsymbol{X}$ and $m$-dimensional vectors $\boldsymbol{\alpha}$, $\boldsymbol{\beta}$, and $\boldsymbol{\theta}$ so as to minimize a loss function defined by

$$\phi(\boldsymbol{X}, \boldsymbol{\alpha}, \boldsymbol{\beta}, \boldsymbol{\theta}) = \sum_{i=1}^{n} \sum_{j \neq i}^{n} \sum_{k=1}^{m} (\omega_{ijk} - o_{ijk})^2 \tag{4.219}$$

The problem is handled by an alternating least squares approach. The algorithm consists of two major phases, the first phase to obtain $\boldsymbol{X}$ and the second phase to obtain $\boldsymbol{\alpha}$, $\boldsymbol{\beta}$, and $\boldsymbol{\theta}$. In the second phase, the problem leads to $m$ linear problems of least squares with non-negative constraints on parameters. TSCALE is provided with a variety of linear contrast models and many options for data analysis. For applications of TSCALE to fast food restaurant data and soft drinks data, refer to DeSarbo et al. (1992).

# 5

# Cluster Analysis of Asymmetric Data

## 5.1. OVERVIEW AND PRELIMINARIES

Clustering and classification are fundamental across many sciences or practical data processing in business fields. A variety of methods and procedures have been studied for that purpose. As mentioned in Chapter 1, many clustering algorithms were proposed on the basis of proximity data, to which multidimensional scaling is also applicable. In this close connection, cluster analysis has been developed for asymmetric data under two major approaches.

In one approach, asymmetric proximity data are treated in accordance with the original form of the data, that is, one-mode two-way data, and analyzed in view of the asymmetry. Hubert (1973) proposed *min and max clustering* for the asymmetric similarity, which was generalized by Fujiwara (1980). These algorithms are extensions of the single linkage algorithm and the complete linkage algorithm for asymmetric clustering. Okada and Iwamoto (1996) proposed the weighted average algorithm for asymmetric clustering. Yadohisa (2002) proposed an updating formula for representing the asymmetric hierarchical clustering in a unified treatment. To represent the clustering results, extensions of the dendrogram were developed by Okada and Iwamoto

(1995, 1996) and Yadohisa (2002). Algorithms and dendrograms mentioned so far are described in Sec. 5.2. Noting different points from those algorithms, we present the algorithm suggested by Ozawa (1983) in Sec. 5.3, and the algorithm by Brossier (1982) in Sec. 5.4. In a special context from these algorithms, Ferligoj and Batagelj (1983) proposed a clustering algorithm for dissimilarity data with asymmetric relational constraints.

In the other approach, the asymmetric proximity data are regarded as a special case of two-mode two-way data, in which the row and column elements are considered two different modes. Let us classify the algorithms in this approach into three classes, referring to Eckes and Orlik (1993).

The algorithms in the first class generally aim at fitting tree structures, that is, ultrametric and/or additive trees to the data. Least squares procedures for estimating ultrametric and additive trees were proposed by DeSarbo and De Soete (1984), DeSarbo et al. (1990) and De Soete et al. (1984). Espejo and Gaul (1986) developed a two-mode variant of the classical average linkage algorithm. Many algorithms belonging to this class were summarized in De Soete and Carroll (1996). We take up the algorithm by De Soete et al. (1984) in Sec. 5.5.

The second class is called *direct clustering*. Common to the algorithms in this class, we point out reordering of rows and columns of the data and determination of clusters. As algorithms involved in the class, we describe in Sec. 5.6 the *Bond energy* algorithm (McCormick et al., 1972), which was improved by Arabie et al. (1988), and turn to the centroid effect algorithm (Eckes and Orlik, 1993) in Sec. 5.7.

The third class contains GENNCLUS by DeSarbo (1982) and PENCLUS by Both and Gaul (1986). These algorithms are a generalization of ADCLUS by Shepard and Arabie (1979) for the case of asymmetric or two-mode data. In these algorithms, the proximity is represented in terms of discrete and possibly overlapping properties. In Sec. 5.8, we treat GENNCLUS (DeSarbo, 1982).

## 5.2. HUBERT ALGORITHMS AND THEIR EXTENSIONS (ONE-MODE APPROACH)

First, we introduce asymmetric hierarchical clustering algorithms (Hubert, 1973; Fujiwara, 1980; Okada and Iwamoto, 1996). Then we set out a new updating formula for an asymmetric clustering algorithm proposed by Yadohisa (2002). In addition, we illustrate extended dendrograms for representing the results. A pioneering study of asymmetric clustering is given in Hubert

(1973). Here we follow the formulation and notation by Fujiwara (1980), including the Hubert's algorithms.

### 5.2.1. Algorithms

There are two types of algorithms in Fujiwara (1980), with or without symmetrization of the original data $\boldsymbol{O} = (o_{ij})$. Both types of algorithms consist of two phases: the first phase for selection of the objects for merging and the second phase for updating the (dis)similarity between clusters. Note that the definitions of the two phases lead to formulation of the clustering algorithms.

#### Algorithms with Symmetrization (Hubert, 1973; Fujiwara, 1980)

For phase (i), we transform an asymmetric similarity matrix $\boldsymbol{O} = (o_{ij})$ to symmetric similarity matrix $\boldsymbol{O}^* = (o_{ij}{}^*)$ by either of the following formulas:

$$\text{(A)} \quad o_{ij}^* = o_{ji}^* = \max(o_{ij}, o_{ji}) \tag{5.1}$$

$$\text{(B)} \quad o_{ij}^* = o_{ji}^* = \min(o_{ij}, o_{ji}) \tag{5.2}$$

Next, we choose clusters $C_I$ and $C_J$ to be combined, so that objects in these clusters should satisfy

$$o_{pq}^* = \max_{i<j}(o_{ij}^*), \qquad (p \in C_I,\, q \in C_J) \tag{5.3}$$

For phase (ii), we first merge $C_I$ and $C_J$, and denote the resulting cluster by $C_{IJ}$. Next, we update the similarity between clusters $C_{IJ}$ and other cluster $C_K$, using either of the following formulas:

$$\text{(a)} \quad o_{rt}^* = o_{tr}^* = \max(o_{pt}^*, o_{qt}^*) \tag{5.4}$$

$$\text{(b)} \quad o_{rt}^* = o_{tr}^* = \min(o_{pt}^*, o_{qt}^*) \tag{5.5}$$

where $r \in C_{IJ}$ and $t \in C_K$.

Combining phase (i) and phase (ii) gives rise to four algorithms. By (A − a) we denote the algorithm combining formula (A) for phase (i) and formula (a) for phase (ii). Then algorithms proposed by Hubert (1973) are represented by (A − a), (A − b), and (B − b).

## Algorithms without Symmetrization I (Fujiwara, 1980)

For phase (i), we choose clusters $C_I$ and $C_J$ to be combined, so that object $p \in C_I$ and object $q \in C_J$ should satisfy either of the following formulas:

$$\text{(C)} \quad \max(o_{pq}, o_{qp}) = \max_{i<j}(\max(o_{ij}, o_{ji})) \tag{5.6}$$

$$\text{(D)} \quad \min(o_{pq}, o_{qp}) = \max_{i<j}(\min(o_{ij}, o_{ji})) \tag{5.7}$$

For phase (ii), first we merge $C_I$ and $C_J$, and denote the resulting cluster by $C_{IJ}$. Next, we update the similarity between cluster $C_{IJ}$ and other cluster $C_K$, using either of the following formulas:

$$\text{(c)} \quad o_{rt} = \max(o_{pt}, o_{qt}), \qquad o_{tr} = \max(o_{tp}, o_{tq}) \tag{5.8}$$

$$\text{(d)} \quad o_{rt} = \min(s_{pt}, s_{qt}), \qquad o_{tr} = \min(o_{tp}, o_{tq}) \tag{5.9}$$

where $r \in C_{IJ}$ and $t \in C_K$. Combination of phase (i) and phase (ii) gives four algorithms.

## Algorithms without Symmetrization II (Okada and Iwamoto, 1995)

For phase (i), we choose the clusters in the same way as the algorithms without symmetrization I.

For phase (ii), we update the similarity between cluster $C_{IJ}$, which is made from clusters $C_I$ and $C_J$, and another cluster $C_K$, using the following formula:

$$\text{(e)} \quad o_{rt} = \tfrac{1}{2}(o_{pt} + o_{qt}), \qquad o_{tr} = \tfrac{1}{2}(o_{tp} + o_{tq}) \tag{5.10}$$

where $r \in C_{IJ}$ and $t \in C_K$. Selecting options of phase (i) gives two algorithms.

## 5.2.2. Extended Updating Formula and Clustering Algorithm

There are several formulations for the clustering algorithm. For extension of algorithms mentioned above, we like to revise Fujiwara's formulation with some differences. Such revisions are useful, and we contrast the algorithms and may implement them into computer programs. For symmetric clustering, we facilitate comparisons by using the updating formula proposed by Lance and Williams (1967).

Here we introduce an extension of this formula by Yadohisa (2002) and formulate the asymmetric hierarchical clustering algorithms by using Yadohisa's formula. Without loss of generality, dissimilarity is used for convenience of description.

## Lance and Williams Updating Formula

Let us follow the Lance and Williams updating formula. Then we have a formula for dissimilarity $\Delta_{(IJ)K}$ between cluster $C_{IJ}$ and another cluster $C_K$ as

$$\Delta_{(IJ)K} = \alpha_I \Delta_{IK} + \alpha_J \Delta_{JK} + \beta \Delta_{IJ} + \gamma |\Delta_{IK} - \Delta_{JK}| \tag{5.11}$$

where $\alpha_I$, $\alpha_J$, $\beta$, and $\gamma$ are either constants or functions. The $\Delta_{IJ}$, $\Delta_{IK}$, and $\Delta_{JK}$ denote the dissimilarities between clusters designated by suffix.

## Definition: Extended Updating Formula (Yadohisa, 2002)

Let $C_{IJ}$ be the cluster formed by merging clusters $C_I$ and $C_J$. We consider updating formula for dissimilarity $\Delta_{(IJ)K}$ from cluster $C_{IJ}$ to another cluster $C_K$. In view of asymmetry, it is defined as:

$$\begin{aligned}\Delta_{(IJ)K} = {} & \alpha_{IJ}^1 f^1(\Delta_{IK}, \Delta_{KI}) + \alpha_{JJ}^1 f^1(\Delta_{JK}, \Delta_{KJ}) + \beta^1 g^1(\Delta_{IJ}, \Delta_{JI}) \\ & + \gamma^1 |f^1(\Delta_{IK}, \Delta_{KI}) - f^1(\Delta_{JK}, \Delta_{KJ})|\end{aligned} \tag{5.12}$$

The updating formula for dissimilarity $\Delta_{K(IJ)}$ from cluster $C_K$ to cluster $C_{IJ}$ is written as:

$$\begin{aligned}\Delta_{K(IJ)} = {} & \alpha_{IJ}^2 f^2(\Delta_{IK}, \Delta_{KI}) + \alpha_{JJ}^2 f^2(\Delta_{JK}, \Delta_{KJ}) + \beta^2 g^2(\Delta_{IJ}, \Delta_{JI}) \\ & + \gamma^2 |f^2(\Delta_{IK}, \Delta_{KI}) - f^2(\Delta_{JK}, \Delta_{KJ})|\end{aligned} \tag{5.13}$$

Here $\Delta_{IJ}$ indicates the asymmetric dissimilarity from cluster $C_I$ to cluster $C_J$; $\alpha_I^1$, $\alpha_J^1$, $\alpha_I^2$, $\alpha_J^2$, $\beta^1$, $\beta^2$, $\gamma^1$, $\gamma^2$ are constants specified prior to analysis, and $f^1, f^2, g^1, g^2$ are functions of two dissimilarities determined prior to analysis. We call this pair of formulas extended updating formula hereafter.

In the same way as the symmetric case, we can update the dissimilarity by using this formula. Thus, we can calculate the dissimilarities between clusters at stage $m + 1$ from the dissimilarities at stage $m$. Unlike the symmetric case, these formulas update $\Delta_{(IJ)K}$ and $\Delta_{K(IJ)}$ separately. Note that the constants and the functions determine the clustering algorithm uniquely.

In the same way as the Lance and Williams formula, parameters such as $\alpha_I^1$ control the combination of the clusters. In addition, functions $f^1$, $f^2$, $g^1$, and $g^2$ control some weighting of the asymmetric relationship between a pair of clusters. Figure 5.1 represents asymmetric dissimilarities between clusters.

## Definition: Asymmetric Hierarchical Clustering Algorithm (Yadohisa, 2002)

Let us define an asymmetric hierarchical clustering algorithm, according to the extended updating formula. The algorithm consists of two phases, the object selection phase and the updating phase. Suppose that an asymmetric dissimilarity matrix $\boldsymbol{O} = (\Delta_{ST})$ is given.

First phase: We form $C_{IJ}$ by merging clusters $C_I$ and $C_J$, of which dissimilarity satisfies

$$\Delta_{IJ} = \min_{S<T} D(\Delta_{ST}, \Delta_{TS}) \tag{5.14}$$

where $D$ indicate a combining criterion, which may be max, min, mean, and so on.

Second phase: We update $\Delta_{(IJ)K}$ and $\Delta_{K(IJ)}$ by the extended updating formulas.

We iterate the first and second phases until all the objects will be one cluster.

Let $C_S$ and $C_T$ be arbitrary clusters. Let $\Delta_{ST}$ and $\Delta_{TS}$ denote asymmetric dissimilarities between them, and $C_{IJ}$ and $C_K$ denote particular clusters merged at a stage. As in the symmetric case, without loss of generality, it is

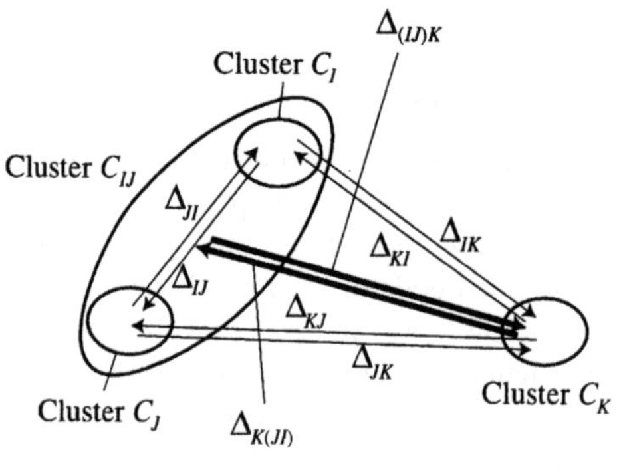


**FIGURE 5.1** Asymmetric dissimilarities between clusters

assumed that cluster $C_K$ may be a singleton, or was formed before cluster $C_{IJ}$ has been formed. In addition, we assume

$$D(\Delta_{IJ}, \Delta_{JI}) \leq D(\Delta_{IK}, \Delta_{KI}) < D(\Delta_{JK}, \Delta_{KJ}) \tag{5.15}$$

We summarize the algorithms so far provided in terms of extended updating formula in Table 5.1. We denote (A − a) when (A) is selected in the first phase and (a) is selected in the second one. Algorithm (A − a) and algorithm (B − b) are equivalent to algorithms (C − c) and (D − d), respectively (Fujiwara, 1980). In the representation by the extended updating formula, the clustering results by (A − a), (A − b), and (B − b) should coincide with Hubert's result; however, we do not symmetrize the data prior to analysis.

The algorithms in Hubert (1973) and Fujiwara (1980) correspond to either a single linkage algorithm or a complete linkage algorithm for the symmetric case. The algorithm in Okada and Iwamoto (1996) is extended from the weighted average algorithm for the asymmetric case. Note that it is possible to develop various new algorithms by specifying the functions $f$, $g$, criterion $D$, and the parameters in the extended updating formula.

### 5.2.3. Extended Dendrograms

Several ways are conceivable for representing the results of asymmetric clustering. Let us differentiate two types. In the first type, information about the asymmetry is shown separately from information about the symmetry. In the second type, all information derived from the analysis is shown simultaneously in a figure. Now we introduce some proposed methods with focus on the second type.

**TABLE 5.1** Parameters of the Extended Updating Formula for Existing Algorithms

| Algorithm | $D$ | $\alpha_I^1(=\alpha_I^2)$ | $\alpha_J^1(=\alpha_J^2)$ | $\beta^1(=\beta^2)$ | $\gamma^1(=\gamma^2)$ | $f^1(=g^1)$ | $f^2(=g^2)$ |
|---|---|---|---|---|---|---|---|
| (A − a) | min | 1/2 | 1/2 | 0 | −1/2 | min($x$, $y$) | min($x$, $y$) |
| (A − b) | min | 1/2 | 1/2 | 0 | 1/2 | min($x$, $y$) | min($x$, $y$) |
| (B − a) | max | 1/2 | 1/2 | 0 | −1/2 | max($x$, $y$) | max($x$, $y$) |
| (B − b) | max | 1/2 | 1/2 | 0 | 1/2 | max($x$, $y$) | max($x$, $y$) |
| (C − c) | min | 1/2 | 1/2 | 0 | −1/2 | $x$ | $y$ |
| (C − d) | min | 1/2 | 1/2 | 0 | 1/2 | $x$ | $y$ |
| (C − e) | min | 1/2 | 1/2 | 0 | 0 | $x$ | $y$ |
| (D − c) | max | 1/2 | 1/2 | 0 | −1/2 | $x$ | $y$ |
| (D − d) | max | 1/2 | 1/2 | 0 | 1/2 | $x$ | $y$ |
| (D − e) | max | 1/2 | 1/2 | 0 | 0 | $x$ | $y$ |

Okada and Iwamoto (1995) show the asymmetry as a ratio between $o_{pq}$ and $o_{qp}$ at the combining stage, where $o_{pq}$ is a similarity from object $q$ to object $p$ and $o_{qp}$ is a similarity from object $p$ to object $q$ (Fig. 5.2).

Okada and Iwamoto (1996) show the asymmetry by direction of the combination at the stage. When $o_{pq} > o_{qp}$ cluster $q$ absorbs cluster $p$, and when $o_{qp} > o_{pq}$ cluster $p$ absorbs cluster $q$ (Fig. 5.3).

In the extended dendrogram of Yadohisa (2002), some distance information is provided to the customary dendrogram. The height of bars in the dendrogram represents combined distance $D(\Delta_{(IJ)K}, \Delta_{K(IJ)})$, which is defined as a combining criterion. In addition, the maximum and the minimum distances between two combining clusters are provided at each combining stages. Thus, three distances $D(\Delta_{(IJ)K}, \Delta_{K(IJ)})$, $\max(\Delta_{(IJ)K}, \Delta_{K(IJ)})$, and $\min(\Delta_{(IJ)K}, \Delta_{K(IJ)})$ considered at each combining stage are represented in the extended dendrogram (Fig. 5.4). It serves to examine the asymmetry by comparing the value of $D$ with the max and the min. Finally, it should be noted that all the information about the asymmetry in Okada and Iwamoto (1995, 1996) is represented in the extended dendrogram by Yadohisa (2002).

### 5.2.4. Example

We analyzed a set of cross-citation data about statistics journals (Stigler, 1994). The frequency of citations is considered to indicate the similarity

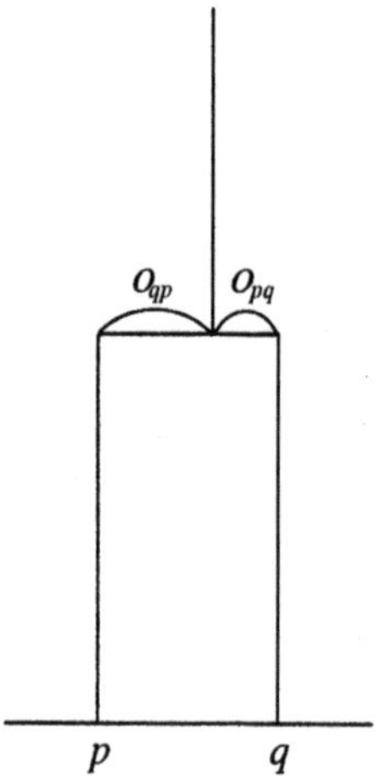


**FIGURE 5.2** Dendrogram by Okada and Iwamoto (1995)

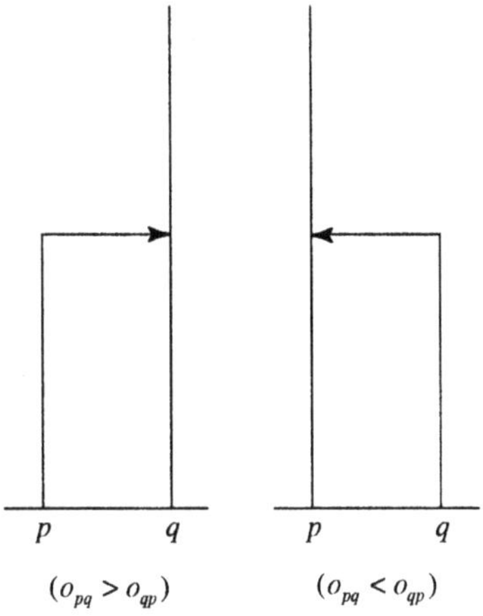


**FIGURE 5.3** Dendrogram by Okada and Iwamoto (1996)

between journals. Table 5.2 shows the bilateral references among eight journals, during years 1987–1989, where the rows indicate journals giving citation, the columns represent citations received. The journals are "The Annals of Statistics," "Biometrics," "Biometrika," "Communications in Statistics,"

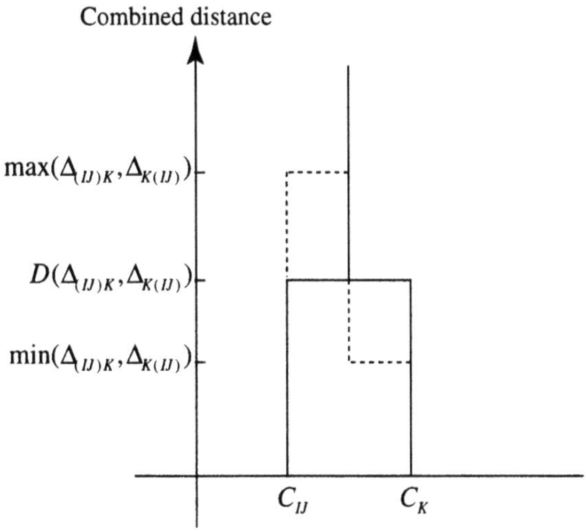


**FIGURE 5.4** Extended dendrogram by Yadohisa (2002)

**TABLE 5.2** Citation Among Statistical Journals

| | Cited journals | | | | | | | |
|---|---|---|---|---|---|---|---|---|
| Citing journals | A | B | C | D | E | F | G | H |
| A : AnnSt | 1623 | 42 | 275 | 47 | 340 | 179 | 28 | 57 |
| B : Biocs | 155 | 770 | 419 | 37 | 348 | 163 | 85 | 66 |
| C : Bioka | 466 | 141 | 714 | 33 | 320 | 284 | 68 | 81 |
| D : ComSt | 1025 | 237 | 730 | 425 | 813 | 276 | 94 | 418 |
| E : JASA | 739 | 264 | 498 | 68 | 1072 | 325 | 104 | 117 |
| F : JRSSB | 182 | 60 | 221 | 17 | 142 | 188 | 43 | 27 |
| G : JRSSC | 88 | 134 | 163 | 19 | 145 | 104 | 211 | 62 |
| H : Tech | 112 | 45 | 147 | 27 | 181 | 116 | 41 | 386 |

AnnSt, The Annals of Statistics; Biocs, Biometrics; Bioka, Biometrika; ComSt, Communications in Statistics; JASA, The Journal of the American Statistical Association; JRSSB, The Journal of the Royal Statistical Society (Series B); JRSSC, The Journal of the Royal Statistical Society (Series C); Tech, Technometrics.

"The Journal of the American Statistical Association," "The Journal of the Royal Statistical Society (Series B)," "The Journal of the Royal Statistical Society (Series C)," and "Technometrics," which are abbreviated in that table to "AnnSt," "Biocs," "Bioka," "ComSt," "JASA," "JRSSB," "JRSSC," and "Tech," respectively.

A total of 155 papers, for example, in Biometrics published in 1987–1989 have referred to those in the Annals of Statistics, whereas the reverse reference is only 42.

Figures 5.5, 5.6, and 5.7 show the results of asymmetric single linkage algorithms, asymmetric complete linkage algorithms, and asymmetric weighted average algorithms, respectively. We observe much difference in the results, which arises from differences in treating the asymmetry. This means we must not ignore the asymmetry. We would like to consign a detailed review of these results to the reader.

## 5.3. CLASSIC (ONE-MODE APPROACH)

Given an asymmetric similarity matrix, Ozawa (1983) proposed a hierarchical clustering algorithm, called CLASSIC, by using the nearest neighbors relation (NNR).

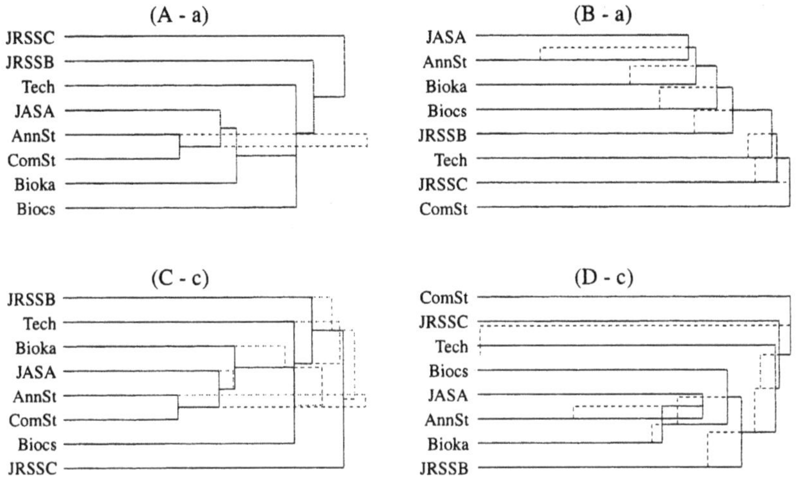


**FIGURE 5.5** Results of asymmetric single linkage algorithms

## 5.3.1. Preparation

Let $X$ be a set of $n$ objects. Denote the similarity for a pair of objects $(j, k)$ by $\lambda_{ij}$. The similarity is assumed to satisfy

$$\lambda_{ii} = 1 \qquad (i \in X) \tag{5.16}$$

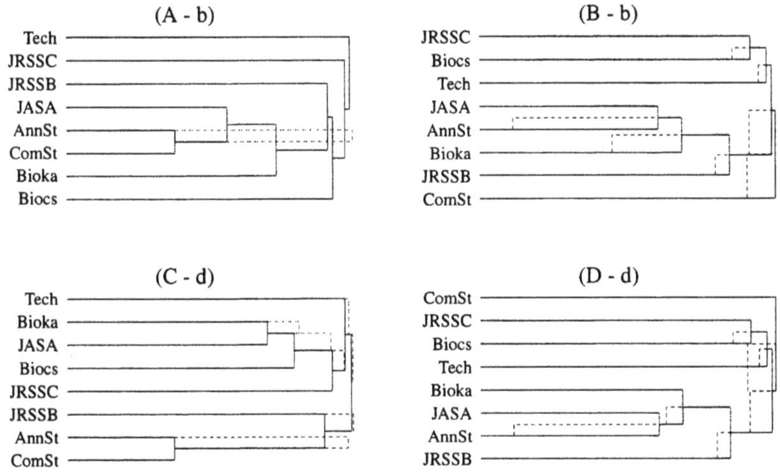


**FIGURE 5.6** Results of asymmetric complete linkage algorithms

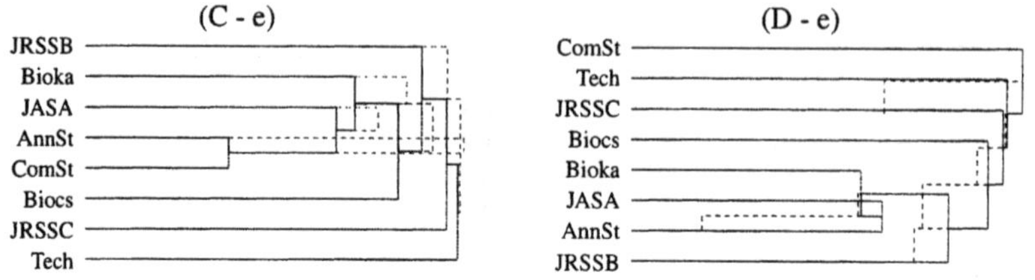


**FIGURE 5.7** Results of asymmetric weighted average algorithms

and

$$0 \leq \lambda_{ij} < 1 \qquad (i \neq j;\ i, j \in X) \tag{5.17}$$

Write an $n \times n$ similarity matrix as

$$\boldsymbol{\Lambda} = (\lambda_{ij}) \tag{5.18}$$

For development to follow, we provide four definitions based on binary relations: composition, union, inclusion, and transitive closure. To begin with, for each object $i$, we rank all objects $j$ in descending order of similarity $\lambda_{ij}$. The order might be such that

$$i, m_1, m_2, \ldots, m_k, \ldots, m_{n-1} \tag{5.19}$$

which means that object $i$ is most similar to itself, next most similar to object $m_1$, next most similar to object $m_2$, and so on. If object $i$ is equally similar to object $m_k$ and object $m_{k+1}$, their rank should be same. Here we assign non-negative integers in ascending order to all the pairs, such that

$$\begin{aligned} o_{ii} &= 0 \\ o_{im_1} &= 1 \\ &\vdots \\ o_{im_{n-1}} &= M \end{aligned} \tag{5.20}$$

where the same integers are assigned to all the pairs of which similarities are equal, so that $M \leq n - 1$. Then we have the *similarity order matrix* $\boldsymbol{O} = (o_{ij})$.

Given a set of objects $X = \{1, 2, \ldots, n\}$ and its similarity order matrix $\boldsymbol{O}$, we define a binary relation $R$ between a pair of objects $i$ and $j$ as

$$iRj \Longleftrightarrow o_{ij} = o_{ji} \leq 1 \tag{5.21}$$

$R$ is reflexive and symmetric, thus

$$iRi \qquad (i \in X) \tag{5.22}$$

and

$$iRj \Longrightarrow jRi \qquad (i \neq j;\ i, j \in X) \tag{5.23}$$

However, $R$ is not transitive, that is, $R$ is not an equivalence relation. Based on $R$, we consider fundamental relations as follows.

## Composition of Relations

For two binary relations $R_1$ and $R_2$ on $X$, we define the composition of $R_1$ and $R_2$, written by $R_1R_2$, as

$$i(R_1R_2)j \Longrightarrow iR_1u \quad \text{and} \quad uR_2j \text{ for some } u \in X \tag{5.24}$$

## Union of Relations

For two binary relations $R_1$ and $R_2$, we define the union of $R_1$ and $R_2$, written by $R_1 \cup R_2$, as

$$i(R_1 \cup R_2)j \Longrightarrow iR_1j \quad \text{or} \quad iR_2j \tag{5.25}$$

## Inclusion of Relations

For two binary relations $R_1$ and $R_2$, we define the inclusion of relations, written by $R_1 \subset R_2$, as

$$R_l \subset R_2 \Longleftrightarrow (iR_1j \Longrightarrow iR_2j) \tag{5.26}$$

## Transitive Closure

Let us define the transitive closure $\hat{R}$ as

$$\hat{R} = \bigcup_{m=1}^{\infty} R^m = R \cup RR \cup RRR \cup \ldots \tag{5.27}$$

$\hat{R}$ is in general written as an infinite union. If the set $X$ is finite, then $\hat{R}$ is finite and for some non-negative integer $h$,

$$\hat{R} = \bigcup_{m=1}^{h} R^m \tag{5.28}$$

When $X$ contains $n$ objects,

$$h \leq n - 1 \tag{5.29}$$

$\hat{R}$ is transitive, reflexive, and symmetric by (5.22) and (5.23), thus $\hat{R}$ is an equivalence relation. Here we call $\hat{R}$ the nearest neighbors relation (NNR), because it detects the two objects that are mutually nearest neighbors, and so it partitions them into subsets of $X$.

As an example of NNR, we assume a set of objects $X = \{1, 2, 3, 4, 5, 6, 7, 8\}$ with an asymmetric similarity matrix $\mathbf{\Lambda}$ given by

$$\mathbf{\Lambda} = \begin{pmatrix} 1.0 & 1.0 & 0.8 & 0.2 & 1.0 & 1.0 & 0.4 & 1.0 \\ 1.0 & 1.0 & 0.2 & 0.4 & 0.6 & 1.0 & 0.8 & 0.0 \\ 0.8 & 0.2 & 1.0 & 0.2 & 1.0 & 1.0 & 0.4 & 0.2 \\ 0.8 & 0.4 & 0.8 & 1.0 & 0.4 & 0.6 & 0.8 & 1.0 \\ 0.6 & 0.0 & 1.0 & 0.2 & 1.0 & 0.2 & 0.8 & 0.4 \\ 0.4 & 0.6 & 0.0 & 0.0 & 0.4 & 1.0 & 1.0 & 0.8 \\ 0.4 & 0.6 & 0.6 & 1.0 & 0.8 & 1.0 & 1.0 & 0.0 \\ 0.0 & 0.2 & 0.2 & 0.8 & 0.6 & 1.0 & 0.4 & 1.0 \end{pmatrix} \tag{5.30}$$

Since the similarity satisfies (5.16) and (5.17), we obtain such a similarity order matrix as

$$\boldsymbol{O} = \begin{pmatrix} 0 & 1 & 2 & 4 & 1 & 1 & 3 & 1 \\ 1 & 0 & 5 & 4 & 3 & 1 & 2 & 6 \\ 2 & 4 & 0 & 4 & 1 & 1 & 3 & 4 \\ 2 & 4 & 2 & 0 & 4 & 3 & 2 & 1 \\ 3 & 6 & 1 & 5 & 0 & 5 & 2 & 4 \\ 4 & 3 & 5 & 5 & 4 & 0 & 1 & 2 \\ 4 & 3 & 3 & 1 & 2 & 1 & 0 & 5 \\ 6 & 5 & 5 & 2 & 3 & 1 & 4 & 0 \end{pmatrix} \tag{5.31}$$

Based on (5.31), NNR, that is, $\hat{R}$, is given by the following table:

| $i \backslash j$ | 1 | 2 | 3 | 4 | 5 | 6 | 7 | 8 |
|---|---|---|---|---|---|---|---|---|
| 1 | $R$ | $R$ | | | | | | |
| 2 | $R$ | $R$ | | | | | | |
| 3 | | | $R$ | | $R$ | | | |
| 4 | | | | $R$ | | | | |
| 5 | | | $R$ | | $R$ | | | |
| 6 | | | | | | $R$ | $R$ | |
| 7 | | | | | | $R$ | $R$ | |
| 8 | | | | | | | | $R$ |

It is found that NNR partitions $X$ into five subsets; {1, 2}, {3, 5}, {4}, {6, 7}, and {8}. Subsets {l, 2}, {3, 5}, and {6, 7} are, respectively, three clusters composed of nearest neighbors. The NNR thus induces some clusters of nearest neighbors from a given similarity order matrix. However, since NNR groups only the highest ranking pairs of objects, almost all objects are left ungrouped. We describe an iterative procedure, called RANKOR, to induce a hierarchy of combining clusters based on the nested sequence of NNRs.

## 5.3.2. Algorithm

On the basis of a similarity order matrix $\boldsymbol{O}$, NNR partitions a set $X$ into the subsets, almost all of which contain single objects, except for a few clusters consisting of a small number of nearest neighbors.

Let us consider a nested sequence of NNRs $\hat{R}_0 \subset \hat{R}_1 \subset \hat{R}_2 \subset \ldots$, which are, respectively, given by a sequence of similarity order matrices $\boldsymbol{O}_0$, $\boldsymbol{O}_1$, $\boldsymbol{O}_2, \ldots$. Then we can construct a hierarchy of clusters.

Rewrite the original similarity order matrix $\boldsymbol{O}$ and $\hat{R}$

$$\boldsymbol{O}_0 = \boldsymbol{O} \quad \text{and} \quad \hat{R}_0 = \hat{R} \tag{5.32}$$

Let $\boldsymbol{O}_k = (o_{ij}^k)$ and $\hat{R}_k$ $(i, j = 1, 2, \ldots, n; k = 0, 1, \ldots)$ be the similarity order matrix and NNR at the $k$th stage, respectively. The following algorithm recurrently defines $\boldsymbol{O}_{k+1}$ in terms of $\boldsymbol{O}_k$. We note that this algorithm cannot be described in the form of the Lance and Williams updating formula. Here we state it in a usual algorithmic form.

### Algorithm RANKOR (Ozawa, 1983)

for $r \leftarrow 1$ to $n$ do
    for $s \leftarrow 1$ to $n$ do
        if $o_{rs}^k = o_{sr}^k \leq 1$
            then $o_{rs}^{k+1} = 0$
            else $o_{rs}^{k+1} = o_{rs}^k$
    end
    SECMIN $\leftarrow 2$
    for $s \leftarrow 1$ to $n$ do
        if $o_{rs}^{k+1} = 1$ then SECMIN $\leftarrow 1$
    end
    for $s \leftarrow 1$ to $n$ do
        if $o_{rs}^{k+1} = 0$

then $o_{rs}^{k+1} = 0$

else $o_{rs}^{k+1} = o_{rs}^{k+1} - \text{SECMIN} + 1$

end

end

$\hat{R}$ is defined by (5.21) and (5.28); the binary relation $R_k$ is defined by

$$iR_k j \iff o_{ij}^k = o_{ji}^k \leq 1 \tag{5.33}$$

thereafter, the transitive closure $\hat{R}_k$ can be defined.

In this combining process, two clusters $C$ and $C'$ at the $k$th stage are merged through the intermediation of two objects $x\,(\in C)$ and $x'(\in C')$ satisfying

$$o_{xx'}^k = o_{x'x}^k = 1 \tag{5.34}$$

Both $x$ and $x'$ are "nearest" in the sense of the mutually highest rank among all pairs of two objects contained in $C$ and $C'$, respectively. In this case, although not explicitly defined, $o_{xx'}^k$ implies the rank of the similarity from cluster $C$ to $C'$ rather than that from object $x$ to $x'$.

The sequence of NNRs $\hat{R}_0, \hat{R}_1, \hat{R}_2, \ldots$ given by (5.21), (5.28), (5.33), and RANKOR induces a hierarchical partition of a set $X$, that is,

$$\hat{R}_k \subset \hat{R}_{k'} \quad \text{for any} \quad k < k' \tag{5.35}$$

### 5.3.3. Example

To illustrate CLASSIC, we assume again the set of objects $X = \{1, 2, 3, 4, 5, 6, 7, 8\}$ with matrix $\boldsymbol{\Lambda}$ given by (5.30). At stage 1, NNR $\hat{R}_1$ defined from $\boldsymbol{O}_1$, which is the same as $\boldsymbol{O}$ in (5.31), induces five clusters {1, 2}, {3, 5}, {4}, {6, 7}, and {8}, as previously stated. The result of combining stages is shown as follows:

$\hat{R}_1$ Stage 1: {1, 2}, {3, 5}, {4}, {6, 7}, {8}
$\hat{R}_2$ Stage 2: {1, 2}, {3, 5}, {4}, {6, 7, 8}
$\hat{R}_3$ Stage 3: {1, 2, 4, 6, 7, 8}, {3, 5}
$\hat{R}_4$ Stage 4: {1, 2, 4, 6, 7, 8}, {3, 5}
$\hat{R}_5$ Stage 5: {1, 2, 3, 4, 5, 6, 7, 8}

Figure 5.8 gives a graphical representation of the result by dendrogram. Note that $\boldsymbol{O}_3 \neq \boldsymbol{O}_4$ in contrast with $\hat{R}_3 = \hat{R}_4$. The combining process thus terminates when all objects in the original set are merged into the single

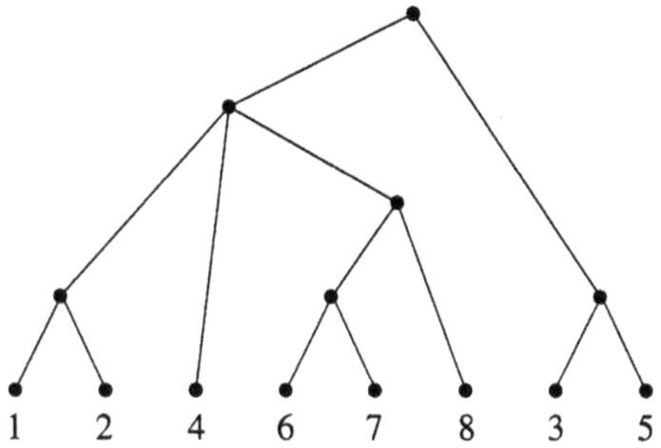


**FIGURE 5.8** The entire results of CLASSIC

cluster. In practice, if necessary, we can optionally stop the combining process when a prescribed condition is satisfied, for example, at the earliest stage where the number of clusters is less than a specified value.

## 5.4. BROSSIER ALGORITHM (ONE-MODE APPROACH)

Here we introduce a matrix decomposition approach proposed by Brossier (1982).

### 5.4.1. Preparation

Let $\boldsymbol{O}$ be an $n \times n$ asymmetric matrix. We decomposed it as $\boldsymbol{O} = \boldsymbol{S} + \boldsymbol{A}$ using (3.1) and (3.2) to (3.3). The skew-symmetric matrix $\boldsymbol{A}$ is the main target for the present analysis. Let us consider the singular value decomposition of $\boldsymbol{A}$ mentioned in Sec. 3.2, and thus utilize the decomposition (3.28). Now we suppose a case that $\boldsymbol{A}$ is of rank 2 and is expressed as

$$\boldsymbol{A} = \boldsymbol{b}\boldsymbol{1}' - \boldsymbol{1}\boldsymbol{b}' \tag{5.36}$$

where $\boldsymbol{1}$ is the vector of unities. Let $\boldsymbol{A} = (a_{ij})$ and $\boldsymbol{b} = (b_j)$, then $a_{ij} = b_i - b_j$. Referring to the argument around Theorem 3.1, it is interesting to remember that $\boldsymbol{A}$ of (5.36) involves one-dimensional structure.

### 5.4.2. Method

In Brossier's formulation, we approximate the symmetric matrix $\boldsymbol{S}$ with an ultrametric matrix $\boldsymbol{U} = u_{ij}$, which is obtained from a hierarchical tree representation of $\boldsymbol{S}$. Regarding the skew-symmetric matrix $\boldsymbol{A}$, we approximate it with a skew-symmetric matrix of rank 2 such that

$$\boldsymbol{E} = (e_{ij}) = \boldsymbol{e}\mathbf{1}' - \mathbf{1}\boldsymbol{e}' \tag{5.37}$$

Then we aim to determine $\boldsymbol{e} = (e_i)$ to meet a least squares criterion.

For the sake of explanation, let us consider the case of a social exchange, which is represented by an exchange matrix $\boldsymbol{O} = (o_{ij})$ with resource $i$ received and resource $j$ returned. Then $s_{ij}$ indicates the average level of exchanges between resources $i$ and $j$, and $a_{ij}$ indicates the surplus or the deficit in the exchanges. In order to approximate matrix $\boldsymbol{A}$ by matrix $\boldsymbol{E}$, we associate a surplus (or a deficit) $e_i$ to resource $i$. Thus we approximate $a_{ij}$ by the difference $e_i - e_j$. The difference means the disequilibrium of exchanges between $i$ and $j$. By adding the quantity $e_i - e_j$ to the ultrametric distance $u_{ij}$, we superpose the two representations in order to approximate $o_{ij}$ in such a way that

$$o_{ij} = s_{ij} + a_{ij} \simeq u_{ij} + e_i - e_j \tag{5.38}$$

The superposition is achieved by increasing or reducing the length of the tree branches at each terminal node $i$ as much as $e_i$. We thus have a simultaneous representation of the symmetric part and the skew-symmetric part, showing both the average level of exchanges and the disequilibriums between them (see Fig. 5.9).

## Approximation of the Symmetric Part

Given a symmetric similarity matrix $\boldsymbol{S} = (s_{ij})$, we consider approximating it by an ultrametric matrix $\boldsymbol{U} = (u_{ij})$. As is known, there is no efficient algorithm to find out the ultrametric in the least squares approach. Here we suggest applying hierarchical agglomerative clustering, and in particular, using the average linkage algorithm, for two reasons. First, the result is interpreted as an average level of exchange. Secondly, this algorithm generally provides a good approximation of $\boldsymbol{S}$ in the sense of least squares.

## Approximation of the Skew-Symmetric Part

Given an $n \times n$ skew-symmetric matrix $\boldsymbol{A}$, we aim to approximate it by an $n \times n$ matrix $\boldsymbol{E} = (e_{ij})$ in the form of (5.37). It is easily found that $\|\boldsymbol{A} - \boldsymbol{E}\|$ attains its minimum when

$$e_i = \alpha_i = \sum_{j=1}^{n} a_{ij} \tag{5.39}$$

Thus we have a least squares approximation of $\boldsymbol{A}$, and we may interpret $e_i$ as indicating the average surplus.

For reading the final representation of matrix $\boldsymbol{O}$, it is suggested to use quantity $u_{ij} + e_i$ rather than $u_{ij} + (e_i - e_j)$. That means that we consider only the surplus of resource received. This treatment can be understood by the following.

Let $\boldsymbol{A}$ be an $n \times n$ matrix and $\boldsymbol{F}$ be an $n \times n$ matrix in the form $f_{ij} = f_i$. Then, $\|\boldsymbol{A} - \boldsymbol{F}\|$ is minimized by taking $f_i = \alpha_i/n$. The minimum is equal to

$$\|\boldsymbol{A}\|^2 - \frac{1}{n}\boldsymbol{\alpha}'\boldsymbol{\alpha}, \qquad \text{where} \quad \boldsymbol{\alpha} = (\alpha_i) \qquad (i = 1, \ldots, n) \tag{5.40}$$

## Computation

The entire algorithm involves the following steps.

Step 1: Decomposition of $\boldsymbol{O}$ into symmetric $\boldsymbol{S}$ and skew-symmetric $\boldsymbol{A}$.
Step 2: Hierarchical clustering based on $\boldsymbol{S}$ by a certain algorithm.
Step 3: Calculation of vector $\boldsymbol{e}$.
Step 4: Change of the length of branches by adding $e_i$.

## 5.4.3. Example

We analyzed the data of trade exchanges between nine countries of the European Community (Belgium and Luxemburg being grouped into one country), which are given in Table 5.3. The data are taken from database statistics of the community (Brossier, 1982), showing the trade for the year 1980.

In a similar way to the structural analysis in international economics, one may normalize the volume of trade by the Gross National Product (GNP) of the respective countries. Denote the volume of exchanges from country $i$ to country $j$ by $o_{ij}$ and let $\boldsymbol{O} = (o_{ij})$. Table 5.4 shows the Gross Domestic Product (GDP) of the countries. Let $p_i$ be the GDP of country $i$,

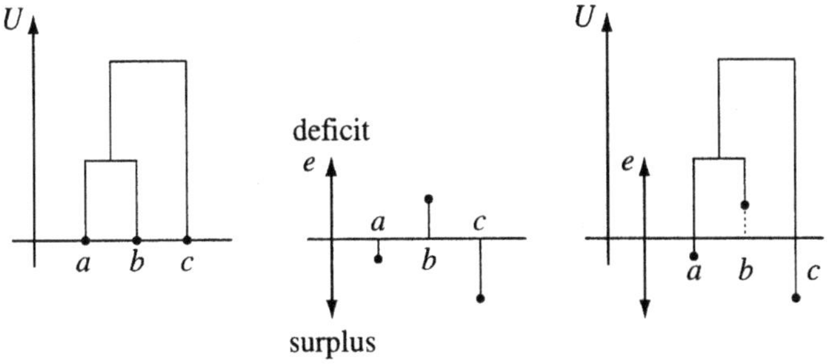


**FIGURE 5.9** Illustration of the hierarchy with asymmetry

**TABLE 5.3** World Trading Data in Europe

| | Import countries | | | | | | | | |
|---|---|---|---|---|---|---|---|---|---|
| Export countries | A | B | C | D | E | F | G | H | I |
| A : Germany | | 15708 | 11923 | 12297 | 10152 | 9209 | 486 | 2645 | 1064 |
| B : France | 14542 | | 9971 | 3707 | 7459 | 6290 | 396 | 564 | 472 |
| C : Italy | 10707 | 9121 | | 1746 | 1853 | 3834 | 157 | 385 | 625 |
| D : Netherlands | 16225 | 5262 | 2993 | | 8454 | 4826 | 285 | 1091 | 295 |
| E : Belgiusn | 9962 | 8117 | 2553 | 6449 | | 4039 | 161 | 503 | 150 |
| F : Great Britain | 9162 | 5246 | 3182 | 4522 | 4169 | | 4417 | 1688 | 349 |
| G : Ireland | 599 | 544 | 190 | 285 | 193 | 2893 | | 41 | 28 |
| H : Denmark | 2327 | 621 | 631 | 480 | 239 | 1809 | 52 | | 46 |
| I : Greece | 1086 | 331 | 350 | 216 | 78 | 206 | 6 | 27 | |

and convert $o_{ij}$ into $o_{ij}^*$ by

$$o_{ij}^* = \frac{o_{ij}}{p_i p_j} \tag{5.41}$$

Table 5.5 shows the matrix of $\boldsymbol{O}^* = (o_{ij}^*)$.

## Analysis of Matrix *O*

The result of the analysis of matrix $\boldsymbol{O}$ is shown in Fig. 5.10. First we realize the degree of exchanges between France and Germany, which is a very noticeable finding in terms of volume for European exchanges. To these central two are added Italy and Ireland, thus forming the group of large exchanges. Next to them appears the natural class of The Netherlands and Belgium, with a fair amount of exchanges. The remaining countries then join the group at very low levels corresponding to the low trade flow.

The analysis of asymmetry reveals the special role played by The Netherlands, although known in the previous analyses, and also the surplus of relations Germany–France and France–Italy.

**TABLE 5.4** GDP in 1979

| | | | |
|---|---|---|---|
| Germany | 557 | United Kingdom | 292 |
| France | 417 | Ireland | 10 |
| Italy | 236 | Denmark | 48 |
| Netherlands | 108 | Greece | 28 |
| Belgium | 82 | | |

**TABLE 5.5** World Trading Data in Europe

| Export countries | Import Countries A | B | C | D | E | F | G | H | I |
|---|---|---|---|---|---|---|---|---|---|
| A : Germany | | 0.0676 | 0.0907 | 0.2044 | 0.2223 | 0.0566 | 0.0873 | 0.0989 | 0.0682 |
| B : France | 0.0626 | | 0.1013 | 0.0823 | 0.2181 | 0.0517 | 0.0950 | 0.0282 | 0.0404 |
| C : Italy | 0.0815 | 0.0927 | | 0.0685 | 0.0958 | 0.0556 | 0.0665 | 0.0340 | 0.0946 |
| D : Netherlands | 0.2697 | 0.1168 | 0.1174 | | 0.9546 | 0.1530 | 0.2639 | 0.2105 | 0.0976 |
| E : Belgium | 0.2181 | 0.2374 | 0.1319 | 0.7282 | | 0.1687 | 0.1963 | 0.1278 | 0.0653 |
| F : Great Britain | 0.0563 | 0.0431 | 0.0462 | 0.1434 | 0.1741 | | 1.5127 | 0.1204 | 0.0427 |
| G : Ireland | 0.1075 | 0.1305 | 0.0805 | 0.2639 | 0.2354 | 0.9908 | | 0.0854 | 0.1000 |
| H : Denmark | 0.0870 | 0.0310 | 0.0557 | 0.0926 | 0.0607 | 0.1291 | 0.1083 | | 0.0342 |
| I : Greece | 0.0696 | 0.0283 | 0.0530 | 0.0714 | 0.0340 | 0.0252 | 0.0214 | 0.0201 | |

## Analysis of Structural Matrix $O^*$

We present analysis of matrix $O^*$. The result is represented in Fig. 5.11. Here we notice two clusters, The Netherlands–Belgium, and Great Britain–Ireland, and note the asymmetrical direction of exchanges. Unlike the preceding

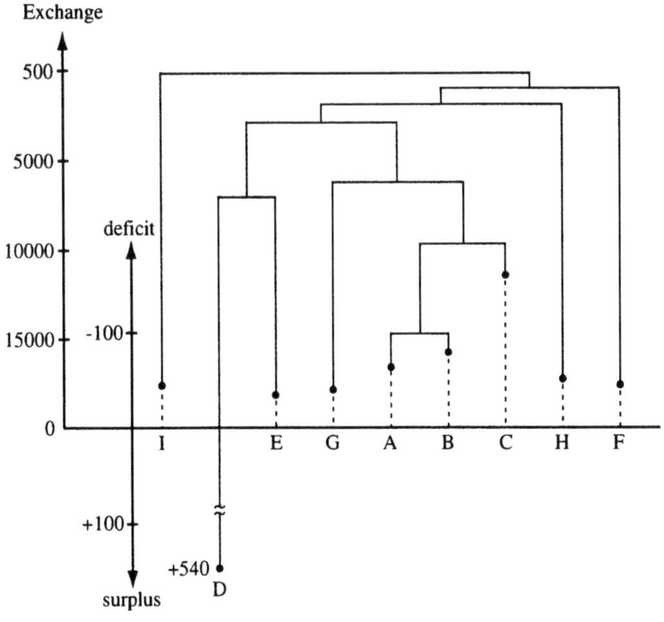


**FIGURE 5.10** Result of the analysis of matrix $O$

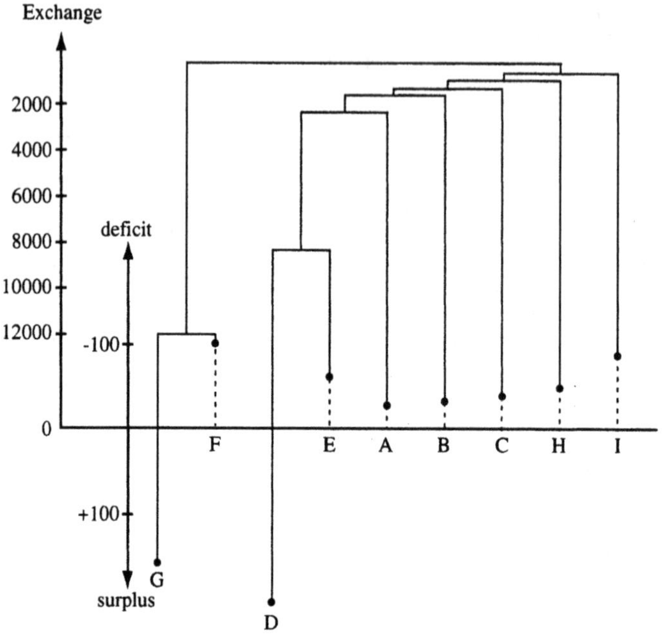


**FIGURE 5.11** Result of the analysis of structural matrix $O^*$

analysis, the other countries do not form a class in themselves. However, they join the Netherlands–Belgium cluster at a relatively low level.

## 5.5. DE SOETE ET AL. ALGORITHM (TWO-MODE APPROACH)

Since asymmetric one-mode two-way data are regarded as a special case of two-mode two-way data, it is possible to consider a two-mode approach to the data. Furnas (1980) proposed representing the proximity relations, using an ultrametric or additive tree. Here we introduce the De Soete et al. algorithm, which is based on a least squares approach, for fitting ultrametric trees to two-mode, two-way data.

### 5.5.1. Preparation

Let $d_{ij}$ denote the ultrametric distance between the nodes $i$ and $j$ on a tree representation. Then $d_{ii} = 0$ and $d_{ij} = d_{ji}$ for all pairs $(i, j)$. For the distance matrix

$D = (d_{ij})$, the following inequality is satisfied for all triples $(i, j, k)$:

$$d_{ij} \leq \max(d_{ik}, d_{ki}) \qquad (5.42)$$

This inequality is known as the *ultrametric inequality*. It means that for any three objects $i$, $j$, and $k$, the largest two of the distances $d_{ij}$, $d_{ik}$, and $d_{jk}$ are equal. Alternatively, it states that in an ultrametric tree, all triangles are either acute isosceles or equilateral. Combining that $d_{ii} = 0$ (for all $i$) with (5.42) yields the triangular inequality and the non-negativity in such a way that

$$d_{ij} + d_{jk} \geq d_{ik} \quad \text{and} \quad d_{ij} \geq 0 \quad (i \neq j) \qquad (5.43)$$

Each ultrametric tree uniquely defines a set of distances $D$ that satisfy (5.42). Conversely, whenever a set of distances satisfies the ultrametric inequality, it uniquely defines an ultrametric tree. Hence, when dissimilarity data $O$ satisfy the ultrametric inequality for all triples $(i, j, k)$, they can be uniquely represented by an ultrametric tree whose distances exactly correspond to dissimilarities. In most cases, $O$ does not satisfy the ultrametric inequality perfectly, and we aim to find an ultrametric tree representation whose ultrametric distances $D$ are as close as possible to the observed dissimilarities $O$.

In the case of rectangular distance matrices, however, it is not possible to test the ultrametric inequality since one of the three distances might be missing for every triple. To indicate two modes, let $A = \{1, 2, \ldots, j, \ldots, n\}$ be the set of row elements, and $B = \{1, 2, \ldots, k, \ldots, m\}$ be the set of row elements. According to Furnas, for a rectangular matrix $D = (d_{jk})$ where $j \in A$ and $k \in B$, it is necessary and sufficient to obtain an ultrametric tree representation that the following two-class ultrametric condition holds

$$d_{i\ell} \leq \max(d_{ik}, d_{jk}, d_{j\ell}), \quad \text{where} \quad i, j \in A \quad \text{and} \quad k, \ell \in B \qquad (5.44)$$

When this inequality is satisfied, the representation is unique up to the internal structure of one-class subtrees.

### 5.5.2. Algorithm for Estimating an Ultrametric Tree

The algorithm consists of the following phases:

#### Phase I

Given an $n \times m$ data matrix $O$, we like to find an $n \times m$ matrix $T$ approximating $O$ best in a least squares sense. Here $T$ should satisfy the two-class ultrametric inequality:

$$t_{i\ell} \leq \max(t_{ik}, t_{jk}, t_{j\ell}) \quad \text{where} \quad i, j \in A \quad \text{and} \quad k, \ell \in B \qquad (5.45)$$

Equivalently, it states that for every quadruple of elements comprised of two elements from each class, the two largest of those four distances must be equal. The purpose leads to solving the optimization problem. We minimize

$$L(\boldsymbol{T}) = \sum_{i=1}^{n}\sum_{k=1}^{m}(o_{ik} - t_{ik})^2 \tag{5.46}$$

with respect to $\boldsymbol{T}$ under the condition that $\boldsymbol{T}$ satisfies (5.45). For this purpose, an exterior penalty function approach is utilized to convert the constrained problem into a series of unconstrained ones. We minimize the augmented function:

$$\Phi(\boldsymbol{T}, \rho) = L(\boldsymbol{T}) + \rho P(\boldsymbol{T}) \tag{5.47}$$

with $\rho > 0$, for an increasing sequence of values of $\rho$. The penalty function is defined as

$$P(\boldsymbol{T}) = \sum_{i=2}^{n}\sum_{j=1}^{i-1}\sum_{k=2}^{m}\sum_{\ell=1}^{k-1}(u_{ijk\ell} - v_{ijk\ell})^2 \tag{5.48}$$

where

$$u_{ijk\ell} = \max(t_{i\ell}, t_{ik}, t_{jk}, t_{j\ell}) \tag{5.49}$$

and

$$v_{ijk\ell} = \begin{cases} \max(t_{i\ell}, t_{jk}, t_{j\ell}) & \text{if} \quad u_{ijk\ell} = t_{ik} \\ \max(t_{ik}, t_{jk}, t_{j\ell}) & \text{if} \quad u_{ijk\ell} = t_{i\ell} \\ \max(t_{i\ell}, t_{ik}, t_{j\ell}) & \text{if} \quad u_{ijk\ell} = t_{jk} \\ \max(t_{i\ell}, t_{ik}, t_{jk}) & \text{if} \quad u_{ijk\ell} = t_{j\ell} \end{cases} \tag{5.50}$$

## Phase II

From $\boldsymbol{T}$, we construct a square $(m+n) \times (m+n)$ matrix $\boldsymbol{D} = (d_{ab})$ which satisfies the ordinary one-class ultrametric inequality (5.42). $\boldsymbol{D}$ is symmetric $d_{ab} = d_{ba}$ and is defined for $a \neq b$. Because of symmetry, we need only determine $d_{ab}$ for $a > b$. For the $n \times m$ submatrix consisting of the last $n$ rows and the first $m$ columns of $\boldsymbol{D}$, we may substitute $\boldsymbol{T}$. Then, the problem is to fill in the (lower half of the symmetric) $m \times m$ and $n \times n$ submatrices comprising the first $m$ rows and columns and the last $n$ rows and columns, respectively. This is accomplished by using the following equations

(Furnas, 1980):

$$d_{ab} = \begin{cases} t_{(a-m)b} & \text{if } m+1 \le a \le m+n, \\ & \text{and } 1 \le b \le m, \\ \min\limits_{i=1,\ldots,n} (\max(t_{ia}, t_{ib})) & \text{if } 1 \le a \le m, \\ & \text{and } 1 \le b \le m, \\ \min\limits_{k=1,\ldots,m} (\max(t_{(a-m)k}, t_{(b-m)k})) & \text{if } m+1 \le a \le m+n, \\ & \text{and } m+1 \le b \le m+n. \end{cases} \tag{5.51}$$

If necessary, a positive constant is added to the $d_{ab}$ so that they satisfy the triangular inequality.

### Phase III

Using a certain hierarchical clustering method, the ultrametric tree representation of both row and column elements is obtained from $\boldsymbol{D}$.

## 5.5.3. Example

We applied the De Soete et al. algorithm to the soft drinks brand switching data provided in Table 3.3. The data in Table 3.3 was normalized by dividing each cell by the product of the respective row and column averages as in DeSarbo (1982).

Figure 5.12 shows the estimated ultrametric tree, in which the brands with (R) correspond to row objects in period $t$ and the brands with (C) corresponds to column objects in period $t+1$. We find a diet cluster (Diet Pepsi, Tab, Like, and Fresca) and a nondiet cluster (Sprite, Pepsi, 7-Up, and Coke). Only the Tab brand merges to different brands before it merges to itself of the other period.

# 5.6. BOND ENERGY ALGORITHM (TWO-MODE APPROACH)

McCormick et al. (1972) proposed a general approach *Bond energy algorithm* for clustering the rows and/or columns of data matrices. This is one of the most well-known algorithms in *direct clustering* and is adaptive for two-mode or asymmetric matrices. The algorithm detects overlapping or nonoverlapping blocks of submatrices in the data matrix. Those blocks are determined to be a maximal subset of objects and a maximal subset of attributes, such that the corresponding objects/attributes entries fulfill a certain homogeneity criterion.

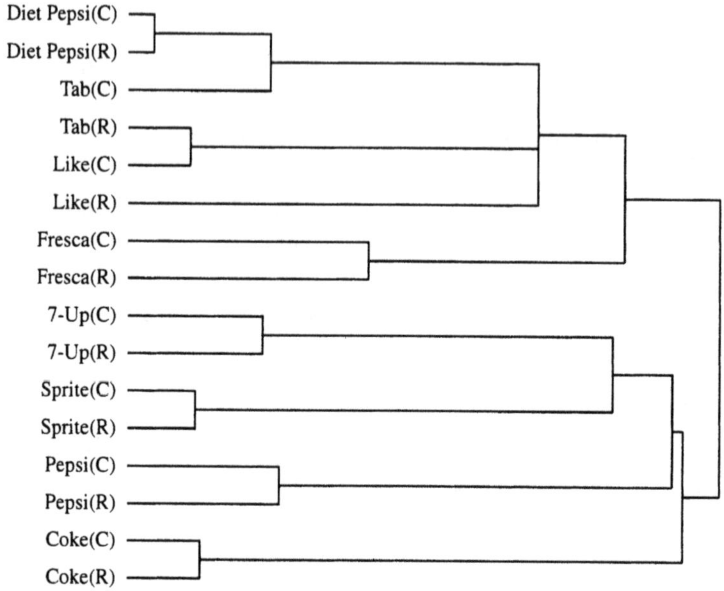


**FIGURE 5.12** Estimated ultrametric tree

## 5.6.1. Preparation

Let $\boldsymbol{O} = (o_{j,k})$ be an $M \times N$ data matrix with non-negative entries. In the bond energy approach, we maximize

$$BE(\boldsymbol{O}) = \sum_j \sum_k o_{j,k}(o_{j,k-1} + o_{j,k+1} + o_{j-1,k} + o_{j+1,k}) \tag{5.52}$$

over all $M!N!$ possible arrays by permuting elements of $\boldsymbol{O}$, where we define

$$o_{j,0} = o_{0,k} = o_{j,N+1} = o_{M+1,k} = 0 \tag{5.53}$$

Because this problem is $NP$-complete, many heuristic algorithms have been proposed.

## 5.6.2. Algorithm

Here we introduce an algorithm proposed by McCormick et al. (1972). First, place one arbitrary column in the first position of the columns. Then, each of the remaining $N-1$ columns is tested in the second positions and placed in the one giving the largest incremental contribution by the respective column to the expression (5.52). We repeat this for the remaining $N-2$ columns. Then the same procedure applies to the rows. Thus, we obtain the matrix that maximizes the bond energy.

Arabie et al. (1988) approved the heuristic approach by use of additional iteration, while McCormick et al. proposed only one overall iteration for the column and another for the row (Arabie et al., 1988, for details).

For the asymmetric matrix that attains $BE(\boldsymbol{O})$ maximum, we divide the matrix between the two columns/(or two rows) attaining the least to the overall bond energy, obtaining a bipartition. The process continues iteratively to form a partition with more clusters. This treatment corresponds to clustering, that is, these submatrices obtained by partitions represent clusters.

### 5.6.3. Example

Here we show the result of applying the bond energy algorithm to the soft drinks brand switching data described in Sec. 5.5.

The bond energy of the original matrix is 3,553,196, the attained maximum is 7,555,994. Using the Arabie et al. algorithm, we obtained the result as shown in Table 5.6. The number of clusters, three for rows and four for columns, is specified as in Arabie et al. (1988). For a detailed interpretation, refer to Arabie et al. (1988, p. 220) in comparison with DeSarbo (1982, p. 465).

## 5.7. CENTROID EFFECT ALGORITHM (TWO-MODE APPROACH)

The two-mode clustering procedure proposed by Eckes and Orlik (1990) makes use of the advantages of the methods belonging to the first and the

**TABLE 5.6** Result of the Bond Energy Algorithm

| | Period $[t+1]$ | | | | | | | |
|---|---|---|---|---|---|---|---|---|
| Period $[t]$ | G | C | D | B | A | E | F | H |
| G : Diet Pepsi | 256 | 186 | 93 | 47 | 93 | 116 | 93 | 116 |
| C : Tab | 80 | 160 | 360 | 120 | 80 | 80 | 40 | 80 |
| D : Like | 131 | 87 | 152 | 152 | 87 | 239 | 43 | 109 |
| E : Pepsi | 26 | 8 | 30 | 132 | 177 | 515 | 75 | 37 |
| H : Fresca | 67 | 53 | 107 | 93 | 226 | 147 | 107 | 200 |
| A : Coke | 13 | 10 | 33 | 107 | 612 | 134 | 55 | 36 |
| B : 7-Up | 12 | 5 | 64 | 448 | 186 | 140 | 99 | 46 |
| F : Sprite | 29 | 29 | 71 | 185 | 114 | 157 | 329 | 86 |

third class in the two-mode approach (see Sec. 5.1). In a similar way to the direct clustering methods, the procedure performs a reorganization of the two-mode data matrix, yielding clusters. Like the tree-fitting methods, it constructs an ultrametric tree representation of the two-mode data. It is based on an agglomerative clustering criterion that defines maximally cohesive elements of two-mode clusters.

## 5.7.1. Preparation

Let $A$ and $B$ represent modes. Let $i$ denote $n$ objects of $A$ (row elements), and $j$ denote $m$ variables of $B$ (column elements). A two-mode array is defined as the Cartesian product $A \times B$ with cells $(i, j)$. Given a two-mode data matrix $\boldsymbol{O}$, $o_{ij}$ indicates an observation to the elements $(i, j)$ of a two-mode array. Let $A'$ and $B'$ be subsets of $A$ and $B$, respectively. A *two-mode cluster* or *bi-cluster* $C_r$ is defined as the union of the two sets $A'$ and $B'$: $C_r = A' \cup B'$. A *two-mode submatrix* is an assignment of scale values $o_{i'j'}$ to elements of $A' \times B' \subset A \times B$. The two-mode submatrix corresponding to $A' \times B'$ is $\boldsymbol{O}_r = (o_{i'j'})$ with $n_r m_r$ elements, where $n_r$ is the number of entities in $A'$ and $m_r$ is the number of entities in $B'$.

Given two bi-clusters, $C_p = A' \cup B'$ and $C_q = A'' \cup B''$, where $A'' \subset A, B'' \subset B, A' \cap A'' = \varnothing, B' \cap B'' = \varnothing$, we define the union as

$$C_t = C_p \cup C_q = (A' \cup B') \cup (A'' \cup B''). \tag{5.54}$$

## 5.7.2. Algorithm

Suppose that two bi-clusters $C_p$ and $C_q$ exist at a certain hierarchical level. The assignment of scale values to the elements of a two-mode array $A^+ \times B^+$ with $A^+ = A' \cup A''$ and $B^+ = B' \cup B''$ yields a submatrix $\boldsymbol{O}_t$. This submatrix is decomposed into four submatrices, two of which correspond to the bi-clusters

$$C_p = A' \cup B' \quad \text{and} \quad C_q = A'' \cup B'' \tag{5.55}$$

and $\boldsymbol{O}_p = (o_{i'j'})$ and $\boldsymbol{O}_q = (o_{i''j''})$, respectively; the other two submatrices correspond to the sets

$$R_\alpha = A' \cup B'' \quad \text{and} \quad R_\beta = A'' \cup B' \tag{5.56}$$

with $\boldsymbol{O}_\alpha = (o_{i'j''})$ containing $n_p m_q$ elements, and $\boldsymbol{O}_\beta = (o_{i''j'})$ containing $n_q m_p$ elements, respectively.

Now we specify the strategy for constructing a two-mode hierarchical clustering. At each step in the analysis, every possible pair of bi-clusters is considered and the two bi-clusters $C_p$ and $C_q$ are merged into a bi-cluster $C_t$ so that the merger results in the minimum increase in an internal heterogeneity measure. The heterogeneity measure of a bi-cluster $C_t$ is given by

$$\text{MSD}_t = \frac{1}{n_t m_t} \sum_{i^+ \in A^+} \sum_{j^+ \in B^+} (o_{i^+j^+} - \mu)^2 \tag{5.57}$$

where $\mu = \max_{i,j}(o_{ij})$, that is, $\mu$ is the maximum entry in the original matrix $\boldsymbol{O}$. Thus, $\text{MSD}_t$ indicates the mean-squared deviation of entries $o_{i^+j^+}$ in the corresponding submatrix $\boldsymbol{O}$ from the maximum entry $\mu$.

Since

$$\frac{1}{n_t m_t} \sum_{i^+ \in A^+} \sum_{j^+ \in B^+} o_{i^+j^+}^2 = s_t^2 + \bar{o}_t^2 \tag{5.58}$$

where $s_t^2$ is the variance of entries in $\boldsymbol{O}_t$, and $\bar{o}_t^2$ is the corresponding squared mean, $\text{MSD}_t$ can be written as

$$\text{MSD}_t = s_t^2 + (\bar{o}_t - \mu)^2 \tag{5.59}$$

The squared difference between $\bar{o}_t$ and $\mu$ is called the *centroid effect* of cluster $C_t$. It can be seen that the problem of minimizing $\text{MSD}_t$ is equivalent to finding a cluster $C_t$ for which the sum of the variance and the centroid effect is minimum. Hence, the MSD index may be interpreted as a two-mode error variance, analogous to the variance criterion in traditional techniques of hierarchical or nonhierarchical (partitioning) one-mode clustering.

The merging rule is specified as follows. At each step of the agglomerative process, a subset of mode $A$ is merged with a subset of mode $B$ so that the increase in the internal heterogeneity measure of the resulting two-mode cluster is as small as possible.

To accomplish this objective, several heuristic criteria are implemented to minimize the MSD index. Which criterion to use at any particular step in the process depends on the subsets considered. Three general cases are distinguished as follows. In each case, those two subsets to yield the smallest criterion value will be merged.

## Case I

A single-element subset $\{i'\}$ is to be merged with another single-element subset $\{j'\}$:

$$\text{MSD}_{i'j'} = (o_{i'j'} - \mu)^2 \tag{5.60}$$

## Case IIa

A single-element subset $\{j'\}$ is to be merged with an existing two-mode cluster $C_r$:

$$\mathrm{MSD}_\alpha = \frac{1}{n_r} \sum_{i' \in A'} (o_{i'j'} - \mu)^2 \tag{5.61}$$

## Case IIb

A single-element subset $\{i'\}$ is to be merged with an existing two-mode cluster $C_r$:

$$\mathrm{MSD}_\beta = \frac{1}{m_r} \sum_{j' \in B'} (o_{i'j'} - \mu)^2 \tag{5.62}$$

Now let us consider a case for merging two existing two-mode clusters. According to the definition of the union of two clusters given above and schematically illustrated in Fig. 5.13, the squared deviations are computed only for those entities of mode $A$ and only for those entities of mode $B$ that belong to the sets $R_\alpha$ and $R_\beta$, respectively. This case may be formally expressed as follows.

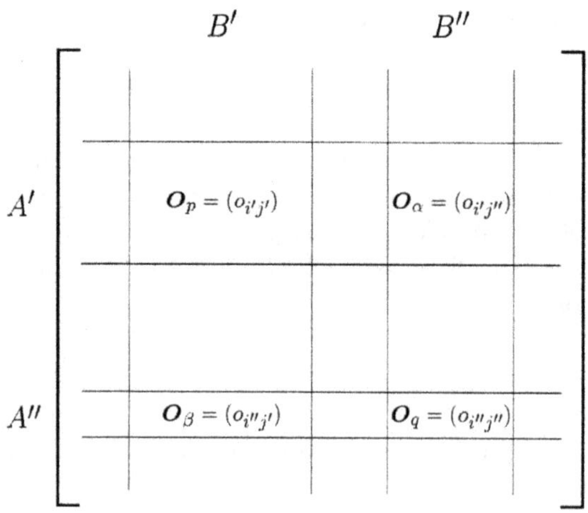


**FIGURE 5.13** Four submatrices

### Case III

A two-mode cluster $C_p = A' \cup B'$ and another two-mode cluster $C_q = A'' \cup B''$ are merged into a new cluster $C_t$:

$$\mathrm{MSD}_{\alpha\beta} = \frac{1}{n_p m_q + n_q m_p} \cdot \left( \sum_{i' \in A'} \sum_{j'' \in B''} (o_{i'j''} - \mu)^2 + \sum_{i'' \in A''} \sum_{j' \in B'} (o_{i''j'} - \mu)^2 \right) \quad (5.63)$$

For determining the number of two-mode clusters, we regard a large increase in MSD values as indicative of the formation of a relatively heterogeneous cluster. Thus, we can decide the number of clusters in an analogous manner to the one-mode clustering method (Ward's method) based on the error-sum-of-squares.

## 5.7.3. Example

We show the results of applying the centroid effect algorithm to the soft drinks brand switching data described in Sec. 5.5.3.

Figure 5.14 represents the dendrogram derived by the centroid effect algorithm. The brands with (R) correspond to row objects in period $t$ and the brands with (C) to column objects in period $t+1$. The major clusters are obtained: one with nondiet objects including Pepsi, Coke, 7-Up, and Sprite, and the other with mostly diet objects such as Diet Pepsi, Tab, Like, and Fresca. The tree is in perfect agreement with the results of DeSarbo and De Soete (1984).

# 5.8. GENNCLUS (ALTERNATING LEAST SQUARES APPROACH)

The ADCLUS (ADditive CLUStering) model proposed by Shepard and Arabie (1979) represents (dis)similarities as combination of discrete and possibly overlapping properties. DeSarbo (1982) proposed GENNCLUS as a generalization of ADCLUS for either overlapping or nonoverlapping properties; symmetric or nonsymmetric (two-mode) clustering. In this section we deal with GENNCLUS as an asymmetric clustering algorithm.

The GENNCLUS model generalizes the ADCLUS model to allow for: either overlapping or nonoverlapping clusters; either a symmetric or

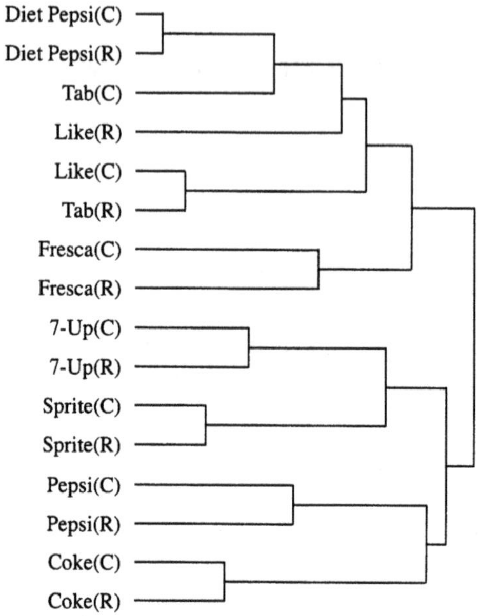


**FIGURE 5.14** The result of the centroid effect algorithm

nonsymmetric unconditional input matrix; and either a diagonal matrix $\boldsymbol{W}$ or an arbitrary symmetric matrix $\boldsymbol{W}$.

The primary advantage of such a generalization is that GENNCLUS will allow one to analyze nonsymmetric data, especially two-mode data. This model is applicable for overlapping clusters, whereas the other models described in this chapter do not allow overlapping.

## 5.8.1. Model

The model of GENNCLUS is stated as:

$$\boldsymbol{O} = \boldsymbol{PWQ}' + \boldsymbol{C} + \boldsymbol{E} \tag{5.64}$$

where $\boldsymbol{O}$ is an $n \times l$ matrix of observed similarities; $\boldsymbol{P}$ is an $n \times m$ matrix indicating the membership of the $n$ row objects in $m$ clusters; $\boldsymbol{W}$ is an $m \times m$ symmetric matrix representing the association between the $m$ derived clusters; $\boldsymbol{Q}$ is an $l \times m$ matrix indicating the membership of the $l$ column objects in $m$

clusters; $\boldsymbol{C}$ is an $n \times l$ matrix where every element is an identical constant; and $\boldsymbol{E}$ is an $n \times l$ error matrix.

Given $\boldsymbol{O}$, in the GENNCLUS algorithm we attempt to estimate $\boldsymbol{P}$, $\boldsymbol{W}$, $\boldsymbol{Q}$, and $\boldsymbol{C}$. Note, if $\boldsymbol{O}$ is symmetric, then $\boldsymbol{P} = \boldsymbol{Q}$, and the model becomes symmetric. One can also specify whether overlapping or nonoverlapping clusters are desired, that is, whether

$$\sum_{k=1}^{m} p_{ik} = \sum_{k=1}^{m} q_{jk} = 1 \quad (i, j = 1, \ldots, n) \tag{5.65}$$

or

$$\sum_{k=1}^{m} p_{ik} \geq 1 \quad (i = 1, \ldots, n) \tag{5.66}$$

and

$$\sum_{k=1}^{m} q_{jk} \geq 1 \quad (j = 1, \ldots, n) \tag{5.67}$$

Unlike MAPCLUS, which is incorporated in the ADCLUS model, GENNCLUS specifies a "complete covering," that is, each object belongs to at least one cluster. In addition, it is provided with an option to constrain $\boldsymbol{W}$ to be diagonal, such that

$$w_{kh} = 0 \quad (k \neq h; \quad k, h = 1, \ldots, m) \tag{5.68}$$

All of these variants can be applied to interval or ratio scaled data. Thus, there are 16 variants of the algorithm by specifying type of clustering (overlapping vs. nonoverlapping), type of data (symmetric vs. asymmetric), type of scale (interval vs. ratio), and type of constraint ($\boldsymbol{W}$ unconstrained vs. $\boldsymbol{W}$ constrained).

### 5.8.2. Algorithm

The 16 variants of the GENNCLUS algorithm are based on the alternating least squares approach. The objective is to minimize

$$\sum_{i=1}^{n} \sum_{j=1}^{l} (o_{ij} - \hat{o}_{ij})^2 = \sum_{i=1}^{n} \sum_{j=1}^{l} \left( o_{ij} - \sum_{k=1}^{m} \sum_{k'=1}^{m} p_{ik} w_{kk'} q_{jk'} - c \right)^2 \tag{5.69}$$

with respect to $\boldsymbol{P}$, $\boldsymbol{W}$, $\boldsymbol{Q}$, and $c$. Here $p_{jh}$ and $q_{jk}$ represent indicator matrices,

$$p_{ik} \in \{0, 1\} \quad (i = 1, \ldots, n; \quad k = 1, \ldots, m) \tag{5.70}$$

$$q_{jk'} \in \{0, 1\} \quad (j = 1, \ldots, l; \quad k' = 1, \ldots, m) \tag{5.71}$$

and $w_{kk'}$ is symmetric, that is,

$$w_{kk'} = w_{k'k} \quad {\scriptstyle (k,k'=1,\ldots,m)} \tag{5.72}$$

Constraints are imposed as

$$\sum_k p_{ik} \geq 1 (= 1 \text{ if nonoverlapping}) \tag{5.73}$$

$$\sum_k q_{jk} \geq 1 (= 1 \text{ if nonoverlapping}) \tag{5.74}$$

$$w_{kh} = 0 \quad (k \neq h; \quad k, h = 1, \ldots, m) \quad \text{if } \boldsymbol{W} \text{ diagonal} \tag{5.75}$$

$$p_{ik} = q_{jk} \quad (i, j = 1, \ldots, n(= l); \quad k = 1, \ldots, m) \quad \text{if } \boldsymbol{O} \text{ symmetric} \tag{5.76}$$

Note that the number of parameters to be estimated is written in general as

$$\left( n(m - \delta_1) + l\delta_2(m - \delta_1) + \delta_3 m + (1 - \delta_3)\left(\frac{m^2 + m}{2}\right) + \delta_4 \right) \tag{5.77}$$

**TABLE 5.7** Three Cluster Overlapping Solution for Brand Switching Data

| | ***P*** (Period $t$) | | | ***Q*** (Period $t+1$) | | |
|---|---|---|---|---|---|---|
| Coke | 1 | 0 | 0 | 1 | 0 | 0 |
| 7-Up | 0 | 0 | 1 | 0 | 0 | 1 |
| Tab | 1 | 1 | 0 | 1 | 1 | 0 |
| Like | 1 | 1 | 0 | 1 | 1 | 0 |
| Pepsi | 1 | 0 | 0 | 1 | 0 | 0 |
| Sprite | 0 | 0 | 1 | 0 | 0 | 1 |
| Diet Pepsi | 1 | 1 | 0 | 1 | 1 | 0 |
| Fresca | 0 | 0 | 1 | 0 | 1 | 1 |

where

$$\delta_1 = \begin{cases} 1 & \text{if nonoverlapping clustering} \\ 0 & \text{if overlapping clustering} \end{cases} \tag{5.78}$$

$$\delta_2 = \begin{cases} 1 & \text{if } \boldsymbol{P} \neq \boldsymbol{Q} \\ 0 & \text{if } \boldsymbol{P} = \boldsymbol{Q} \end{cases} \tag{5.79}$$

$$\delta_3 = \begin{cases} 1 & \text{if } \boldsymbol{W} \text{ is diagonal} \\ 0 & \text{else} \end{cases} \tag{5.80}$$

$$\delta_4 = \begin{cases} 1 & \text{if interval scaled } s_{ij} \\ 0 & \text{if ratio scaled } s_{ij} \end{cases} \tag{5.81}$$

There is a possibility of estimating quite a few parameters for a set of data ($nl$ observations). Obviously the maximum of $m$ is conditioned by the degrees of freedom of a specified model.

## 5.8.3. Example

Table 5.7 shows the result of applying the GENNCLUS algorithm to the soft drinks brand switching data described in Sec. 5.5. From this result, one can see that $\boldsymbol{P}$ and $\boldsymbol{Q}$ are not so different, that is, $\boldsymbol{O}$ does not involve a large degree of asymmetry. In fact, both $\boldsymbol{P}$ and $\boldsymbol{Q}$ are identical except for Fresca.

# 6

# Network Analysis of Asymmetric Data

## 6.1. OVERVIEW AND PRELIMINARIES

### 6.1.1. Asymmetry in Network Analysis

In this chapter we describe some methods and procedures for analyzing asymmetric relationships regarding networks. Network analysis is not only concerned with social phenomena but also with psychological phenomena. In psychological studies, network theories have been used in a variety of ways, such as neural nets, associative networks, and semantic networks among others (Hutchinson, 1989). In most cases the concept of a network served as the model or the frame of reference to account for the phenomena and the network itself was not observed directly in experiments. Turning to social network studies, in contrast, the network data are collected directly through observation, interview, questionnaire, or archival records in most cases (except for cases such as non-network relational data). From this point of view, it seems natural that network analysis has developed in social studies more than in psychological studies. Asymmetry is observed in social network data more prominently than in psychological data for network analysis. Hence, analysis of social networks is of principal concern. For a moment, let us review several terms of social networks and graph theories.

### 6.1.2. Preliminaries

#### Social Network

A social network consists of a set of $n$ social actors and a set of relational ties among actors, and the ties specify how these actors are linked together. A relation refers to the set of ties. Let $N$ denote a set of actors $N = \{1, 2, \ldots, n\}$. Let $u_{ij}$ represent the relational tie of an ordered pair of actors $(i, j)$. Define an $n \times n$ matrix $\boldsymbol{U} = (u_{ij})$, which is called the *relational matrix* or *sociomatrix*. Thus the matrix describes a relation on the set of actors.

Regarding the direction of the relation, there are two cases. One is the directed case in which $u_{ij} \neq u_{ji}$ generally, then $\boldsymbol{U}$ is asymmetric. The other is the nondirectional case in which $u_{ij} = u_{ji}$ for all pairs, then $\boldsymbol{U}$ is symmetric.

When the relation is dichotomous, it is represented as $u_{ij} = 1$ if the relational tie (simply, tie) from $i$ to $j$ exists, and as $u_{ij} = 0$ if it does not. When the relation is valued, $u_{ij}$ indicates the degree of relational ties, taking a real value defined on a measurement scale.

For the actor set, there may be multiple relations, that is, $r$ kinds of relations, giving rise to $r$ sociomatrices. For example, we can think of relationships in terms of friendship, getting on, and working together for a set of students, which will be indicated by three sociomatrices. If a network is defined for two sets of actors, it is called a two-mode network in contrast to a one-mode network for a single set of actors as so far mentioned.

#### Graph Theoretic Terms

A directed graph (*digraph*) is a set of vertices, together with a set of directed arcs connecting certain pairs of vertices. A network is a directed graph in which lengths (positive real numbers) are associated with the arcs. A walk is a nonempty alternating sequence of vertices and arcs, beginning and ending with vertices. A walk in which no vertex occurs more than once is called a path. The length of a path is the sum of the lengths of the arcs.

The distance from vertex $j$ to vertex $k$ is some function of the lengths of the paths connecting $j$ and $k$. The distance in terms of the minimum pathlength is called the minimum pathlength metric. Structures with the metric are called *geodesic*. The distance matrix of a network contains the distances defined for each ordered pair of vertices. If there is no path connecting one vertex to another, the corresponding distance is considered to be infinite.

Let us explain correspondence of a graph representation and a directional network. Regarding a social network, a vertex corresponds to an actor (generally, a social entity), an arc to a social tie between a pair of actors, and the arc length to the degree of the tie. Regarding a network derived from psychological data, a vertex corresponds to a stimulus (generally, a psychological object or a state). An arc and the arc length are considered in connection with the proximity between a pair of stimuli.

### 6.1.3. Subjects of the Chapter

Putting our focus on general terms, we are concerned with the analysis of a one-mode, directed network of a single relation. As a note, it is to be remembered that we provided a brief comment in Chapter 4 and fair amount of description in Chapter 5 on analysis of two-mode asymmetric data. We take up three subjects in what follows.

In Sec. 6.2 we deal with detecting cohesive groups in a social network of individuals that may be a dichotomous or valued one (Fershtman, 1997). The concept of cluster is closely related to the concept of cohesiveness. In Sec. 6.3 we consider constructing a valued network, given a set of proximity data (Hutchinson, 1989). The proximity may represent relationship among social entities or proximity among psychological stimuli (cognitive similarity, word association). In Sec. 6.4 we treat analysis of a dichotomous network by statistical models by introducing random variables in the formulation (Strauss and Ikeda, 1990; Wasserman and Pattison, 1996).

## 6.2. DETECTION OF COHESIVE GROUPS

It is often of interest to detect or identify a cohesive group in a social network. A cohesive group of actors is a subset of actors, among whom the relationships are relatively frequent, high, intense, strong, or important in comparison with the relationships between the group members and the non-group members. For a practical analysis of network data, cohesiveness would be specified formally, depending on the purpose of analysis.

For a review, we would like to refer to Wasserman and Faust (1994). The term clique is a fundamental concept to study cohesive groups. For a non-directional dichotomous network, a clique is precisely defined in terms of graph theory, accordingly a procedure is formulated. For a directional dichotomous network, one performs a symmetrization of the sociomatrix, from which follows a new non-directional dichotomous relation. Then one can apply the procedure for detecting cliques.

They also suggested possible utilization of MDS for a study of cohesive groups in a network. Constructing some appropriate measure of proximity (e.g., dissimilarity) based on the network data, one can apply MDS to derive a configuration of actors. Then one would examine what subsets of actors are close in the spatial configuration and compare the closeness with graph theoretic distance (e.g., geodesic distance). Fershtman (1997) proposed a procedure to detect cohesive groups in a directional network by using the segregation matrix index. In this section we describe the procedure and its application to real data of a social network.

### 6.2.1. Segregation Matrix Index

We consider a situation in which $n$ individuals choose each other for friendship or liking. That is, we regard the people as an actor set and a choice as a directional tie between a pair of individuals.

#### The Case of a Dichotomous Network

Let $A$ be a group of $n_A$ actors. Let $B$ be the rest of $n_B$ actors ($n_B = n - n_A$). Define $u_{ij} = 1$ if actor $i$ chooses actor $j$, and $u_{ij} = 0$ if $i$ does not choose $j$. Let $u_{ii} = 0$. Denote the numbers of choices given by actor $i$ toward members of $A$ and toward those of $B$ by

$$u_{iA} = \sum_{j \in A} u_{ij} \quad \text{and} \quad u_{iB} = \sum_{j \in B} u_{ij} \tag{6.1}$$

respectively. Let $u_{AA}$ be the number of mutual choices among $A$, and $u_{AB}$ the number of choices made by members of $A$ toward $B$, in such a way that

$$u_{AA} = \sum_{i \in A} u_{iA} = \sum_{i \in A} \sum_{j \in A} u_{ij} \tag{6.2}$$

$$u_{AB} = \sum_{i \in A} u_{iB} = \sum_{i \in A} \sum_{j \in B} u_{ij} \tag{6.3}$$

Let us express the density of choices among $A$ and that between $A$ and $B$ in terms of the average number of choices as

$$d_{AA} = \frac{u_{AA}}{n_A(n_A - 1)} \quad \text{and} \quad d_{AB} = \frac{u_{AB}}{n_A n_B} \tag{6.4}$$

respectively. It is noted that the density measure is the same as the average linkage in cluster analysis. Fershtman proposed an index to measure the tendency of group $A$ toward segregated attitude by

$$r(A) = \frac{d_{AA}}{d_{AB}} \tag{6.5}$$

It takes a value in the range of (0, ∞). For normalization, we define the segregation matrix index (SMI) of group $A$ as,

$$\psi(A) = \frac{r(A) - 1}{r(A) + 1} = \frac{d_{AA} - d_{AB}}{d_{AA} + d_{AB}} \tag{6.6}$$

The value of $\psi$ is bounded as

$$-1 \leq \psi(A) \leq 1 \tag{6.7}$$

## The Case of a Valued Network

The index is easily extended to the case of a valued network. Denote the intensity of interaction of $i$ toward $j$ by $v_{ij}$. Let $\boldsymbol{V} = (v_{ij})$, which is usually asymmetric. We can redefine $u_{iA}$ and $u_{iB}$ by putting $v_{ij}$ instead of $u_{ij}$ in (6.1). Then (6.2) to (6.4) are restated in terms of $v_{ij}$. Thus we have the segregation matrix index $\psi(A)$ for a valued graph.

# 6.2.2. Algorithm for Detecting S-Cliques

## Operational Definition of S-Clique

Given a directed social network, which is a binary graph or a valued one, Fershtman defined the $S$-clique (a clique based on the segregation matrix index) operationally. A group of actors $A$ of size $n_A$ is an $S$-clique, if the following inequalities hold for any group $A'$ that includes at least one member of $A$ ($A \cap A' \neq \phi$) and whose size $n_{A'}$ ranges from $n_A - 1$ to $n_A + 1$:

1. $\psi(A) > \psi(A')$ if $n_{A'} = n_A + 1$
2. $\psi(A) \geq \psi(A')$ if $n_{A'} = n_A$ or $n_A - 1$

## The Algorithm

On the basis of the definition above, the algorithm for detecting $S$-cliques is stated as follows.

Step 1: Set threshold values $\gamma_j$ ($j = 2, \ldots, \ell$)
Step 2: Compute SMI for all possible two-member groups $A_2$. Store all groups such that $\psi(A_2) > \gamma_2$.
Step 3: Add an actor to each $A_2$ stored above, generating all three-member groups $A_3$. Compute $\psi$ for those groups, and store all groups such that $\psi(A_3) > \gamma_3$.
Step 4: Set $j = 3$.

Step 5: Add an actor to each $j$-member group (stored in the preceding step), generating all $(j+1)$-member groups $A_{j+1}$. Compute $\psi$ for those groups and store all groups such that $\psi(A_{j+1}) > \gamma_{j+1}$.

Step 6: Check the following inequalities for each $A_j$:

(a) $\psi(A_j) > \psi(A_{j+1})$ for each $A_{j+1}$ that has at least one common member with $A_j$.

(b) $\psi(A_j) \geq \psi(A_j')$ for each $A_j'$ that has at least one common member with $A_j$.

(c) $\psi(A_j) \geq \psi(A_{j-1})$ for each $A_{j-1}$ that has at least one common member with $A_j$.

Step 7: If all the three inequalities hold, $A_j$ is an $S$-clique.

Step 8: If $j < \ell$, then set $j = j + 1$ and go to Step 5.

The number of $j$-member groups in a network of $n$ actors is $n!/(j!(n-j)!)$. It takes an extremely long time to check all the possible groups. If we set high threshold values, we can decrease the number of groups to be checked. However, by setting a very high value, we might fail to find $S$-cliques.

## 6.2.3. Example

Fershtman (1997) examined and showed the adequacy of the operational definition to detect $S$-cliques through two social networks: a social network in a monastery and a social network in a karate club. Here we describe the first.

According to Fershtman, S. F. Sampson originally collected the data of a social network among 18 monks in a monastery. Sampson found four groups in the network: Young Turks (monks 1, 2, 7, 12, 14, 15, and 16), Loyal Opposition (monks 4, 5, 6, 9, and 11), Outcast (monks 3, 17, 18), Waverers (monks 8, 10, 13).

Reitz (1988) analyzed the network data, constructing a valued network $v_{ij}$ that represents the choice interaction intensity of actor $i$ toward actor $j$ ($v_{ij}$ was given in four graded scores; 0.25, 0.5, 0.75, 1). The derived network is shown in Table 6.1 and illustrated in Fig. 6.1. In the figure, directed arcs are drawn only for choices such that $v_{ij} > 0.25$.

Fershtman applied the algorithm above to the sociomatrix $V = (v_{ij})$, finding three $S$-cliques, $A$, $B$, and $C$. Clique $A$ ($\psi = 0.918$) consists of a subclique of monks (1, 2, 7, 12) with $\psi = 0.847$ and three other monks. This clique $A$ coincides with the Young Turks group. Clique $B$ ($\psi = 0.818$) corresponds to the Loyal Opposition group plus a Waverer (monk 8). Clique $C$ ($\psi = 0.787$) coincides with the Outcasts group. Monks 10 and 13 do not belong to a clique.

These results indicate that the detected $S$-cliques correspond to Sampson's findings quite well, and the algorithm even revealed an inner structure

**TABLE 6.1** Choice Intensity Among 18 Monks

| Monk | Chosen monks | | | | | | | | | | | | | | | | | |
|---|---|---|---|---|---|---|---|---|---|---|---|---|---|---|---|---|---|---|
| no. | 1 | 2 | 3 | 4 | 5 | 6 | 7 | 8 | 9 | 10 | 11 | 12 | 13 | 14 | 15 | 16 | 17 | 18 |
| 1 | | 0.25 | 0.25 | | 0.25 | | 0.25 | 0.25 | | | | 1.0 | | 0.50 | | | 0.25 | |
| 2 | 1.0 | | | | 0.25 | | 0.75 | 0.25 | | | | 0.75 | | | | | | |
| 3 | 1.0 | | | | | | | | | | | | 1.0 | | | | 1.0 | 0.50 |
| 4 | | | | | 0.25 | 1.0 | | | | 0.75 | 1.0 | | | | | | | |
| 5 | | 0.50 | | 1.0 | | | | | 0.50 | | 0.75 | | | | | | | |
| 6 | | 0.25 | | 1.0 | 0.25 | | | | 0.75 | | 0.50 | | | | | | | |
| 7 | | 1.0 | | | | | | | | | | 0.75 | | | 0.50 | 0.25 | | |
| 8 | 0.25 | | | 1.0 | 0.25 | 0.75 | | | 0.50 | | 0.50 | | | | | | | |
| 9 | 0.50 | 0.25 | | 0.50 | 0.25 | | | 0.50 | | | | 0.25 | | | | | | |
| 10 | | | | 0.25 | 0.25 | | 0.25 | | 0.25 | | | | 0.25 | | | | | |
| 11 | | | | 0.50 | 0.75 | | | 0.50 | 0.50 | | | | | 0.25 | | 0.25 | | |
| 12 | 0.75 | 0.75 | | | | | 0.50 | | | | | | | 0.25 | | | | |
| 13 | | | | | 1.0 | | 0.75 | | | | 0.75 | | | | | | | 0.25 |
| 14 | 1.0 | 0.75 | | | | | | | | | | 0.75 | | | 1.0 | 0.25 | | |
| 15 | 0.50 | 1.0 | | | | | 0.50 | | | | | 0.25 | | 0.75 | | | | |
| 16 | | 1.0 | | | | | 1.0 | | | | | 0.25 | | | 1.0 | | | |
| 17 | | 0.75 | 0.75 | | | | | | | | | | 0.25 | | | | | 0.75 |
| 18 | 0.25 | 1.0 | 0.75 | | | | | | | | | | | | | | 1.0 | |

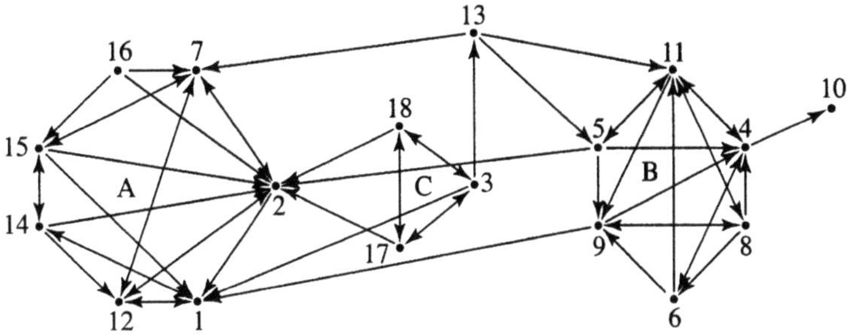


**FIGURE 6.1** Graph representation of monk data

in clique $A$. In passing, it may be of interest to compare these cliques with those groups (Loyal Opposition + Waverers, Young Turks, Outcasts) that Reitz (1988) represented with group density by using his algorithm.

## 6.3. NETWORK SCALING

We are concerned with a structure of asymmetric relationship by constructing a network from empirical data. For scaling of an asymmetric network, Hutchinson (1989) developed a procedure (NETSCAL) based on a set of asymmetric proximity data. Let us describe the procedure, referring to the graph theoretic terms.

### 6.3.1. Construction of a Network from Data

A matrix $\boldsymbol{O} = (o_{jk})$ of proximity is called realizable if there is a network $G$ such that the values of $\boldsymbol{O}$ are equal to the corresponding distances in $G$.

*Theorem 6.1.* Let $\boldsymbol{O} = (o_{jk})$ be a matrix of a distance-like measure among $n$ objects. The necessary and sufficient conditions for $\boldsymbol{O}$ to be realizable are

$$o_{jj} = 0 \tag{6.8}$$

$$o_{jk} > 0 \tag{6.9}$$

$$o_{ij} + o_{jk} \geq o_{ik} \tag{6.10}$$

This theorem means that the complete graph, in which all vertices are reciprocally connected, is a realization of any matrix satisfying (6.8) through (6.10). In such a case, each distance is equivalent to an arc length. It is noted that the complete graph may not be the only realization of matrix $\boldsymbol{O}$. The network (digraph) may contain redundant arcs.

An arc is called redundant if its deletion changes no distances in the network. A network is irreducible if it contains no redundant arcs.

*Theorem 6.2.* Let $G$ be an irreducible representation of some matrix $\boldsymbol{O}$ satisfying (6.8) through (6.10). The arc $(j, k)$ is present in $G$ if and only if

$$j \neq k \quad \text{and} \quad o_{jk} < \min\,(o_{ji} + o_{jk};\, i \neq j,\, k) \tag{6.11}$$

We see that the addition in the right-hand side is a rather strict condition, considering that the measurement of $\boldsymbol{O}$ on a ratio scale is usually difficult. The addition is weakened to the maximum function that gives an invariant result to the monotonic transformation of values of $\boldsymbol{O}$. The following theorem gives the sufficient condition for the presence of arc.

*Theorem 6.3.* Let $\boldsymbol{O}$ be a matrix satisfying (6.8) through (6.10). If

$$o_{jk} \leq \min\,(\max\,(o_{ji},\ o_{ik});\, i \neq j,\ k) \tag{6.12}$$

then $(j, k)$ is an arc in $G$, where $G$ is an irreducible representation of matrix $\boldsymbol{O}$.

*Theorem 6.4.* For any network $G$, and for any two vertices $x$ and $y$ in $G$, if there is a path in $G$ from $x$ to $y$, then there is a path in $G$ from $x$ to $y$ containing only arcs that satisfy Theorem 6.3.

## NETSCAL Algorithm

Given a matrix of proximity $\boldsymbol{O} = (o_{jk})$, we aim to construct a network based on $\boldsymbol{O}$ and determine distances on the network. The algorithm to meet the requirements is outlined in two steps as follows.

1. Digraph is constructed by checking each ordered pair to see if the assumption of Theorem 6.3 is satisfied. If it is, the corresponding arc is added to the structure.
2. Specific arc lengths are estimated and the resulting network distances can be compared to the data. Only distances between distinct vertices are computed.

There are many ways to compute arc lengths. Hutchinson suggested an iterative procedure to estimate arc lengths, from which distances $d_{jk}$ are computed according to the minimum pathlength metric.

### 6.3.2. Example

An example is taken from Hutchinson (1989), which is based on the data regarding recalled confusion of grammatical transformations collected by Mehler (1963). The data matrix indicates the frequency with which one type of sentence was transformed in memory by subjects to another. Table 6.2 lists the grammatical types and sentences. Table 6.3 shows the frequency matrix on the kernel sentence and its seven transformations. Applying the NETSCAL procedure to the data yielded the network illustrated in Fig. 6.2. In the figure, thickness of the arcs indicates the degree of arc length. A thick segment indicates short distance. The representation reveals that virtually all of the arcs correspond to a single transformation. The squared correlation between $o_{jk}$ and distances on the network was 0.73. Prior to the analysis, it was known that the eight sentences constitute a three-dimensional binary space (Miller–Chomsky cube). Comparing with it, the application of NETSCAL provided arc lengths and revealed two diagonal arcs.

## 6.4. STATISTICAL MODELS FOR SOCIAL NETWORK

Let us consider a dichotomous directed network for a set $n$ actors. Suppose that we are given a sociomatrix $\boldsymbol{U} = (u_{kl})$, which represents a dichotomous relation among the set of actors. Researchers have developed many models to account for each relational tie as a function of actor properties or network statistics (Wasserman & Pattison, 1996). Here our description is confined to probabilistic models in terms of network statistics. Let us explain some typical statistics.

**TABLE 6.2** Grammatical Types and Sentences

| Abb. | Grammatical type | Sentence |
|---|---|---|
| K | Karnel | "The man opened the door." |
| N | Negative | "The man did not open the door." |
| Q | Question | "Did the man open the door?" |
| P | Passive | "The door was opened by the man." |
| NQ | Negative Question | "Didn't the man open the door?" |
| NP | Negative Passive | "The door was not opened by the man." |
| PQ | Passive Question | "Was the door opened by the man?" |
| NPQ | Negative Passive Question | "Wasn't the door opened by the man?." |

**TABLE 6.3** Errors in Recall

| | Responses | | | | | | | |
|---|---|---|---|---|---|---|---|---|
| Stimulus | K | N | Q | P | NQ | NP | PQ | NPQ |
| K | 300 | 14 | 12 | 14 | 8 | 4 | 1 | 3 |
| N | 36 | 234 | 20 | 3 | 29 | 11 | 6 | 2 |
| Q | 31 | 16 | 210 | 1 | 72 | 2 | 8 | 12 |
| P | 43 | 3 | 8 | 243 | 15 | 10 | 30 | 13 |
| NQ | 29 | 15 | 31 | 3 | 221 | 3 | 7 | 23 |
| NP | 6 | 49 | 9 | 18 | 16 | 191 | 16 | 31 |
| QP | 13 | 5 | 32 | 27 | 29 | 15 | 145 | 60 |
| NPQ | 2 | 2 | 14 | 16 | 44 | 5 | 38 | 182 |

Let $L$ be the number of ties and $M$ be the number of mutual dyads (number of reciprocity). Denote the directional tie from $i$ to $j$ by $i \rightarrow j$. For statistics of three members, we consider $S_1$ the number of 2-in stars (triple of nodes such that $j \rightarrow i$ and $k \rightarrow i$), $S_2$ the number of 2-out stars (triple of nodes such that $i \rightarrow j$ and $i \rightarrow k$), $S_3$ the 2-mixed stars ($j \rightarrow i$ and $i \rightarrow k$). Further, let $T_C$ be the number of cyclic triads (triple of nodes such that $i \rightarrow j$ and $j \rightarrow k$), $T_T$ the number of transitive triads and $T_I$ the number of intransitive triads. These statistics are defined as follows:

$$L = \sum_{i,j} u_{ij}, \qquad M = \sum_{i<j} u_{ij}u_{ji} \tag{6.13}$$

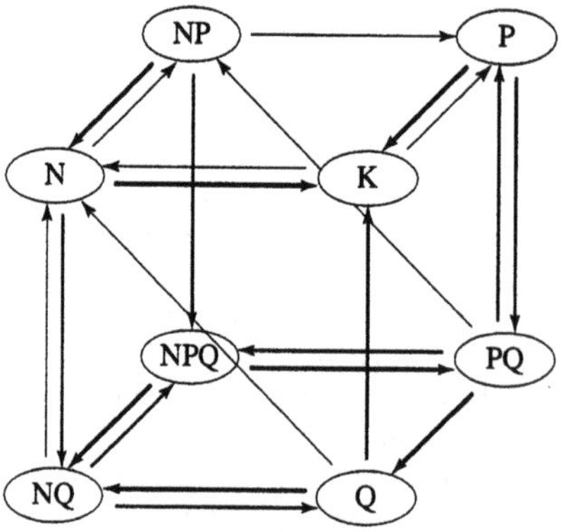


**FIGURE 6.2** NETSCAL result of grammatical transformations data

$$S_1 = \sum_{i \neq j \neq k} u_{ji}u_{ki}, \qquad S_2 = \sum_{i \neq j \neq k} u_{ij}u_{ik}, \qquad S_3 = \sum_{i \neq j \neq k} u_{ji}u_{ik} \tag{6.14}$$

$$T_T = \sum_{i \neq j \neq k} u_{ij}u_{jk}u_{ik}, \qquad T_I = \sum_{i \neq j \neq k} u_{ij}u_{jk}(1 - u_{ik}),$$

$$T_C = \sum_{i \neq j \neq k} u_{ij}u_{jk}u_{ki} \tag{6.15}$$

In developing the probabilistic models, we provide $n \times n$ matrices $\boldsymbol{U}_{ij}^{+}$, $\boldsymbol{U}_{ij}^{-}$, and $\boldsymbol{U}_{ij}^{c}$ follows:

$$\boldsymbol{U}_{ij}^{+} = (w_{kl}) \qquad \text{where} \begin{cases} w_{kl} = u_{kl} \text{ for } (k, l) \neq (i, j) \\ w_{kl} = 1 \text{ for } (k, l) = (i, j) \end{cases} \tag{6.16}$$

$$\boldsymbol{U}_{ij}^{-} = (w_{kl}) \qquad \text{where} \begin{cases} w_{kl} = u_{kl} \text{ for } (k, l) \neq (i, j) \\ w_{kl} = 0 \text{ for } (k, l) = (i, j) \end{cases} \tag{6.17}$$

$$\boldsymbol{U}_{ij}^{c} = (w_{kl}) \qquad \text{where} \begin{cases} w_{kl} = u_{kl} \text{ for } (k, l) \neq (i, j) \\ (w_{kl}) = \text{undefined for } (k, l) = (i, j) \end{cases} \tag{6.18}$$

Define $\boldsymbol{X} = (x_{kl})$. If

$$x_{kl} = u_{kl} \quad \text{for} \quad (k, l) \neq (i, j) \tag{6.19}$$

we denote

$$\boldsymbol{X} = \boldsymbol{U}_{ij}^{c} \tag{6.20}$$

In view of the dichotomous nature of random variable $u_{ij}$, we state the conditional probability as

$$\Pr(u_{ij} = 1 | \boldsymbol{X} = \boldsymbol{U}_{ij}^{c}) = \frac{\Pr(\boldsymbol{X} = \boldsymbol{U}_{ij}^{+})}{\Pr(\boldsymbol{X} = \boldsymbol{U}_{ij}^{+}) + \Pr(\boldsymbol{X} = \boldsymbol{U}_{ij}^{-})} \tag{6.21}$$

### 6.4.1. Logit Models

Given a sociomatrix $\boldsymbol{U}$, we regard it as an observed matrix of random variables, which is a function of network statistics. Let $z(\boldsymbol{U})$ be a vector of network statistics. Let $\boldsymbol{\theta}$ be a vector of model parameters, indicating weights of the linear combination of those statistics. The family of $p^*$ models are stated as

$$\Pr(\boldsymbol{X} = \boldsymbol{U}) = \frac{\exp(\boldsymbol{\theta}' z(\boldsymbol{U}))}{\kappa(\boldsymbol{\theta})} \tag{6.22}$$

(Strauss and Ikeda, 1990). For example, we may write a very simple case using a scalar $\theta$, as

$$\Pr(\boldsymbol{X} = \boldsymbol{U}) = \frac{\exp(\theta L)}{\kappa(\theta)} \tag{6.23}$$

Constraints might be imposed on $\boldsymbol{\theta}$. The function $\kappa(\boldsymbol{\theta})$ is a normalizing factor. For analysis of network data, we need to estimate $\boldsymbol{\theta}$. The sociomatrix is usually asymmetric. Setting aside such a formal asymmetry, one would realize asymmetry at structural levels by examining the weight $\boldsymbol{\theta}$ to network statistics. As examples of those statistics, we may consider the number of mutual dyads $M$, the number of ties $L$, and so on (Table 6.4).

From (6.21) and (6.22), the conditional probability is stated as

$$\Pr(u_{ij} = 1 | \boldsymbol{X} = \boldsymbol{U}_{ij}^{c}) = \frac{\exp(\boldsymbol{\theta}' z(\boldsymbol{U}_{ij}^{+}))}{\exp(\boldsymbol{\theta}' z(\boldsymbol{U}_{ij}^{+})) + \exp(\boldsymbol{\theta}' z(\boldsymbol{U}_{ij}^{-}))} \tag{6.24}$$

Accordingly, the odds ratio is given in such a way that

$$\begin{aligned} \frac{\Pr(u_{ij} = 1 | \boldsymbol{X} = \boldsymbol{U}_{ij}^{c})}{\Pr(u_{ij} = 0 | \boldsymbol{X} = \boldsymbol{U}_{ij}^{c})} &= \frac{\exp(\boldsymbol{\theta}' z(\boldsymbol{U}_{ij}^{+}))}{\exp(\boldsymbol{\theta}' z(\boldsymbol{U}_{ij}^{-}))} \\ &= \exp(\boldsymbol{\theta}'(z(\boldsymbol{U}_{ij}^{+}) - z(\boldsymbol{U}_{ij}^{-}))) \end{aligned} \tag{6.25}$$

Denote the log odds ratio (the logit) by $\omega_{ij}$. Then the logit $p^*$ models are generally expressed as

$$\omega_{ij} = \boldsymbol{\theta}'(z(\boldsymbol{U}_{ij}^{+}) - z(\boldsymbol{U}_{ij}^{-})) \tag{6.26}$$

To specify a logit $p^*$ model, one should consider what network statistics to involve in the model, which are supposed to affect the logit value. That is, the model specification depends on the hypothesized features of a network structure.

## 6.4.2. Parameter Estimation

For directed relations, Table 6.4 lists parameters and graph statistics for modeling logit models (Wasserman and Pattison, 1996). Some of the parameters measure the effects reflected by triads. It gives parameters for homogeneous effects in the upper side and those for prominence homogeneity in the lower side, where $d_{ij}$ indicates geodesic on the graph. The parameters will be interpreted in connection with the network, by examining the estimates and their contribution to the likelihood.

**TABLE 6.4** Homogeneous Parameters and Graph Statistics

| Label | Parameter | Graph statistic $z(\boldsymbol{U})$ |
|---|---|---|
| Homogeneous effects | | |
| Choice | $\theta$ | $L = u_{++}$ |
| Reciprocity | $\rho$ | $M = \sum_{i,j} u_{ij}u_{ji}$ |
| Transitivity | $\tau_T$ | $T_T = \sum_{i,j,k} u_{ij}u_{jk}u_{ik}$ |
| Intransitivity | $\tau_I$ | $T_I = \sum_{i,j,k} u_{ij}u_{jk}(1 - u_{ik})$ |
| Cyclicity | $\tau_C$ | $T_C = \sum_{i,j,k} u_{ij}u_{jk}u_{ki}$ |
| 2-in stars | $\sigma_1$ | $S_1 = \sum_{i,j,k} u_{ji}u_{ki}$ |
| 2-out stars | $\sigma_2$ | $S_2 = \sum_{i,j,k} u_{ij}u_{ik}$ |
| 2-mixed stars | $\sigma_3$ | $S_3 = \sum_{i,j,k} u_{ji}u_{ik}$ |
| ⋮ | ⋮ | ⋮ |
| $k$-in stars | $\upsilon_{kI}$ | $U_{kI} = \sum_{i,j_1,j_2,\ldots,j_k} u_{j_1i}u_{j_2i}\cdots u_{j_ki}$ |
| $k$-out stars | $\upsilon_{kO}$ | $U_{kO} = \sum_{i,j_1,j_2,\ldots,j_k} u_{ij_1}u_{ij_2}\cdots u_{ij_k}$ |
| ⋮ | ⋮ | ⋮ |
| 3-paths | $\pi_3$ | $P_3 = \sum_{i,j,k,l} u_{ij}u_{jk}u_{kl}$ |
| ⋮ | ⋮ | ⋮ |
| $k$-paths | $\pi_k$ | $P_k = \sum_{i,j_1,j_2,\ldots,j_k} u_{ij_1}u_{j_1j_2}\cdots u_{j_{k-1}j_k}$ |
| Connectivity | $\nu$ | $N =$ minimum number of edges whose removal disconnects the digraph (edge connectivity) |
| Indices of prominence homogeneity (centralization/prestige) | | |
| | $\phi_1$ | $C_{G_1} =$ sum of lengths of geodesics ($d_{ij}$) |
| | $\phi_2$ | $C_{G_2} =$ variance of geodesic lengths ($d_{ij}$) |
| | $\phi_3$ | $C_E =$ maximum of geodesic length (eccentricity) |
| | $\phi_4$ | $C_C =$ variance of geodesic length sums ($d_{i+}$) (closeness centralization) |
| | $\phi_5$ | $C_D =$ variance of $u_{i+}$ (degree centralization) |
| | $\phi_6$ | $C_B =$ variance of numbers of geodesics containing $i$ (betweeness centralization) |
| | $\phi_7$ | $P_P =$ variance of geodesic length sums ($d_{+i}$) (proximity group prestige) |
| | $\phi_8$ | $P_D =$ variance of $u_{+i}$ (degree group prestige) |

Actor-level parameters and graph statistics are also of interest in the study of social networks (Wasserman and Pattison, 1996). Such parameters include, for example, those to measure centrality and prestige, and so on.

We consider the likelihood function for the $p^*$ model of (6.22). Taking an approximate estimation approach (Strauss and Ikeda, 1990), one assumes

**TABLE 6.5** Vickers and Chan Data: "Get-on with" Relation

| | 1 | 2 | 3 | 4 | 5 | 6 | 7 | 8 | 9 | 10 | 11 | 12 | 13 | 14 | 15 | 16 | 17 | 18 | 19 | 20 | 21 | 22 | 23 | 24 | 25 | 26 | 27 | 28 | 29 |
|---|---|---|---|---|---|---|---|---|---|---|---|---|---|---|---|---|---|---|---|---|---|---|---|---|---|---|---|---|---|
| 1 | 0 | 0 | 0 | 0 | 0 | 1 | 0 | 1 | 0 | 0 | 1 | 1 | 0 | 1 | 0 | 1 | 1 | 0 | 1 | 0 | 1 | 1 | 0 | 0 | 0 | 1 | 1 | 0 | 0 |
| 2 | 1 | 0 | 1 | 1 | 1 | 1 | 1 | 1 | 0 | 1 | 1 | 0 | 0 | 1 | 1 | 1 | 0 | 0 | 1 | 1 | 1 | 0 | 0 | 0 | 0 | 0 | 0 | 0 | 0 |
| 3 | 0 | 0 | 0 | 1 | 0 | 0 | 1 | 1 | 1 | 0 | 1 | 0 | 0 | 0 | 0 | 0 | 0 | 0 | 0 | 0 | 0 | 0 | 0 | 0 | 0 | 0 | 0 | 0 | 0 |
| 4 | 0 | 1 | 1 | 0 | 1 | 1 | 1 | 0 | 0 | 1 | 0 | 0 | 0 | 0 | 0 | 0 | 0 | 0 | 0 | 0 | 0 | 0 | 0 | 0 | 0 | 0 | 0 | 0 | 0 |
| 5 | 1 | 1 | 0 | 0 | 0 | 1 | 0 | 1 | 0 | 1 | 1 | 0 | 0 | 1 | 1 | 1 | 0 | 0 | 1 | 1 | 1 | 1 | 1 | 0 | 0 | 0 | 1 | 1 | 1 |
| 6 | 1 | 1 | 1 | 1 | 1 | 0 | 1 | 1 | 0 | 1 | 1 | 1 | 0 | 0 | 0 | 0 | 0 | 0 | 0 | 1 | 1 | 1 | 1 | 1 | 0 | 1 | 1 | 0 | 1 |
| 7 | 0 | 0 | 1 | 0 | 0 | 0 | 0 | 0 | 0 | 0 | 0 | 0 | 0 | 0 | 0 | 0 | 0 | 0 | 0 | 0 | 0 | 0 | 0 | 0 | 0 | 0 | 0 | 0 | 0 |
| 8 | 1 | 1 | 1 | 1 | 1 | 1 | 1 | 0 | 0 | 1 | 1 | 1 | 1 | 1 | 1 | 1 | 1 | 1 | 1 | 1 | 1 | 1 | 1 | 1 | 1 | 1 | 1 | 1 | 1 |
| 9 | 0 | 1 | 1 | 1 | 0 | 0 | 1 | 1 | 0 | 0 | 1 | 1 | 0 | 1 | 0 | 1 | 1 | 1 | 1 | 0 | 0 | 0 | 1 | 1 | 1 | 1 | 0 | 1 | 1 |
| 10 | 0 | 1 | 0 | 0 | 1 | 0 | 0 | 0 | 1 | 0 | 0 | 1 | 0 | 0 | 0 | 0 | 0 | 0 | 0 | 0 | 0 | 0 | 0 | 0 | 0 | 0 | 0 | 0 | 0 |
| 11 | 1 | 1 | 1 | 1 | 1 | 1 | 1 | 1 | 0 | 1 | 0 | 1 | 1 | 1 | 1 | 1 | 1 | 1 | 1 | 1 | 1 | 1 | 1 | 1 | 1 | 1 | 1 | 1 | 1 |
| 12 | 1 | 0 | 0 | 0 | 1 | 1 | 0 | 1 | 0 | 0 | 1 | 0 | 0 | 1 | 0 | 1 | 1 | 0 | 0 | 0 | 0 | 1 | 0 | 0 | 0 | 0 | 0 | 0 | 0 |
| 13 | 0 | 0 | 0 | 0 | 0 | 0 | 0 | 0 | 0 | 0 | 0 | 0 | 0 | 1 | 1 | 1 | 1 | 1 | 1 | 1 | 1 | 1 | 1 | 1 | 1 | 1 | 1 | 0 | 1 |
| 14 | 0 | 1 | 0 | 0 | 1 | 0 | 0 | 0 | 0 | 0 | 1 | 0 | 0 | 0 | 1 | 1 | 1 | 0 | 1 | 1 | 1 | 1 | 1 | 1 | 0 | 1 | 1 | 0 | 0 |
| 15 | 0 | 0 | 0 | 0 | 1 | 0 | 0 | 0 | 0 | 0 | 0 | 0 | 0 | 1 | 0 | 1 | 0 | 0 | 1 | 1 | 1 | 1 | 1 | 1 | 0 | 1 | 1 | 1 | 1 |
| 16 | 1 | 1 | 0 | 0 | 1 | 0 | 0 | 0 | 0 | 1 | 1 | 1 | 0 | 1 | 1 | 0 | 1 | 0 | 1 | 1 | 1 | 1 | 1 | 1 | 0 | 1 | 0 | 0 | 0 |
| 17 | 1 | 0 | 0 | 0 | 0 | 0 | 0 | 0 | 0 | 0 | 1 | 1 | 1 | 1 | 1 | 1 | 0 | 0 | 1 | 0 | 0 | 0 | 0 | 1 | 0 | 1 | 0 | 1 | 1 |
| 18 | 0 | 0 | 0 | 0 | 0 | 0 | 0 | 1 | 1 | 0 | 0 | 0 | 1 | 0 | 0 | 1 | 0 | 0 | 0 | 0 | 0 | 0 | 0 | 0 | 1 | 0 | 0 | 0 | 0 |
| 19 | 0 | 0 | 0 | 0 | 0 | 0 | 0 | 0 | 0 | 0 | 0 | 0 | 0 | 1 | 1 | 1 | 1 | 1 | 0 | 1 | 1 | 1 | 1 | 1 | 0 | 1 | 1 | 0 | 0 |
| 20 | 0 | 1 | 0 | 0 | 1 | 0 | 0 | 0 | 0 | 0 | 1 | 0 | 1 | 1 | 0 | 1 | 0 | 0 | 1 | 0 | 1 | 1 | 1 | 1 | 1 | 1 | 1 | 0 | 0 |
| 21 | 1 | 1 | 0 | 0 | 1 | 0 | 0 | 0 | 0 | 0 | 1 | 0 | 0 | 1 | 0 | 1 | 0 | 0 | 1 | 1 | 0 | 1 | 1 | 0 | 0 | 0 | 0 | 0 | 0 |
| 22 | 0 | 1 | 0 | 0 | 1 | 0 | 0 | 0 | 0 | 0 | 1 | 0 | 0 | 1 | 1 | 1 | 0 | 0 | 1 | 1 | 1 | 0 | 1 | 0 | 0 | 1 | 1 | 0 | 1 |
| 23 | 0 | 0 | 0 | 0 | 0 | 0 | 0 | 0 | 0 | 0 | 0 | 0 | 0 | 1 | 1 | 1 | 0 | 0 | 1 | 1 | 1 | 1 | 0 | 0 | 1 | 1 | 1 | 0 | 1 |
| 24 | 0 | 0 | 0 | 0 | 0 | 0 | 0 | 0 | 0 | 0 | 0 | 0 | 0 | 1 | 1 | 1 | 1 | 1 | 1 | 1 | 1 | 1 | 1 | 1 | 0 | 1 | 1 | 1 | 1 |
| 25 | 0 | 0 | 0 | 0 | 0 | 0 | 0 | 1 | 1 | 0 | 1 | 0 | 1 | 0 | 0 | 0 | 0 | 1 | 0 | 0 | 0 | 0 | 0 | 0 | 0 | 0 | 0 | 0 | 0 |
| 26 | 0 | 0 | 0 | 0 | 0 | 0 | 0 | 0 | 0 | 0 | 0 | 0 | 0 | 1 | 1 | 1 | 0 | 0 | 1 | 1 | 0 | 1 | 1 | 0 | 0 | 0 | 1 | 0 | 0 |
| 27 | 0 | 0 | 0 | 0 | 0 | 0 | 0 | 0 | 0 | 0 | 0 | 0 | 0 | 1 | 0 | 0 | 1 | 0 | 1 | 1 | 1 | 1 | 0 | 0 | 0 | 1 | 0 | 1 | 1 |
| 28 | 0 | 1 | 0 | 1 | 1 | 0 | 0 | 1 | 1 | 0 | 0 | 1 | 1 | 0 | 1 | 0 | 1 | 0 | 0 | 1 | 1 | 0 | 1 | 1 | 0 | 1 | 1 | 0 | 1 |
| 29 | 0 | 1 | 0 | 1 | 0 | 0 | 0 | 1 | 1 | 0 | 0 | 0 | 1 | 0 | 0 | 0 | 1 | 0 | 0 | 0 | 0 | 0 | 1 | 1 | 0 | 1 | 1 | 1 | 0 |

conditional independence over ties (6.21) and defines the pseudo-likelihood function as

$$PL(\boldsymbol{\theta}) = \prod_{i \neq j} \Pr(u_{ij} = 1 | \boldsymbol{X} = \boldsymbol{U}^c_{ij})^{u_{ij}} \Pr(u_{ij} = 0 | \boldsymbol{X} = \boldsymbol{U}^c_{ij})^{(1-u_{ij})} \tag{6.27}$$

Maximizing it by nonlinear optimization, one obtains a maximum pseudo-likelihood estimate (MP estimate) of $\boldsymbol{\theta}$. According to a theorem due to Strauss

**TABLE 6.6** Parameter Estimates for a Model

| Effect | Parameter | Estimate |
|---|---|---|
| Choice | $L$ | −1.18 |
| Mutuality | $M$ | 1.98 |
| Transitivity | $T_T$ | 0.26 |
| Cyclicity | $T_C$ | −0.20 |
| 2-in stars | $S_1$ | −0.01 |
| 2-out stars | $S_2$ | −0.15 |
| 2-mixed stars | $S_3$ | −0.08 |
| Degree-centralization | $C_D$ | 1.29 |
| Degree-prestige | $P_D$ | −0.49 |

and Ikeda (1990), the estimation of $p^*$ models can be performed by using the logit $p^*$ form under the assumption that the relational variables are actually independent.

## 6.4.3. Example

An example is taken from Wasserman and Pattison (1996), which is based on the data collected by Vickers and Chan (1981). The original data consisted of multirelational network records obtained from 29 students. The data of the "get-on with" relation is shown in Table 6.5. Among a variety of homogeneous models that the authors applied to the data for illustration of the logit models, here is presented only the result of analysis by a model in terms of nine parameters. Table 6.6 shows the parameter estimates. For the model, the likelihood ratio was 681.6 and the sum of absolute residuals was 217.5. Large effects are noticed for mutuality, degree-centralization, and transivity, and thus those are likely present in the network.

# 7

# Multivariate Analysis of Asymmetry Between Data Sets

## 7.1. OVERVIEW AND PRELIMINARIES

### 7.1.1. Overview

In this chapter we are concerned with asymmetry between two sets of variables. The asymmetry is focused on the directional analysis that one set of variables accounts for the other set in terms of dependence. As a simplest case, we think of two variables, $x$ and $y$. When it is possible to assume a customary regression model, setting a random variable being dependent on a specified non-random variable, one considers such a directional analysis as dependence of $y$ on $x$.

However, there are situations in some fields, social sciences and behavioral sciences, in which one cannot differentiate the roles of variables so clearly. In such a case it is meaningful to consider two directional analyses of regression. We may set two models of linear regression, one to account for $y$ by $x$ and the other to account for $x$ by $y$. The situation becomes prominent for cases in which a phenomenon is observed by two sets of variables, say $\boldsymbol{x}$ and $\boldsymbol{y}$, where those variables are mutually correlated within each set and between the sets. In this viewpoint, we focus on the asymmetrical relationship in terms of dependence between two variable sets.

Among approaches on this standpoint, we take up redundancy analysis. The concept of redundancy is different from that under the same name in information theory. The concept was first suggested in connection with canonical correlation analysis to examine asymmetry in terms of dependence (Stewart and Love, 1968). On the basis of the concept, redundancy analysis has been developed in a variety of ways, mainly in the field of psychometrics (van den Woollenberg, 1977).

In the process of development, researchers found that redundancy analysis is closely related to regression in terms of composite variates for two variable sets, such as canonical regression analysis, the principal component analysis (PCA) of instrumental variables (Rao, 1964). Furthermore, some procedures of reduced rank regression have been found to be identical or closely related to the procedures mentioned above. For extensions of the subject of this chapter, nonsymmetrical data analysis (Lauro and Esposito, 2000) has been proposed from a geometrical point of view. An approach of different type has been offered for analysis of multivariate data in connection with asymmetric data (Saito, 1993b).

## 7.1.2. Preliminaries

### Multivariate Data

In this chapter, we deal with descriptive data analysis in which random variables are not of concern. Consider variable vectors $\boldsymbol{x} = (x_1, x_2, \ldots, x_p)'$ and $\boldsymbol{y} = (y_1, y_2, \ldots, y_q)'$. Denote observation vector of $x_j$ by $\boldsymbol{x}_j = (x_{1j}, x_{2j}, \ldots, x_{nj})'$ and observation vector of $y_j$ by $\boldsymbol{y}_j = (y_{1j}, y_{2j}, \ldots, y_{nj})'$ Write data matrices of $n$ observations in terms of $\boldsymbol{x}$ or $\boldsymbol{y}$ as

$$\boldsymbol{X} = (\boldsymbol{x}_1, \boldsymbol{x}_2, \ldots, \boldsymbol{x}_p), \qquad \boldsymbol{Y} = (\boldsymbol{y}_1, \boldsymbol{y}_2, \ldots, \boldsymbol{y}_p) \tag{7.1}$$

We use the simple term covariance matrix instead of variance–covariance matrix. Denote the covariance matrix of $\boldsymbol{x}$ and $\boldsymbol{y}$ by $\boldsymbol{S}_{xy}$, and $\boldsymbol{S}_{xx}$, $S_{yx}$, $\boldsymbol{S}_{yy}$ conforming to the same notation.

Throughout this chapter, it is assumed that the data matrices satisfy the following rank condition:

$$\text{rank}\ (\boldsymbol{Y}) = \text{rank}\ (\boldsymbol{X}'\boldsymbol{Y}) = q \leq p = \text{rank}\ (\boldsymbol{X}) \tag{7.2}$$

Define diagonal matrices, each with standard deviations on the diagonal as

$$\boldsymbol{D}_x = \text{diag}(s_{x_1}, \ldots, s_{x_p}), \qquad \boldsymbol{D}_y = \text{diag}(s_{y1}, \ldots, s_{y_q}) \tag{7.3}$$

Let a centering matrix $\boldsymbol{H}$ be

$$\boldsymbol{H} = \boldsymbol{I}_p - \frac{1}{n}\mathbf{1}\mathbf{1}' \tag{7.4}$$

Define

$$\boldsymbol{X}_* = \frac{1}{\sqrt{n}}\boldsymbol{X}\boldsymbol{H} \quad \text{and} \quad \tilde{\boldsymbol{X}} = \frac{1}{\sqrt{n}}\boldsymbol{X}\boldsymbol{H}\boldsymbol{D}_x^{-1} \tag{7.5}$$

In what follows, $\boldsymbol{X}_*$ will be called the centered data matrix of $\boldsymbol{X}$, and $\tilde{\boldsymbol{X}}$ the normalized data matrix. For simplicity, we use a convention, in most parts of the chapter, that the data matrices denoted by $\boldsymbol{X}$ and $\boldsymbol{Y}$ are centered, so that

$$\boldsymbol{X}'\boldsymbol{X} = \boldsymbol{S}_{xx}, \qquad \boldsymbol{X}'\boldsymbol{Y} = \boldsymbol{S}_{xy}, \qquad \boldsymbol{Y}'\boldsymbol{X} = \boldsymbol{S}_{yx}, \qquad \boldsymbol{Y}'\boldsymbol{Y} = \boldsymbol{S}_{yy} \tag{7.6}$$

Let $\boldsymbol{R}_{xy}$ represent the correlation matrix of $\boldsymbol{x}$ and $\boldsymbol{y}$. For some parts, we treat the data matrices as normalized, so that covariance matrices are equal to correlation matrices:

$$\boldsymbol{S}_{xx} = \boldsymbol{R}_{xx}, \qquad \boldsymbol{S}_{xy} = \boldsymbol{R}_{xy}, \qquad \boldsymbol{S}_{yx} = \boldsymbol{R}_{yx}, \qquad \boldsymbol{S}_{yy} = \boldsymbol{R}_{yy} \tag{7.7}$$

Let $\mathrm{ch}(\cdot)$ represent generally the eigenvalues of a matrix. By $\mathrm{ch}_t(\cdot)$, we denote the $t$th eigenvalue in descending order of magnitude.

## Multivariate Regression Analysis

Let $\boldsymbol{x}$ be a vector of $p$ independent variables and $\boldsymbol{y}$ a vector of $q$ dependent variables. Then the model of multivariate regression analysis (MRA) is represented as

$$\boldsymbol{y} = \boldsymbol{C}'\boldsymbol{x} + \boldsymbol{e} \tag{7.8}$$

where $\boldsymbol{e} = (e_1, e_2, \ldots, e_q)$ is the error vector. Let $\boldsymbol{E}$ be the error matrix of order $n \times q$. Then

$$\boldsymbol{Y} = \boldsymbol{X}\boldsymbol{C} + \boldsymbol{E} \tag{7.9}$$

By assumption (7.2) $\boldsymbol{X}$ is nonsingular. Minimizing the least squares criterion defined by

$$\mathrm{tr}(\boldsymbol{E}'\boldsymbol{E}) = \mathrm{tr}((\boldsymbol{Y} - \boldsymbol{X}\boldsymbol{C})'(\boldsymbol{Y} - \boldsymbol{X}\boldsymbol{C})) = \sum_{j=1}^{q} \|\boldsymbol{y}_i - \mathbf{X}\boldsymbol{c}_j\|^2 \tag{7.10}$$

yields the least squares estimate (LSE) of $\boldsymbol{C}$ as

$$\hat{\boldsymbol{C}} = (\boldsymbol{X}'\boldsymbol{X})^{-1}\boldsymbol{X}'\boldsymbol{Y} \tag{7.11}$$

Note that $X'X$ is positive definite by the assumption. Denote the $j$th column vector of $\hat{C}$ by $\hat{c}_j$. When one considers a multiple regression model by which variable $y_j$ is regressed on $x$ such that

$$y_j = Xc_j + e_j \tag{7.12}$$

where $e_j$ is the error vector, then $\hat{c}_j$ is also the LSE of $c_j$ $(j = 1, 2, \ldots, q)$. Then MRA is regarded as a set expression of multiple regression models.

Define an orthogonal projection operator onto the space spanned by the column vectors of $X$ as

$$\Pi_X = X(X'X)^{-1}X' \tag{7.13}$$

Then estimates are given by

$$\hat{Y} = X\hat{C} = \Pi_X Y \tag{7.14}$$

$$\hat{E} = Y - \hat{Y} = (I - \Pi_X)Y \tag{7.15}$$

The covariance matrix of residuals is given by

$$S_{ee} = \hat{E}'\hat{E} = Y'(I - \Pi)Y \tag{7.16}$$

$$= S_{yy} - S_{yx}S_{xx}^{-1}S_{xy} \tag{7.17}$$

The degree of the fit attained by the model is given by

$$\psi = \frac{\text{tr}(\hat{Y}'\hat{Y})}{\text{tr}(Y'Y)} = \frac{\text{tr}(Y'\Pi_X Y)}{\text{tr}(S_{yy})} \tag{7.18}$$

## Canonical Correlation Analysis

In canonical correlation analysis (CCA) under assumption (7.2), one finds pairs of composite variates $f_k = u_k'x$ and $g_k = v_k'y$ $(k = 1, 2, \ldots, q)$, by solving the following eigenequations:

$$S_{yx}u_k = \lambda_k S_{yy}v_k \tag{7.19}$$

$$S_{xy}v_k = \lambda_k S_{xx}u_k \tag{7.20}$$

where

$$\lambda_1 \geq \lambda_2 \geq \cdots \geq \lambda_q > 0 \tag{7.21}$$

The $(f_k, g_k)$ are called $k$th pair of canonical variates with $\text{cor}(f_k, g_k) = \lambda_k$, each of unit variance and uncorrelated with other pairs of canonical variates. The $\lambda_k$ is called $k$th canonical correlation coefficient, and (7.19) and (7.20) are called canonical equations.

Given $q$ pairs of canonical variates, one can construct furthermore $(p - q)$ canonical variates $f_j = \boldsymbol{u}_j'\boldsymbol{x}$, $(j = q+1, \ldots, p)$, each of unit variance and uncorrelated with other canonical variates. In the end, one obtains canonical variates $\boldsymbol{f} = (f_1, f_2, \ldots, f_p)'$ and $\boldsymbol{g} = (g_1, g_2, \ldots, g_q)'$.

Let $\boldsymbol{U} = (\boldsymbol{u}_1, \boldsymbol{u}_2, \ldots, \boldsymbol{u}_p)$ and $\boldsymbol{V} = (\boldsymbol{v}_1, \boldsymbol{v}_2, \ldots, \boldsymbol{v}_q)$. Define $n \times 1$ score vectors $\boldsymbol{f}_j$ and $\boldsymbol{g}_k$, score matrices $\boldsymbol{F}$ and $\boldsymbol{G}$ as:

$$\boldsymbol{F} = (\boldsymbol{f}_1, \boldsymbol{f}_2, \ldots, \boldsymbol{f}_p) \qquad \text{where} \quad \boldsymbol{f}_j = \boldsymbol{X}\boldsymbol{u}_j \tag{7.22}$$

$$\boldsymbol{G} = (\boldsymbol{g}_1, \boldsymbol{g}_2, \ldots, \boldsymbol{g}_p) \qquad \text{where} \quad \boldsymbol{g}_k = \boldsymbol{Y}\boldsymbol{v}_k \tag{7.23}$$

The relationship of canonical variates is represented as

$$\boldsymbol{F}'\boldsymbol{F} = \boldsymbol{U}'\boldsymbol{S}_{xx}\boldsymbol{U} = \boldsymbol{I}_p \tag{7.24}$$

$$\boldsymbol{G}'\boldsymbol{G} = \boldsymbol{V}'\boldsymbol{S}_{yy}\boldsymbol{V} = \boldsymbol{I}_q \tag{7.25}$$

$$\boldsymbol{G}'\boldsymbol{F} = \boldsymbol{V}'\boldsymbol{S}_{yx}\boldsymbol{U} = (\boldsymbol{\Lambda}\boldsymbol{O}) \tag{7.26}$$

where $\boldsymbol{\Lambda} = \text{diag}(\lambda_1, \lambda_2, \ldots, \lambda_q)$, and $\boldsymbol{O}$ denotes a zero matrix of order $q \times (p - q)$.

For later description, some formulas are provided here. For the first $r(\leq q)$ pairs of eigenvectors, let $\boldsymbol{U}_r = (\boldsymbol{u}_1, \boldsymbol{u}_2, \ldots, \boldsymbol{u})$, $\boldsymbol{V}_r = (\boldsymbol{v}_1, \boldsymbol{v}_2, \ldots, \boldsymbol{v}_r)$ and $\boldsymbol{D}_\lambda = \text{diag}(\lambda_1, \lambda_2, \ldots, \lambda_r)$. Then expressions for (7.19) and (7.20) are shown as

$$\boldsymbol{S}_{yx}\boldsymbol{U}_r = \boldsymbol{S}_{yy}\boldsymbol{V}_r\boldsymbol{D}_\lambda \tag{7.27}$$

$$\boldsymbol{S}_{xy}\boldsymbol{V}_r = \boldsymbol{S}_{xx}\boldsymbol{U}_r\boldsymbol{D}_\lambda \tag{7.28}$$

From (7.19) and (7.20) it follows that

$$\boldsymbol{S}_{xy}\boldsymbol{S}_{yy}^{-1}\boldsymbol{S}_{yx}\boldsymbol{u}_k = \lambda_k^2\boldsymbol{S}_{xx}\boldsymbol{u}_k \tag{7.29}$$

$$\boldsymbol{S}_{yx}\boldsymbol{S}_{xx}^{-1}\boldsymbol{S}_{xy}\boldsymbol{v}_k = \lambda_k^2\boldsymbol{S}_{yy}\boldsymbol{v}_k \tag{7.30}$$

By assumption (7.2), one can represent $\boldsymbol{S}_{yy}$ as a product of two identical non-singular matrices in such a way that

$$\boldsymbol{S}_{yy} = \boldsymbol{S}_{yy}^{1/2}\boldsymbol{S}_{yy}^{1/2} \tag{7.31}$$

Let $\boldsymbol{S}_{yy}^{-1/2}$ be the inverse of $\boldsymbol{S}_{yy}^{1/2}$. Then (7.30) may be rewritten as

$$\boldsymbol{S}_{yy}^{-1/2}\boldsymbol{S}_{yx}\boldsymbol{S}_{xx}^{-1}\boldsymbol{S}_{xy}\boldsymbol{S}_{yy}^{-1/2}\boldsymbol{v}_k^* = \lambda_k^2\boldsymbol{v}_k^* \qquad \text{where} \quad \boldsymbol{v}_k^* = \boldsymbol{S}_{yy}^{1/2}\boldsymbol{v}_k \tag{7.32}$$

## 7.2. REDUNDANCY ANALYSIS

### 7.2.1. Concept of Redundancy

Regarding the interpretation of canonical correlations in educational measurement, Stewart and Love (1968) presented the concept of redundancy as a directional (or asymmetric) index to represent the dependence of one test battery (variable set) on the other battery. Since then, the study of the concept has been followed by many researchers (Israëls, 1984, 1986; Johansson, 1981; Lambert et al., 1988; Tyler, 1982; van der Burg and de Leeuw, 1990).

In view of the customary definition of the concept based on the CCA, we will treat for a moment $\boldsymbol{X}$ and $\boldsymbol{Y}$ that are normalized so as to satisfy (7.7). Then the canonical equations (7.19) and (7.20) are stated as

$$\boldsymbol{R}_{yx}\boldsymbol{u}_k = \lambda_k \boldsymbol{R}_{yy}\boldsymbol{v}_k \tag{7.33}$$

$$\boldsymbol{R}_{xy}\boldsymbol{v}_k = \lambda_k \boldsymbol{R}_{xx}\boldsymbol{u}_k \tag{7.34}$$

In practical data analysis, we are concerned only with canonical variates associated with $\lambda \neq 0$, and so we treat the same number of canonical variates. Now suppose that under assumption (7.2) we have constructed $r(<q)$ pairs of canonical variates. Let matrices of these eigenvectors be $\boldsymbol{U}_r = (\boldsymbol{u}_1, \boldsymbol{u}_2, \ldots, \boldsymbol{u}_r)$ and $\boldsymbol{V}_r = (\boldsymbol{v}_1, \boldsymbol{v}_2, \ldots, \boldsymbol{v}_r)$. Denote the vectors of canonical variates by

$$\boldsymbol{f} = (f_1, f_2, \ldots, f_r) = \boldsymbol{U}_r'\boldsymbol{x} \tag{7.35}$$

$$\boldsymbol{g} = (g_1, g_2, \ldots, g_r)' = \boldsymbol{V}_r'\boldsymbol{y} \tag{7.36}$$

and the score vectors of each canonical variate by

$$\boldsymbol{f}_k = \boldsymbol{X}\boldsymbol{u}_k, \qquad \boldsymbol{g}_k = \boldsymbol{Y}\boldsymbol{v}_k \qquad (k = 1, 2, \ldots, r) \tag{7.37}$$

The correlation coefficient of a canonical variate and an observed variable is called the structural correlation coefficient. Intraset structural correlations mean correlations between canonical variates and observed variables of the same set, which we write as $\rho_{jk} = \mathrm{cor}(x_j, f_k)$ or $\tau_{jk} = \mathrm{cor}(y_j, g_k)$. They are given by

$$(\rho_{1k}, \rho_{2k}, \ldots, \rho_{pk})' = \boldsymbol{X}'\boldsymbol{f}_k = \boldsymbol{R}_{xx}\boldsymbol{u}_k \tag{7.38}$$

$$(\tau_{1k}, \tau_{2k}, \ldots, \tau_{qk})' = \boldsymbol{Y}'\boldsymbol{g}_k = \boldsymbol{R}_{yy}\boldsymbol{v}_k \tag{7.39}$$

Interset structural correlations mean correlations between canonical variates of one set and observed variables of the other set, which we write as $\rho_{jk} =$

$\text{cor}(y_j, f_k)$ or $\tau_{jk} = \text{cor}(x_j, g_k)$. They are given by

$$(\rho_{1k}, \rho_{2k}, \ldots, \rho_{pk})' = \boldsymbol{X}'\boldsymbol{g}_k = \boldsymbol{R}_{xy}\boldsymbol{v}_k \tag{7.40}$$

$$(\tau_{1k}, \tau_{2k}, \ldots, \tau_{qk})' = \boldsymbol{Y}'\boldsymbol{f}_k = \boldsymbol{R}_{yx}\boldsymbol{u}_k \tag{7.41}$$

The proportion of the total variance (=$p$) of $\boldsymbol{x}$ extracted by $f_k$ means the average of the squared intraset correlations, which is represented as

$$\text{Ve}(f_k) = \frac{1}{p}\boldsymbol{u}_k'\boldsymbol{R}_{xx}\boldsymbol{R}_{xx}\boldsymbol{u}_k \tag{7.42}$$

Likewise, the proportion of the total variance (=$q$) of $\boldsymbol{y}$ extracted by $g_k$ is represented as

$$\text{Ve}(g_k) = \frac{1}{p}\boldsymbol{u}_k'\boldsymbol{R}_{yy}\boldsymbol{R}_{yy}\boldsymbol{v}_k \tag{7.43}$$

From the definition, we see that

$$\sum_{k=1}^{r} \text{Ve}(g_k) \leq 1 \tag{7.43}$$

$$\sum_{k=1}^{q} \text{Ve}(f_k) \leq 1 \tag{7.45}$$

The equality holds when $r = q$ in (7.44), and also when $r = p = q$ in (7.45).

*Redundancy* represents the proportion of the total variance of a set explained by a canonical variate of the other set. This directional measure is also defined as the average of the squared interset correlations. Let us denote the redundancy of $f_k$ by $\text{Re}(\boldsymbol{y}|f_k)$ or simply $\text{Re}(f_k)$. It is expressed as

$$\text{Re}(\boldsymbol{y}|f_k) = \text{Re}(f_k) = \frac{1}{q}\boldsymbol{u}_k'\boldsymbol{R}_{xy}\boldsymbol{R}_{yx}\boldsymbol{u}_k = \lambda_k^2\text{Ve}(g_k) \tag{7.46}$$

Noting the last term, we find that the proportion of the total variance of $\boldsymbol{y}$ explained by $f_k$ is further stated by the two-step formula, explained by the canonical correlation between the two sets and the proportion of the total variance of $\boldsymbol{y}$ extracted by $g_k$. The sum of $\text{Re}(f_k)$ is called *cumulative redundancy*, which is given by

$$\text{Re}(\boldsymbol{y}|\boldsymbol{f}) = \sum_{k=1}^{r} \text{Re}(f_k) = \frac{1}{q}\text{tr}(\boldsymbol{U}_r'\boldsymbol{R}_{xy}\boldsymbol{R}_{yx}\boldsymbol{U}_r) \tag{7.47}$$

When $r = q$, it is also called *total redundancy*, and becomes equal to the average of the fit attained by a set of regression analyses (7.12) for $y_j$ $(j = 1, 2, \ldots, q)$. In other expressions

$$\sum_{k=1}^{q} \mathrm{Re}(f_k) = \frac{1}{q}\sum_{j=1}^{q} R_{y_j}^2 = \frac{1}{q}\mathrm{tr}(\boldsymbol{R}_{yx}\boldsymbol{R}_{xx}^{-1}\boldsymbol{R}_{xy}) \tag{7.48}$$

Here $R_{y_j}$ refers to the multiple correlation coefficient attained by regression of $y_j$ on $\boldsymbol{x}$.

If $\boldsymbol{y}$ is accounted for by $\boldsymbol{x}$ perfectly, the total redundancy becomes unity. In educational measurement, $\boldsymbol{x}$ and $\boldsymbol{y}$ correspond to two test sets respectively. In such a case, when the total redundancy is unity, it implies that either test is redundant for the measurement purpose. In this sense, (7.48) is sometimes called the *redundancy index*.

In a similar way, redundancy is defined for $g_k$ and $\boldsymbol{g} = (g_1, g_2, \ldots, g_r)'$ as:

$$\mathrm{Re}(\boldsymbol{x}|g_k) = \mathrm{Re}(g_k) = \frac{1}{p}\boldsymbol{v}_k'\boldsymbol{R}_{yx}\boldsymbol{R}_{xy}\boldsymbol{v}_k = \lambda_k^2 \mathrm{Ve}(f_k) \tag{7.49}$$

$$\mathrm{Re}(\boldsymbol{x}|\boldsymbol{g}) = \mathrm{Re}(\boldsymbol{g}) = \sum_{k=1}^{r} \mathrm{Re}(g_k) \tag{7.50}$$

Write the canonical correlation coefficient between $\boldsymbol{x}$ and $\boldsymbol{y}$ as $\lambda(\boldsymbol{x}, \boldsymbol{y})$. Changing $\boldsymbol{x}$ and $\boldsymbol{y}$ in (7.33) and (7.34) gives the same equations with $\lambda(\boldsymbol{y}, \boldsymbol{x})$. As long as we treat nonzero $\lambda$, it is obvious that $\lambda(\boldsymbol{x}, \boldsymbol{y}) = \lambda(\boldsymbol{y}, \boldsymbol{x})$, the correlation is a symmetric index to measure the relationship between the two sets. In contrast, redundancy is a measure of the asymmetric relationship between the two sets. Given $\boldsymbol{X}$ and $\boldsymbol{Y}$, one can compute such asymmetry based on the result of CCA for $\boldsymbol{X}$ and $\boldsymbol{Y}$, by comparing $\mathrm{Re}(\boldsymbol{y}|\boldsymbol{f})$ with $\mathrm{Re}(\boldsymbol{x}|\boldsymbol{g})$.

### 7.2.2. Development of Redundancy Analysis

The redundancy explained above was defined in terms of canonical variates. On the basis of the concept, van den Woollenberg (1977) suggested a general definition of redundancy in terms of composite variates, and proposed a procedure of redundancy analysis (RDA). It gives composite variates to maximize the redundancy. Thus, in a different way from that mentioned above, one can perform analysis of asymmetry between data sets, apart from CCA. The formulation was given on the basis of correlation matrices. For the purpose of later exposition in this chapter, we describe the formulation in terms of covariance matrices.

Because redundancy is a measure of asymmetric relationship between two multivariate data sets $\boldsymbol{X}$ and $\boldsymbol{Y}$, we can consider two directional analysis based on the concept of redundancy. Let RDA $(\boldsymbol{y}|\boldsymbol{x})$ be the redundancy analysis where $\boldsymbol{x}$ accounts for $\boldsymbol{y}$, and RDA$(\boldsymbol{x}|\boldsymbol{y})$ be the analysis in the opposite direction. For the purpose of exposition, we describe RDA$(\boldsymbol{y}|\boldsymbol{x})$ first. Let us assume that both $\boldsymbol{X}$ and $\boldsymbol{Y}$ are centered.

Noting the orthogonality on the canonical score matrix

$$\boldsymbol{F}_r'\boldsymbol{F}_r = \boldsymbol{I}_r \qquad \text{where} \quad \boldsymbol{F}_r = \boldsymbol{X}\boldsymbol{U}_r \tag{7.51}$$

then we can rewrite cumulative redundancy (7.47) as

$$\begin{aligned} \mathrm{Re}(\boldsymbol{y}|\boldsymbol{f}) &= \frac{\mathrm{tr}(\boldsymbol{R}_{yx}\boldsymbol{U}_r\boldsymbol{U}_r'\boldsymbol{R}_{xy})}{\mathrm{tr}(\boldsymbol{R}_{yy})} \\ &= \frac{\mathrm{tr}(\boldsymbol{Y}'\boldsymbol{F}_r(\boldsymbol{F}_r'\boldsymbol{F}_r)^{-1}\boldsymbol{F}_r'\boldsymbol{Y})}{\mathrm{tr}(\boldsymbol{R}_{yy})} \end{aligned} \tag{7.52}$$

Now we consider treating data sets that satisfy (7.2) and (7.6). Let us define a vector of $r$ composite variates such that

$$\begin{aligned} &z = (z_1, z_2, \ldots, z_r)' \qquad \text{where} \quad \boldsymbol{a}_t'z_t = \alpha_t'\boldsymbol{x}_t \\ &(t = 1, 2, \ldots, r) \end{aligned} \tag{7.53}$$

and a score matrix as

$$\boldsymbol{Z} = \boldsymbol{X}\boldsymbol{A} \tag{7.54}$$

where $\boldsymbol{Z}$, $\boldsymbol{X}$, and $\boldsymbol{A}$ are matrices of order $n \times r$, $n \times p$, and $p \times r$, respectively.

We impose a constraint to normalize $z$ as

$$\mathrm{Var}(z) = \boldsymbol{Z}'\boldsymbol{Z} = \boldsymbol{A}'\boldsymbol{S}_{xx}\boldsymbol{A} = \boldsymbol{I}_r \tag{7.55}$$

Referring to (7.52), we define cumulative redundancy as

$$\mathrm{Re}(z) = \frac{\mathrm{tr}(\boldsymbol{S}_{yz}\boldsymbol{S}_{zz}^{-1}\boldsymbol{S}_{zy})}{\mathrm{tr}(\boldsymbol{S}_{yy})} = \frac{\mathrm{tr}(\boldsymbol{A}'\boldsymbol{S}_{xy}\boldsymbol{S}_{yx}\boldsymbol{A})}{\mathrm{tr}(\boldsymbol{S}_{yy})} \tag{7.56}$$

We like to maximize Re($z$) under condition (7.55). Define a Lagrangean function as

$$\mathcal{L} = \mathrm{tr}(\boldsymbol{A}'\boldsymbol{S}_{xy}\boldsymbol{S}_{yx}\boldsymbol{A}) - \mathrm{tr}((\boldsymbol{A}'\boldsymbol{S}_{xx}\boldsymbol{A} - \boldsymbol{I}_r)\boldsymbol{M}) \tag{7.57}$$

Here $\boldsymbol{M} = (m_{st})$ is a symmetric matrix of Lagrangean multipliers. Through some manipulation, we obtain an eigenequation

$$\boldsymbol{S}_{xy}\boldsymbol{S}_{yx}\boldsymbol{a}_t = \mu_t\boldsymbol{S}_{xx}\boldsymbol{a}_t \qquad (t = 1, 2, \ldots, r \leq q) \tag{7.58}$$

to give $\boldsymbol{a}_t$ $(t = 1, 2, \ldots, r)$. The eigenvectors are normalized to satisfy

$$\boldsymbol{a}_j'\boldsymbol{S}_{xx}\boldsymbol{a}_k = \delta_{jk} \tag{7.59}$$

where $\delta_{jk}$ represents Kronecker's delta. Using eigenvectors $\boldsymbol{A}_r = (\boldsymbol{a}_1, \boldsymbol{a}_2, \ldots, \boldsymbol{a}_r)$ we determine the score matrix of composite variates $\boldsymbol{Z} = \boldsymbol{X}\boldsymbol{A}_r$. Then the redundancy is expressed by

$$\mathrm{Re}(z_t) = \mathrm{Re}(\boldsymbol{y}|z_t) = \frac{\mu_t}{\mathrm{tr}(\boldsymbol{S}_{yy})} \tag{7.60}$$

and the cumulated redundancy is

$$\mathrm{Re}(z) = \mathrm{Re}(\boldsymbol{y}|z) = \frac{\sum_{t=1}^{r} \mu_t}{\mathrm{tr}(\boldsymbol{S}_{yy})} \tag{7.61}$$

Since $\mu_t > 0$ and $r \leq q$, we find that for a fixed $r$ the cumulated redundancy takes a maximum when $r = q$. Write the contribution of $\mu_t$ to the sum as

$$\alpha_t = \frac{\mu_t}{\sum_{t=1}^{q} \mu_t} \tag{7.62}$$

Then we see that

$$\frac{\sum_{t=1}^{r} \mathrm{Re}(z_t)}{\sum_{t=1}^{q} \mathrm{Re}(z_t)} = \frac{\sum_{t=1}^{r} \mu_t}{\sum_{t=1}^{q} \mu_t} = \sum_{t=1}^{r} \alpha_t \tag{7.63}$$

It is suggested to use structural correlations rather than using eigenvectors for interpreting the composite variates. Referring to definition (7.38) to (7.39) based on CCA, we write intraset structural correlations as $\rho_{jk} = \mathrm{cor}(x_j, z_k)$ and interset structural correlations as $\tau_{jk} = \mathrm{cor}(y_j, z_k)$. Then

$$(\rho_{1k}, \rho_{2k}, \ldots, \rho_{pk})' = \boldsymbol{D}_x^{-1}\boldsymbol{S}_{xx}\boldsymbol{a}_k \tag{7.64}$$

$$(\tau_{1k}, \tau_{2k}, \ldots, \tau_{qk})' = \boldsymbol{D}_y^{-1}\boldsymbol{S}_{yx}\boldsymbol{a}_k \tag{7.65}$$

Unlike the case of (7.46), the average of the interset structural correlations is not necessarily equal to the redundancy (7.60). The proportion that $z_t$ accounts for the total variance of $\boldsymbol{x}$ is given by

$$\mathrm{Ve}(z_k) = \frac{\boldsymbol{a}_k'\boldsymbol{S}_{xx}\boldsymbol{S}_{xx}\boldsymbol{a}_k}{\mathrm{tr}(\boldsymbol{S}_{xx})} \tag{7.66}$$

and its cumulated sum is given by

$$\sum_{k=1}^{r} \mathrm{Ve}(z_k) = \frac{\mathrm{tr}(\boldsymbol{A}'\boldsymbol{S}_{xx}\boldsymbol{S}_{xx}\boldsymbol{A})}{\mathrm{tr}(\boldsymbol{S}_{xx})} \leq 1 \tag{7.67}$$

When $r = p$, the equality holds.

Now turn to RDA($\boldsymbol{x}|\boldsymbol{y}$). In this case we construct composite variates $\eta_j = \boldsymbol{\beta}_j'\boldsymbol{y}$, which accounts for $\boldsymbol{x}$. By the same formulation above, we derive

$$\boldsymbol{S}_{xy}\boldsymbol{S}_{yx}\boldsymbol{\beta}_t = \nu_t\boldsymbol{S}_{yy}\boldsymbol{\beta}_t \tag{7.68}$$

$$\boldsymbol{\beta}_j'\boldsymbol{S}_{yy}\boldsymbol{\beta}_r = \delta_{jk} \tag{7.69}$$

$$\mathrm{Re}(\eta_t) = \mathrm{Re}(\boldsymbol{x}|\eta_t) = \frac{\nu_t}{\mathrm{tr}(\boldsymbol{S}_{xx})} \tag{7.70}$$

$$\mathrm{Re}(\eta) = \mathrm{Re}(\boldsymbol{x}|\boldsymbol{\eta}) = \frac{\sum_{t=1}^{r} \nu_t}{\mathrm{tr}(\boldsymbol{S}_{xx})} \tag{7.71}$$

## 7.3. MULTIVARIATE REGRESSION ON COMPOSITE VARIATES

In this section we are concerned with two kinds of MRA where one variable set is explained by the set of composite variates constructed from the other set of variables. Noting that model (7.8) does not incorporate the correlation information between $\boldsymbol{x}$ and $\boldsymbol{y}$, one may consider variants of MRA that utilize $\boldsymbol{S}_{xx}$, $\boldsymbol{S}_{xy}$, and $\boldsymbol{S}_{yy}$. On such a standpoint, one can examine the asymmetrical relationship in terms of dependence between data sets. In what follows, we treat data sets satisfying (7.2) and (7.6), and describe two procedures.

### 7.3.1. Canonical Regression Analysis

In canonical regression analysis (CRA), one performs first CCA($\boldsymbol{x}$, $\boldsymbol{y}$), then derive canonical variates $\boldsymbol{f} = (f_1, f_2, \ldots, f_r)'$ from $\boldsymbol{x}$. Next one performs MRA to account for $\boldsymbol{y}$ by $\boldsymbol{f}$. This process utilizes $\boldsymbol{S}_{yy}$, which involves the information of correlations among $\boldsymbol{y}$. The CRA is viewed as regression by a two-stage least squares procedure, with the regression on the second stage.

In CRA, $\boldsymbol{y}$ is regressed on canonical variates $f_1, f_2, \ldots, f_r$ by a least squares approach. The model is

$$\boldsymbol{Y} = \boldsymbol{F}_r\boldsymbol{B} + \boldsymbol{E} \tag{7.72}$$

Through CCA, remember that (7.24) to (7.26) hold for canonical scores. To state the property of CRA, let $\boldsymbol{\Pi}_F$ be the projection operator onto the space spanned by column vectors of $\boldsymbol{F}$. Consider a nonsingular transformation of $\boldsymbol{X}$ such that $\boldsymbol{F} = \boldsymbol{XU}$. Since

$$\boldsymbol{X}(\boldsymbol{X}'\boldsymbol{X})^{-1}\boldsymbol{X}' = \boldsymbol{F}(\boldsymbol{F}'\boldsymbol{F})^{-1}\boldsymbol{F}' \tag{7.73}$$

we find that $\boldsymbol{\Pi}_F = \boldsymbol{\Pi}_X$ from (7.13). Partition $\boldsymbol{U}$ into two parts in such a way that $\boldsymbol{U} = (\boldsymbol{U}_r\,\boldsymbol{U}_s)$ where $p = r + s$. Matrix $\boldsymbol{U}_r$ consists of the first $r$ columns of $\boldsymbol{U}$ and $\boldsymbol{U}_s$ of the remaining columns. In correspondence to the partition, we divide $\boldsymbol{F}$ as $\boldsymbol{F} = (\boldsymbol{F}_r\,\boldsymbol{F}_s)$. Then we have

$$\boldsymbol{\Pi}_X = \boldsymbol{\Pi}_F = \boldsymbol{\Pi}_{F_r} + \boldsymbol{\Pi}_{F_s} \tag{7.74}$$

From (7.24)

$$\boldsymbol{F}_r'\boldsymbol{F}_r = \boldsymbol{I}_r, \qquad \boldsymbol{\Pi}_{F_r} = \boldsymbol{F}_r\boldsymbol{F}_r' \tag{7.75}$$

Using the relation, we apply the theory of MRA on (7.72). Then we obtain the LSE of $\boldsymbol{B}$ and estimates of $\boldsymbol{Y}$ as:

$$\hat{\boldsymbol{B}} = \boldsymbol{F}_r'\boldsymbol{Y} = \boldsymbol{U}_r'\boldsymbol{S}_{xy} \tag{7.76}$$

$$\hat{\boldsymbol{Y}} = \boldsymbol{F}_r\hat{\boldsymbol{B}} = \boldsymbol{\Pi}_{F_r}\boldsymbol{Y} \tag{7.77}$$

From (7.18), the degree of fit is stated as

$$\phi_r = \frac{\mathrm{tr}(\hat{\boldsymbol{Y}}'\hat{\boldsymbol{Y}})}{\mathrm{tr}(\boldsymbol{Y}'\boldsymbol{Y})} = \frac{\mathrm{tr}(\boldsymbol{Y}'\boldsymbol{\Pi}_{F_r}\boldsymbol{Y})}{\mathrm{tr}(\boldsymbol{S}_{yy})} \tag{7.78}$$

Using (7.75) and $\boldsymbol{F}_r = \boldsymbol{XU}_r$, the above expression is restated as

$$\phi_r = \frac{\mathrm{tr}(\boldsymbol{S}_{yx}\boldsymbol{U}_r\boldsymbol{U}_r'\boldsymbol{S}_{xy})}{\mathrm{tr}(\boldsymbol{S}_{yy})} \tag{7.79}$$

Among the $p$ canonical variates of $\boldsymbol{f}$, we may regard $(f_1, f_2, \ldots, f_q)$ meaningful variates, which correspond to nonzero eigenvalues. Using these $q$ variates for (7.72), we have $\boldsymbol{F}_r = \boldsymbol{F}_q$. Regarding $\boldsymbol{F}_s$, we note that $\boldsymbol{G}'\boldsymbol{F}_s = \boldsymbol{O}$ by (7.26), and $\boldsymbol{Y}'\boldsymbol{F}_s = \boldsymbol{O}$ because $\boldsymbol{V}$ is nonsingular. Thus

$$\boldsymbol{\Pi}_{F_s}\boldsymbol{Y} = \boldsymbol{O}, \qquad \boldsymbol{\Pi}_X\boldsymbol{Y} = \boldsymbol{\Pi}_{F_q}\boldsymbol{Y} \tag{7.80}$$

Putting this into (7.78) gives

$$\phi_q = \frac{\mathrm{tr}(\boldsymbol{Y}'\boldsymbol{\Pi}_X\boldsymbol{Y})}{\mathrm{tr}(\boldsymbol{S}_{yy})} \tag{7.81}$$

Therefore, using all the meaningful canonical variates as regressors, CRA attains the same degree of fit as MRA given by (7.18).

For the case in which $\boldsymbol{X}$ and $\boldsymbol{Y}$ are normalized so that the covariance matrices are equal to correlation matrices, (7.76) becomes

$$\hat{\boldsymbol{B}} = \boldsymbol{U}_r' \boldsymbol{R}_{xy} \tag{7.82}$$

Referring to (7.41), we find that $\hat{\boldsymbol{B}}$ is the matrix of structural correlations between $(f_1, f_2, \ldots, f_r)$ and $\boldsymbol{y}$. Equation (7.79) becomes

$$\phi_r = \frac{1}{q} \operatorname{tr}(\boldsymbol{R}_{yx} \boldsymbol{U}_r \boldsymbol{U}_r' \boldsymbol{R}_{xy}) \tag{7.83}$$

This is the same as (7.47), the cumulative redundancy on the result of CCA. From the description above, it is summarized as follows. For normalized data, examining the degree of fit and the regression coefficients in CRA is equivalent to examining the redundancy and the structural correlation coefficients on the result of CCA.

## 7.3.2. Principal Component Analysis with Instrumental Variables

For a phenomenon observing many variables, we consider some of them are of primary concern and others are of secondary concern. For convenience of description, let us call the first set of variables main variables and denote them by $\boldsymbol{y}$ and the second set of variables instrumental variables and denote them by $\boldsymbol{x}$. Rao (1964) proposed procedures for data analysis for such a situation. Among them we take up the procedure that was named principal component analysis with instrumental variables (IPCA).

Unlike CRA, which is a two-stage procedure, one can formulate directly a variant of MRA in terms of composite variates so that the accountability is maximized. On this standpoint, one considers composite variates in terms of linear combination of $\boldsymbol{x}$

$$z_t = \boldsymbol{\alpha}_t' \boldsymbol{x} \qquad (t = 1, 2, \ldots, r) \tag{7.84}$$

so that they predict $\boldsymbol{y}$ as much as possible. Define

$$z_t = \boldsymbol{X}\boldsymbol{a}_t, \qquad \boldsymbol{Z} = \boldsymbol{X}\boldsymbol{A} \tag{7.85}$$

where $\boldsymbol{Z} = (z_1, z_2, \ldots, z_r)$ and $\boldsymbol{A} = (\boldsymbol{a}_1, \boldsymbol{a}_2, \ldots, \boldsymbol{a}_r)$. Let us perform such a prediction based on a regression model such that

$$\boldsymbol{Y} = \boldsymbol{Z}\boldsymbol{B} + \boldsymbol{E} = \boldsymbol{X}\boldsymbol{A}\boldsymbol{B} + \boldsymbol{E} \tag{7.86}$$

The least squares criterion is stated by

$$h(\boldsymbol{A}, \boldsymbol{B}) = \|\boldsymbol{Y} - \boldsymbol{XAB}\|^2 = \mathrm{tr}(\boldsymbol{EE}') = \mathrm{tr}(\boldsymbol{S}_{ee}) \tag{7.87}$$

For $\boldsymbol{Z}$ given by an arbitrary $\boldsymbol{A}$, one can estimate $\boldsymbol{B}$ by a least squares approach. Then the LSE of $\boldsymbol{B}$ is derived by

$$\boldsymbol{B}_{(A)} = (\boldsymbol{Z}'\boldsymbol{Z})^{-1}\boldsymbol{Z}'\boldsymbol{Y} \tag{7.88}$$

Substituting it into (7.87) yields

$$\begin{aligned} \boldsymbol{S}_{ee} &= \boldsymbol{S}_{yy} - \boldsymbol{S}_{yz}\boldsymbol{S}_{zz}^{-1}\boldsymbol{S}_{zy} \\ &= \boldsymbol{S}_{yy} - \boldsymbol{S}_{yx}\boldsymbol{A}(\boldsymbol{A}'\boldsymbol{S}_{xx}\boldsymbol{A})^{-1}\boldsymbol{A}'\boldsymbol{S}_{xy} \end{aligned} \tag{7.89}$$

In order to measure $\boldsymbol{S}_{ee}$, we like to set a scalar criterion $\mathrm{tr}(\boldsymbol{S}_{ee})$. Let us find $\boldsymbol{A}$ to minimize the criterion. Noting that $\mathrm{tr}(\boldsymbol{S}_{yy})$ is a constant, our purpose leads to maximizing an objective function

$$h(\boldsymbol{A}) = \mathrm{tr}(\boldsymbol{S}_{yx}\boldsymbol{A}(\boldsymbol{A}'\boldsymbol{S}_{xx}\boldsymbol{A})^{-1}\boldsymbol{A}'\boldsymbol{S}_{xy}) \tag{7.90}$$

Now we impose constraints,

$$\boldsymbol{Z}'\boldsymbol{Z} = \boldsymbol{A}'\boldsymbol{S}_{xx}\boldsymbol{A} = \boldsymbol{I}_r \tag{7.91}$$

so that $z_1, z_2, \ldots, z_r$ are mutually uncorrelated. Then we have

$$h(\boldsymbol{A}) = \mathrm{tr}(\boldsymbol{A}'\boldsymbol{S}_{xy}\boldsymbol{S}_{yx}\boldsymbol{A}) \tag{7.92}$$

Maximizing this with respect to $\boldsymbol{A}$ leads to the following eigenequation,

$$\boldsymbol{S}_{xy}\boldsymbol{S}_{yx}\boldsymbol{a}_t = \mu_t\boldsymbol{S}_{xx}\boldsymbol{a}_t \qquad (t = 1, 2, \ldots, p) \tag{7.93}$$

$$\mu_1 \geq \mu_2 \geq \cdots \geq \mu_p \geq 0 \tag{7.94}$$

Taking the first $r$ eigenvectors, we set $\boldsymbol{A} = (\boldsymbol{a}_1, \boldsymbol{a}_2, \ldots, \boldsymbol{a}_r)$. The maximum of $h(\boldsymbol{A})$ is

$$h_{\max}(\boldsymbol{A}) = \sum_{t=1}^{r} \boldsymbol{a}_t'\boldsymbol{S}_{xy}\boldsymbol{S}_{yx}\boldsymbol{a}_t = \sum_{t=1}^{r} \mu_t \tag{7.95}$$

From assumption (7.2), we find that $\mathrm{rank}(\boldsymbol{S}_{xx}^{-1}\boldsymbol{S}_{xy}\boldsymbol{S}_{yx}) = q$. Noting that $\mu = \mathrm{ch}(\boldsymbol{S}_{xx}^{-1}\boldsymbol{S}_{xy}\boldsymbol{S}_{yx})$, for (7.93) we see then that

$$\mu_{q+1} = \cdots = \mu_p = 0 \tag{7.96}$$

Denote the total cumulative proportion of eigenvalues by

$$\theta(r) = \frac{\sum_{t=1}^{r} \mu_t}{\sum_{t=1}^{p} \mu t} = \frac{\sum_{t=1}^{r} \mu_t}{\sum_{t=1}^{q} \mu_t} \tag{7.97}$$

This is an index of data reduction in $r$ dimensions.

Let us examine formulation of IPCA. Equation (7.93) is rewritten as

$$\boldsymbol{S}_{yx}\boldsymbol{S}_{xx}^{-1}\boldsymbol{S}_{xy}\boldsymbol{S}_{yx}\boldsymbol{a} = \mu\boldsymbol{S}_{yx}\boldsymbol{a} \tag{7.98}$$

Let $\boldsymbol{w} = \boldsymbol{S}_{yx}\boldsymbol{a}/\sqrt{\mu}$. Then $\boldsymbol{w}'\boldsymbol{w} = 1$. Using the orthogonal projector $\boldsymbol{\Pi}_X$ defined by (7.13),

$$\boldsymbol{S}_{yx}\boldsymbol{S}_{xx}^{-1}\boldsymbol{S}_{xy}\boldsymbol{w} = \boldsymbol{Y}'\boldsymbol{\Pi}_X\boldsymbol{Y}\boldsymbol{w} = \mu\boldsymbol{w} \tag{7.99}$$

Thus $\boldsymbol{Y}_* = \boldsymbol{\Pi}_X\boldsymbol{Y}$ represents the projection of $\boldsymbol{Y}$ on the column space of $\boldsymbol{X}$. Noting $\boldsymbol{\Pi}_X^2 = \boldsymbol{\Pi}_X$, we find that

$$\boldsymbol{Y}_*'\boldsymbol{Y}_*\boldsymbol{w} = \mu\boldsymbol{w} \tag{7.100}$$

Thus IPCA is equivalent to PCA of the covariance matrix of $\boldsymbol{Y}_*$, for which efficiency of the reduction is shown by (7.97).

Estimates of $\boldsymbol{B}$ and $\boldsymbol{Y}$ are given by

$$\hat{\boldsymbol{B}} = \boldsymbol{A}'\boldsymbol{S}_{xy}, \quad \hat{\boldsymbol{Y}} = \boldsymbol{X}\boldsymbol{A}\boldsymbol{A}'\boldsymbol{S}_{xy} \tag{7.101}$$

The degree of fit is represented as

$$\frac{h(\boldsymbol{A})}{\mathrm{tr}(\boldsymbol{S}_{yy})} = \frac{\sum_{t=1}^{r} \mu_t}{\mathrm{tr}(\boldsymbol{S}_{yy})} \tag{7.102}$$

In the nonsymmetrical data analysis, Lauro and Esposito (2000) suggested extending operator $\boldsymbol{\Pi}_X$ to a more general form in a geometrical context.

## 7.4. COMPARISON OF RELATED PROCEDURES

As pointed out by researchers, the procedure of redundancy analysis has a close relationship with other approaches (Muller, 1981; ten Berge, 1985; Saito, 1995). On the standpoint of the present chapter, we would like to give comparisons of RDA with other procedures, such as CRA and reduced rank regression (RRR).

### 7.4.1. RDA in comparison with IPCA

Both RDA and IPCA lead to identical eigenequations. This relationship is confirmed by rewriting the objective function of IPCA. From (7.87), (7.89), and (7.92), we find that

$$h(\boldsymbol{A}, \boldsymbol{B}) \geq \mathrm{tr}(\boldsymbol{S}_{yy}) - h(\boldsymbol{A}) \tag{7.103}$$

Noting (7.92) and (7.56), we obtain

$$\mathrm{Re}(z) = \frac{h(\boldsymbol{A})}{\mathrm{tr}(\boldsymbol{S}_{yy})} \tag{7.104}$$

Therefore, the regression model is incorporated implicitly in RDA, and so $\mathrm{Re}(z)$ indicates the degree to which the model accounts for the total variance of $\boldsymbol{y}$. The two procedures yield the same composite variates.

From (7.93) and (7.99), we see for $t = 1, 2, \ldots, q$ that

$$\mu_t = \mathrm{ch}_t(\boldsymbol{S}_{xx}^{-1}\boldsymbol{S}_{xy}\boldsymbol{S}_{yx}) = \mathrm{ch}_t(\boldsymbol{S}_{yx}\boldsymbol{S}_{xx}^{-1}\boldsymbol{S}_{xy}) = \mathrm{ch}_t(\boldsymbol{Y}'\boldsymbol{\Pi}_X\boldsymbol{Y}) \tag{7.105}$$

Then it follows from (7.61) that

$$\sum_{t=1}^{q} \mathrm{Re}(z_t) = \frac{\mathrm{tr}(\boldsymbol{Y}'\boldsymbol{\Pi}_X\boldsymbol{Y})}{\mathrm{tr}(\boldsymbol{S}_{yy})} \tag{7.106}$$

When $r = q$, the cumulative redundancy by RDA or the maximum accountability by IPCA is the same as the goodness of fit by MRA. In view of the equivalence of IPCA and RDA, we may write IPCA (RDA).

### 7.4.2. RDA in comparison with CRA

As described above, CRA and IPCA (RDA) are regression analyses in terms of composite variates. Referring to the formulations, the degree of fit by CRA should not be larger than the degree of fit by RDA (Saito, 1995). Let us confirm this point. For the same number of composite variates $r$, we write the fit attained by RDA as $\psi_r$ and the fit attained by CRA as $\phi_r$:

$$\psi_r = \frac{\sum_{t=1}^{r} \mathrm{ch}_t(\boldsymbol{Y}'\boldsymbol{\Pi}_X\boldsymbol{Y})}{\mathrm{tr}(\boldsymbol{S}_{yy})} \tag{7.107}$$

$$\phi_r = \frac{\sum_{t=1}^{r} \mathrm{ch}_t(\boldsymbol{Y}'\boldsymbol{\Pi}_{F_r}\boldsymbol{Y})}{\mathrm{tr}(\boldsymbol{S}_{yy})} \tag{7.108}$$

For a moment we consider two subspaces $\boldsymbol{V}$ and $\boldsymbol{W}$ in $n$-dimensional space.

If $\boldsymbol{W}$ is a subspace of $\boldsymbol{V}$, the following inequality holds:

$$\text{ch}_t(\boldsymbol{Y}'\boldsymbol{\Pi}_V\boldsymbol{Y}) \geq \text{ch}_t(\boldsymbol{Y}'\boldsymbol{\Pi}_W\boldsymbol{Y}) \tag{7.109}$$

Here $\boldsymbol{\Pi}_V$ and $\boldsymbol{\Pi}_W$ are orthogonal projection operators. For the subspace spanned by canonical score vectors $\boldsymbol{F}_r$, we find that $\boldsymbol{F}_r$ is a subspace of $\boldsymbol{X}$. Applying (7.109), we derive

$$\text{ch}_t(\boldsymbol{Y}'\boldsymbol{\Pi}_X\boldsymbol{Y}) \geq \text{ch}_t(\boldsymbol{Y}'\boldsymbol{\Pi}_{F_r}\boldsymbol{Y}) \tag{7.110}$$

from which follows

$$\sum_{t=1}^{r} \text{ch}_t(\boldsymbol{Y}'\boldsymbol{\Pi}_X\boldsymbol{Y}) \geq \sum_{t=1}^{r} \text{ch}_t(\boldsymbol{Y}'\boldsymbol{\Pi}_{F_r}\boldsymbol{Y}) = \text{tr}(\boldsymbol{Y}'\boldsymbol{\Pi}_{F_r}\boldsymbol{Y}) \tag{7.111}$$

Combining (7.107), (7.108), and (7.111), we obtain

$$\psi_r \geq \phi_r \qquad (r = 1, 2, \ldots, q) \tag{7.112}$$

The arguments are summarized by the following theorem (Saito, 1995).

*Theorem 7.1.* For RDA and CRA, inequality (7.112) holds. For the same number of composite variates as regressors, RDA does not attain smaller degree of fit than CRA. When $r = q$, the equality holds in (7.112).

## 7.4.3. Comparison of RDA and CRA with RRR

It is often claimed that redundancy analysis is equivalent to reduced rank regression. However, this is right as long as one thinks of the reduced rank regression in some context. In fact, CRA is regarded as a kind of RRR, so utilizing CRA or RDA for tool of analysis of asymmetry between data sets is an application of RRR. It may be of interest to clarify the relationship of CRA and RDA to RRR.

The model of reduced rank regression is generally stated as

$$\boldsymbol{Y} = \boldsymbol{X}\boldsymbol{C} + \boldsymbol{E} \qquad \text{where rank}(\boldsymbol{C}) = r \tag{7.113}$$

Izenman (1975) proposed a general theorem on RRR, which serves to show the relationship on a theoretical basis. Let us assume that $p$-dimensional vector $\boldsymbol{x}$ and $q$-dimensional vector $\boldsymbol{y}$ accord to the $(p+q)$ normal distribution with expectations $\text{E}(\boldsymbol{x}) = \boldsymbol{0}$, $\text{E}(\boldsymbol{y}) = \boldsymbol{0}$ and covariance matrices $\boldsymbol{\Sigma}_{xx}$, $\boldsymbol{\Sigma}_{xy}$, $\boldsymbol{\Sigma}_{yx}$, $\boldsymbol{\Sigma}_{yy}$. Then the following theorem holds (Izenman, 1975).

*Theorem 7.2.* Given a $q \times q$ positive definite matrix $\boldsymbol{\Gamma}$, and for a specified $r \leq \min(p, q)$, a $p \times r$ matrix $\boldsymbol{A}$ and an $r \times q$ matrix $\boldsymbol{B}$ that minimize the criterion defined by

$$g(\boldsymbol{A}, \boldsymbol{B}) = \mathrm{tr}(\mathrm{E}(\boldsymbol{\Gamma}^{1/2}(\boldsymbol{y} - \boldsymbol{B}'\boldsymbol{A}'\boldsymbol{x})(\boldsymbol{y} - \boldsymbol{B}'\boldsymbol{A}'\boldsymbol{x})'\boldsymbol{\Gamma}^{1/2})) \tag{7.114}$$

are obtained as

$$\boldsymbol{A} = \boldsymbol{\Sigma}_{xx}^{-1}\boldsymbol{\Sigma}_{xy}\boldsymbol{\Gamma}^{1/2}\boldsymbol{W}_r \tag{7.115}$$

$$\boldsymbol{B} = \boldsymbol{W}_r'\boldsymbol{\Gamma}^{-1/2} \tag{7.116}$$

Here $\boldsymbol{W}_r = (\boldsymbol{w}_1, \boldsymbol{w}_2, \ldots, \boldsymbol{w}_r)$ are eigenvectors of the following equation:

$$\boldsymbol{\Gamma}^{1/2}\boldsymbol{\Sigma}_{yx}\boldsymbol{\Sigma}_{xx}^{-1}\boldsymbol{\Sigma}_{xy}\boldsymbol{\Gamma}^{1/2}\boldsymbol{w}_j = \theta_j\boldsymbol{w}_j \qquad (j = 1, 2, \ldots, r) \tag{7.117}$$

which should satisfy orthogonality $\boldsymbol{w}_j'\boldsymbol{w}_k = \delta_{jk}$. The minimum is given by

$$g_{\min} = \mathrm{tr}(\boldsymbol{\Gamma}\boldsymbol{\Sigma}_{yy}) - \sum_{j=1}^{r}\theta_j \tag{7.118}$$

Setting $\boldsymbol{\Gamma} = \boldsymbol{I}_q$ in the theorem and substituting sample statistics in the result, we then have from (7.117) an eigenequation such that

$$\boldsymbol{S}_{yx}\boldsymbol{S}_{xx}^{-1}\boldsymbol{S}_{xy}\boldsymbol{w}_j = \mu_j\boldsymbol{w}_j \qquad (j = 1, 2, \ldots, r) \tag{7.119}$$

Let $\boldsymbol{W}_r = (\boldsymbol{w}_1, \boldsymbol{w}_2, \ldots, \boldsymbol{w}_r)$. Then we have estimates

$$\boldsymbol{A}_{(1)} = \boldsymbol{S}_{xx}^{-1}\boldsymbol{S}_{xy}\boldsymbol{W}_r \tag{7.120}$$

$$\boldsymbol{B}_{(1)} = \boldsymbol{W}_r' \tag{7.121}$$

Write the estimate of $\boldsymbol{Y}$ as $\boldsymbol{Y}_{(1)} = \boldsymbol{X}\boldsymbol{A}_{(1)}\boldsymbol{B}_{(1)}$. Let

$$\boldsymbol{a}_j = \frac{1}{\sqrt{\mu_j}}\boldsymbol{S}_{xx}^{-1}\boldsymbol{S}_{xy}\boldsymbol{w}_j, \qquad \boldsymbol{A}_r = \boldsymbol{S}_{xx}^{-1}\boldsymbol{S}_{xy}\boldsymbol{W}_r\boldsymbol{D}_{\mu}^{-1/2} \tag{7.122}$$

$$\boldsymbol{D}_{\mu}^{1/2} = \mathrm{diag}(\sqrt{\mu_1}, \sqrt{\mu_2}, \ldots, \sqrt{\mu_r}) \tag{7.123}$$

then $\boldsymbol{a}_j$ vectors satisfy (7.91) and (7.93).

Write estimates (7.101), which have been given by IPCA, as $\boldsymbol{A}_{(2)}, \boldsymbol{B}_{(2)}, \boldsymbol{Y}_{(2)}$, and substitute them in (7.120) and (7.121) to yield

$$\boldsymbol{A}_{(2)} = \boldsymbol{A}_{(1)}\boldsymbol{D}_{\mu}^{-1/2} \tag{7.124}$$

$$\boldsymbol{B}_{(2)} = \boldsymbol{D}_{\mu}^{-1/2}\boldsymbol{W}_r'\boldsymbol{S}_{yx}\boldsymbol{S}_{xx}^{-1}\boldsymbol{S}_{xy} = \boldsymbol{D}_{\mu}^{1/2}\boldsymbol{B}_{(1)} \tag{7.125}$$

Thus estimates of IPCA (RDA) are related to those of this RRR by the scale transformation, and the degree of fit is the same as shown by $\boldsymbol{Y}_{(1)} = \boldsymbol{Y}_{(2)}$ for the same $r$. It is also found that the least squares solution of RRR suggested by Davies and Tso (1982) gives the same degree of fit as IPCA (RDA).

Setting $\boldsymbol{\Gamma} = \boldsymbol{\Sigma}_{yy}^{-1}$ in the theorem and substituting sample statistics in the result, we obtain from (7.117) that

$$\boldsymbol{S}_{yy}^{-1/2}\boldsymbol{S}_{yx}\boldsymbol{S}_{xx}^{-1}\boldsymbol{S}_{xy}\boldsymbol{S}_{yy}^{-1/2}\boldsymbol{w}_k = \lambda_k^2\boldsymbol{w}_k \qquad (k = 1, 2, \ldots, r) \tag{7.126}$$

It is remembered that this is the canonical equation (7.32). Let the solution be $\boldsymbol{W}_r = (\boldsymbol{w}_1, \boldsymbol{w}_2, \ldots, \boldsymbol{w}_r)$. Rewriting $\boldsymbol{W}_r = \boldsymbol{S}_{yy}^{-1/2}\boldsymbol{V}_r$ and noting relation (7.28), we then have

$$\boldsymbol{A}_{(3)} = \boldsymbol{S}_{xx}^{-1}\boldsymbol{S}_{xy}\boldsymbol{V}_r = \boldsymbol{U}_r\boldsymbol{D}_\lambda \tag{7.127}$$

$$\boldsymbol{B}_{(3)} = \boldsymbol{V}_r'\boldsymbol{S}_{yy} = \boldsymbol{D}_\lambda^{-1}\boldsymbol{U}_r'\boldsymbol{S}_{xy} \tag{7.128}$$

Let us apply a scale transformation to these two such that

$$\boldsymbol{A}_{(4)} = \boldsymbol{U}_r = \boldsymbol{A}_{(3)}\boldsymbol{D}_\lambda^{-1} \tag{7.129}$$

$$\boldsymbol{B}_{(4)} = \boldsymbol{U}_r'\boldsymbol{S}_{xy} = \boldsymbol{D}_\lambda\boldsymbol{B}_{(3)} \tag{7.130}$$

It should be noticed that these are the estimates of CRA. Thus the estimates are related to those of this RRR by the scale transformation, and both procedures give the same degree of fit for the same $r$. It is also noted that $\boldsymbol{A}_{(4)}$ and $\boldsymbol{B}_{(4)}$ are the estimates given by the maximum likelihood procedure of RRR due to Tso (1981).

Figure 7.1 illustrates the relationship among procedures in reference to Theorems 7.1 and 7.2. Among various formulations of RRR so far developed, Theorem 7.2 is concerned with two sets of random variables. As shown above,

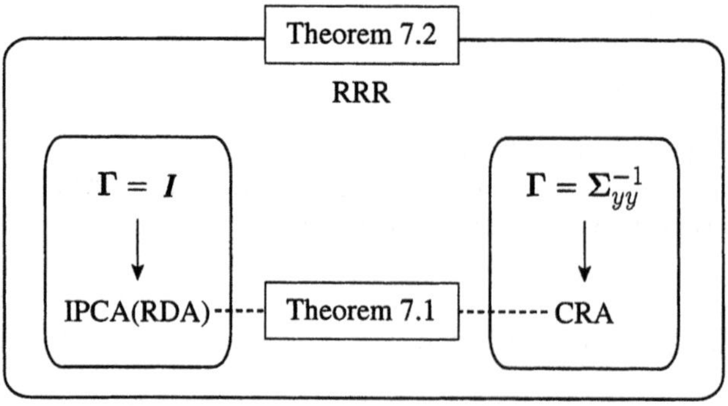


**FIGURE 7.1** Relationship of procedures

substituting sample statistics in the statements of the theorem yields the same equations as derived by the procedures in descriptive data analysis. Then, the analysis of asymmetry by performing two directional analyses of RDA or CRA is regarded as utilization of RRR in two directional ways.

## 7.5. NUMERICAL EXAMPLE

Here we present examples of RDA and CRA with a focus on the asymmetry between two data sets, which were observed on a group of $N$ applicants for admission to a university. One set of variables $\boldsymbol{x}$ describes high school grade point averages of five subjects of the applicants. Another set $\boldsymbol{y}$ describes

**TABLE 7.1** Correlation Matrices

| | $R_{xx}$ | | | | |
|---|---|---|---|---|---|
| | $x_1$ | $x_2$ | $x_3$ | $x_4$ | $x_5$ |
| $x_1$ | 1.000 | 0.326 | 0.179 | 0.279 | 0.089 |
| $x_2$ | 0.326 | 1.000 | 0.293 | 0.386 | 0.384 |
| $x_3$ | 0.179 | 0.293 | 1.000 | 0.301 | 0.321 |
| $x_4$ | 0.279 | 0.386 | 0.301 | 1.000 | 0.276 |
| $x_5$ | 0.089 | 0.384 | 0.321 | 0.276 | 1.000 |
| | $R_{xy}$ | | | | |
| | $y_1$ | $y_2$ | $y_3$ | $y_4$ | $y_5$ |
| $x_1$ | 0.344 | 0.156 | 0.195 | 0.220 | 0.098 |
| $x_2$ | 0.278 | 0.463 | 0.282 | 0.226 | 0.315 |
| $x_3$ | 0.129 | 0.151 | 0.300 | 0.185 | 0.192 |
| $x_4$ | 0.271 | 0.311 | 0.324 | 0.436 | 0.269 |
| $x_5$ | 0.064 | 0.210 | 0.278 | 0.126 | 0.412 |
| | $R_{yy}$ | | | | |
| | $y_1$ | $y_2$ | $y_3$ | $y_4$ | $y_5$ |
| $y_1$ | 1.000 | 0.670 | 0.661 | 0.696 | 0.525 |
| $y_2$ | 0.670 | 1.000 | 0.660 | 0.614 | 0.706 |
| $y_3$ | 0.661 | 0.660 | 1.000 | 0.680 | 0.707 |
| $y_4$ | 0.696 | 0.614 | 0.680 | 1.000 | 0.546 |
| $y_5$ | 0.525 | 0.706 | 0.707 | 0.546 | 1.000 |

**TABLE 7.2** Eigenvalues and Related Quantities

| Dimension $t$ | | 1 | 2 | 3 | 4 | 5 |
|---|---|---|---|---|---|---|
| CCA ($\boldsymbol{x}, \boldsymbol{y}$) | Eigenvalue $\lambda^2$ | 0.249 | 0.213 | 0.166 | 0.082 | 0.038 |
| | Proportion | 0.332 | 0.285 | 0.222 | 0.110 | 0.051 |
| | Cum. prop. | 0.332 | 0.617 | 0.839 | 0.949 | 1.000 |
| | Can. cor. $\lambda$ | 0.499 | 0.461 | 0.407 | 0.287 | 0.196 |
| RDA ($\boldsymbol{y}\|\boldsymbol{x}$) | Eigenvalue $\mu$ | 0.795 | 0.117 | 0.062 | 0.024 | 0.009 |
| | Proportion | 0.790 | 0.116 | 0.061 | 0.024 | 0.009 |
| | Cum. prop. | 0.790 | 0.906 | 0.967 | 0.991 | 1.000 |
| RDA ($\boldsymbol{x}\|\boldsymbol{y}$) | Eigenvalue $\nu$ | 0.484 | 0.192 | 0.103 | 0.058 | 0.025 |
| | Proportion | 0.561 | 0.222 | 0.120 | 0.067 | 0.030 |
| | Cum. prop. | 0.561 | 0.784 | 0.903 | 0.970 | 1.000 |

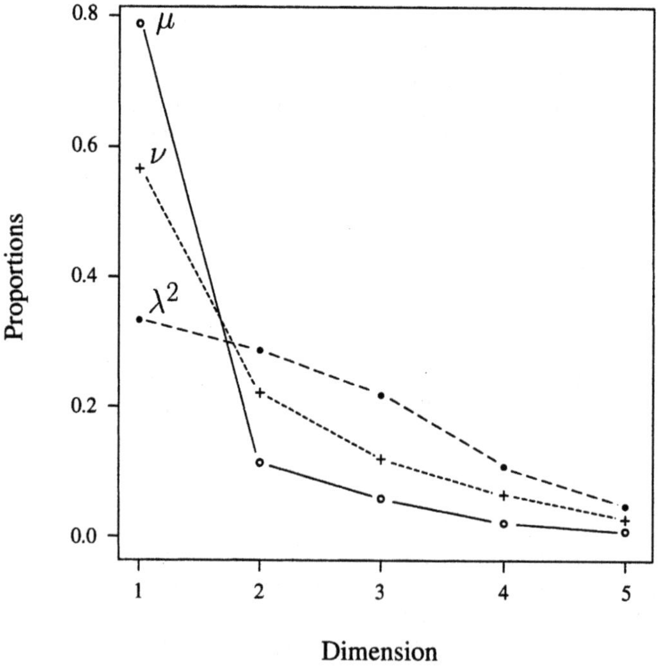


**FIGURE 7.2** Proportion of eigenvalues

**TABLE 7.3** Redundancy Analysis

| | RDA $(y\|x)$ | | | RDA $(x\|y)$ | | |
|---|---|---|---|---|---|---|
| | Composite variate | | | Composite variate | | |
| | $\xi_1$ | $\xi_2$ | $\xi_3$ | $\eta_1$ | $\eta_2$ | $\eta_3$ |
| Variance ext. | 0.413 | 0.180 | 0.126 | 0.666 | 0.137 | 0.077 |
| Cumulative | 0.413 | 0.593 | 0.719 | 0.666 | 0.803 | 0.881 |
| Redundancy | 0.159 | 0.023 | 0.012 | 0.097 | 0.038 | 0.021 |
| Cumulative | 0.159 | 0.182 | 0.195 | 0.097 | 0.135 | 0.156 |

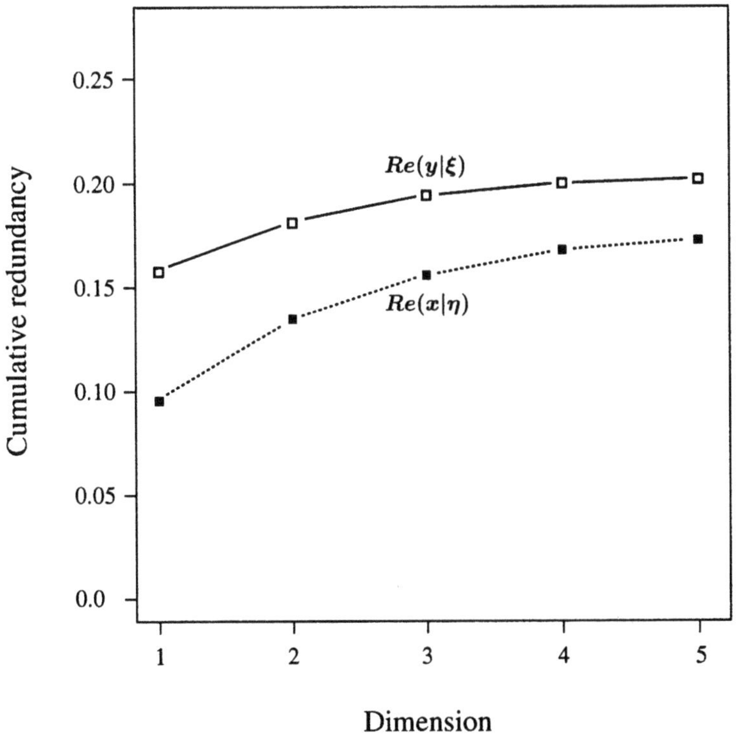


**FIGURE 7.3** Comparison of RDA$(y|x)$ and RDA$(x|y)$ (redundancy)

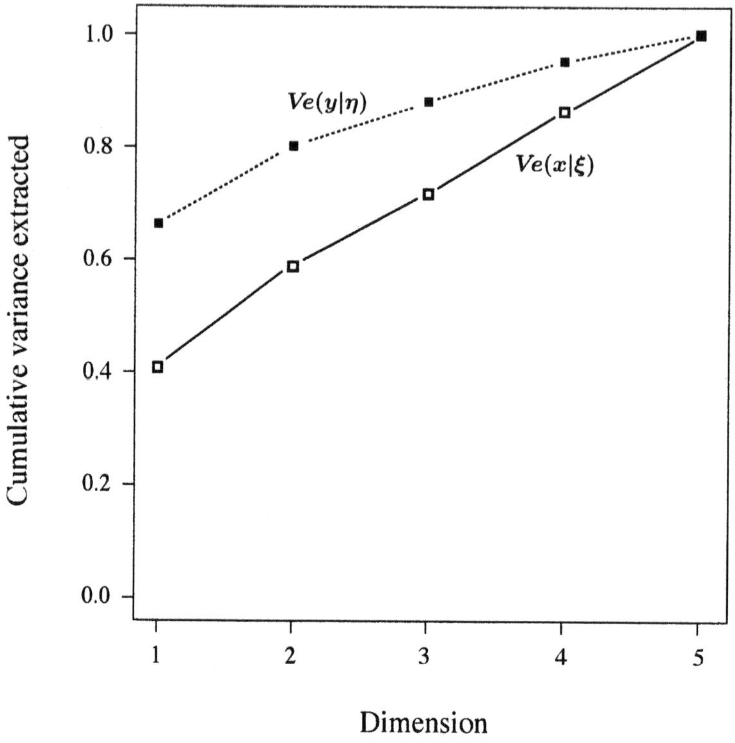


**FIGURE 7.4** Comparison of RDA($\boldsymbol{y}|\boldsymbol{x}$) and RDA($\boldsymbol{x}|\boldsymbol{y}$) (variance extracted)

**TABLE 7.4** Canonical Regression Analysis

| | CRA ($\boldsymbol{x}|\boldsymbol{f}$) Canonical variate | | | RDA ($\boldsymbol{y}|\boldsymbol{g}$) Canonical variate | | |
|---|---|---|---|---|---|---|
| | $f_1$ | $f_2$ | $f_3$ | $g_1$ | $g_2$ | $g_3$ |
| Variance ext. | 0.361 | 0.184 | 0.146 | 0.521 | 0.184 | 0.127 |
| Cumulative | 0.361 | 0.545 | 0.691 | 0.521 | 0.704 | 0.831 |
| Redundancy | 0.129 | 0.039 | 0.021 | 0.090 | 0.039 | 0.024 |
| Cumulative | 0.129 | 0.168 | 0.190 | 0.090 | 0.129 | 0.153 |

achievement test scores of the same five subjects of the applicants. The personal information was originally recorded in the form of $N \times 5$ ($N > 1000$). For the present purposes, we normalize the data. That gives rise to correlation matrices, which are shown in Table 7.1.

We applied redundancy analysis to the correlation matrices in two directions, RDA($\boldsymbol{y}|\boldsymbol{x}$) and RDA($\boldsymbol{x}|\boldsymbol{y}$). In comparison, we performed CCA between $\boldsymbol{x}$ and $\boldsymbol{y}$, which is referred to as CCA($\boldsymbol{x}, \boldsymbol{y}$). Then we computed redundancy based on the result, which means CRA using canonical variates $\boldsymbol{f}$ or $\boldsymbol{g}$ in two directions, CRA($\boldsymbol{y}|\boldsymbol{f}$) and CRA($\boldsymbol{x}|\boldsymbol{g}$). Across those analyses, the number of composite variates, $r$, was changed from 1 to 5.

Table 7.2 shows eigenvalues, proportions of eigenvalues to the sum and cumulated proportions, regarding $\mu_t(t = 1, 2, \ldots, 5)$ of RDA($\boldsymbol{y}|\boldsymbol{x}$), $\nu_t(t = 1, 2, \ldots, 5)$ of RDA($\boldsymbol{x}|\boldsymbol{y}$) and $\lambda_t^2(t = 1, 2, \ldots, 5)$ of CCA($\boldsymbol{x}, \boldsymbol{y}$). Figure 7.2 shows changes of proportions of eigenvalues in each analysis. One may find

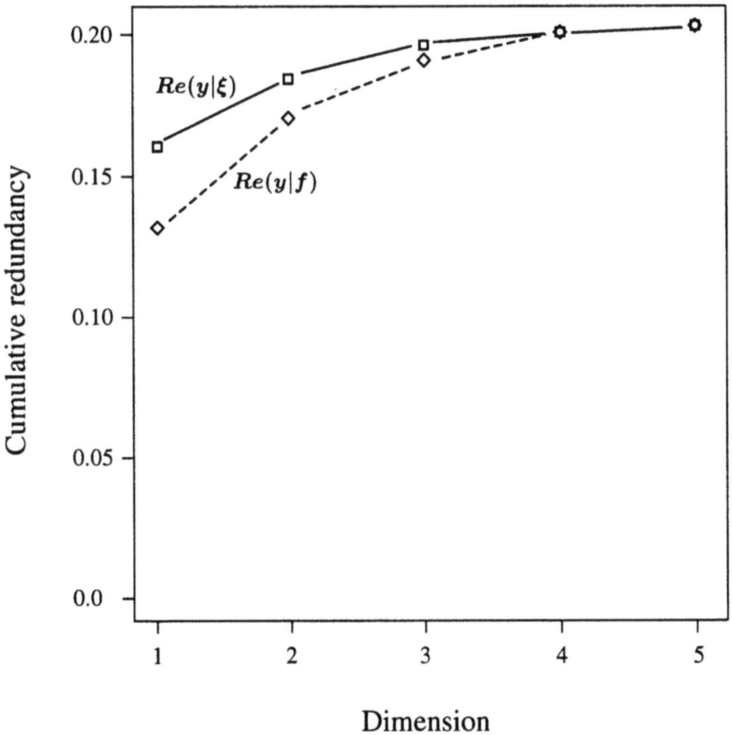


**FIGURE 7.5** Comparison of RDA($\boldsymbol{y}|\boldsymbol{x}$) and CRA($\boldsymbol{y}|\boldsymbol{f}$)

that the data reduction in smaller dimensions is more efficient in RDA than in CCA.

Table 7.3 summarizes the results of RDA($\boldsymbol{y}|\boldsymbol{x}$) on the left side and the results of RDA($\boldsymbol{x}|\boldsymbol{y}$) on the right side. We denote composite variates in terms of $\boldsymbol{x}$ by $\boldsymbol{\xi}$ and those in terms of $\boldsymbol{y}$ by $\boldsymbol{\eta}$. Examining the values of cumulative redundancy ($r = 1, 2, 3$) in two cases, we find a tendency that variates $\boldsymbol{\xi}$ account for $\boldsymbol{y}$ larger than variates $\boldsymbol{\eta}$ do for $\boldsymbol{x}$. Thus, in view of this difference, we observe the asymmetry between the two variable sets. Figure 7.3 reveals the tendency clearly.

The variance extracted by each $\xi_t$ or $\eta_t$ is given in Table 7.3, which is contrasted for the two cases in Fig. 7.4. It is of interest to note that there is more loss of information in constructing $\boldsymbol{\xi}$ from $\boldsymbol{x}$ than in constructing $\boldsymbol{\eta}$ from $\boldsymbol{y}$.

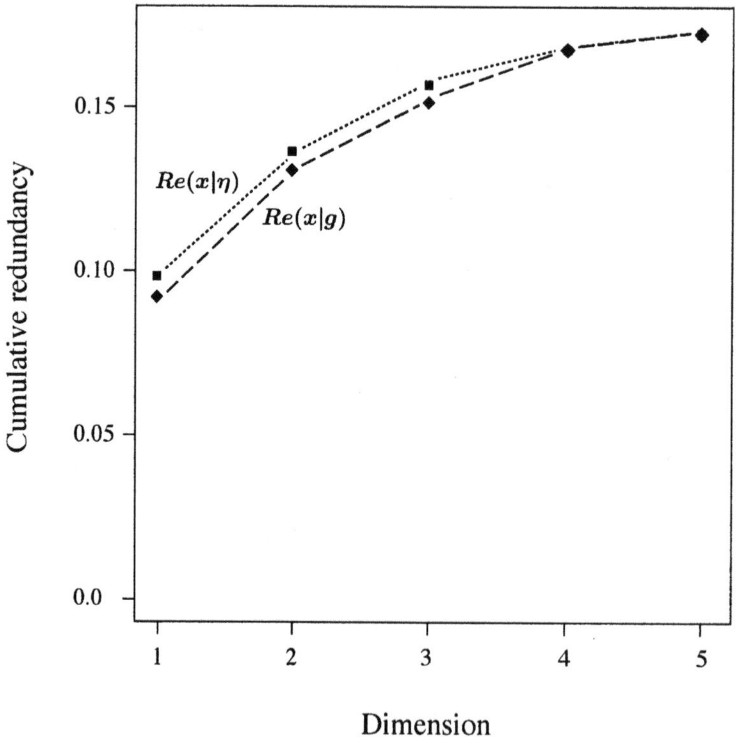


**FIGURE 7.6** Comparison of RDA($\boldsymbol{x}|\boldsymbol{y}$) and CRA($\boldsymbol{x}|\boldsymbol{g}$)

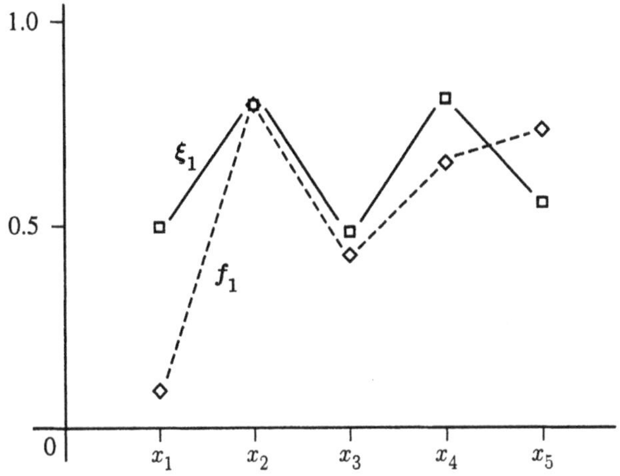


**FIGURE 7.7** Intraset correlation: cor($x_j$, $\xi_1$) and cor($x_j$, $f_1$)

Table 7.4 provides a summary of CRA($\boldsymbol{y}|\boldsymbol{f}$), and CRA($\boldsymbol{x}|\boldsymbol{y}$). One may note that the variance of $\boldsymbol{x}$ extracted by $\xi_1$ is larger than that extracted by $f_1$, and that the variance of $\boldsymbol{y}$ extracted by $\eta_1$ is larger than that extracted by $g_1$. Remembering the ways in which the composite variates are constructed, those results are quite natural. Figure 7.5 reveals the difference in the

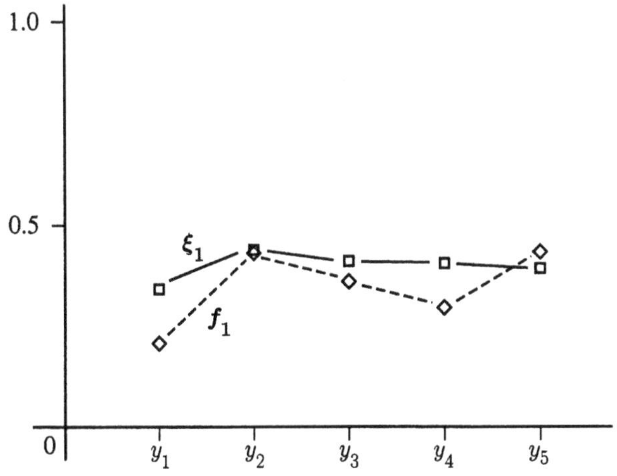


**FIGURE 7.8** Interset correlation: cor($y_j$, $\xi_1$) and cor($y_j$, $f_1$)

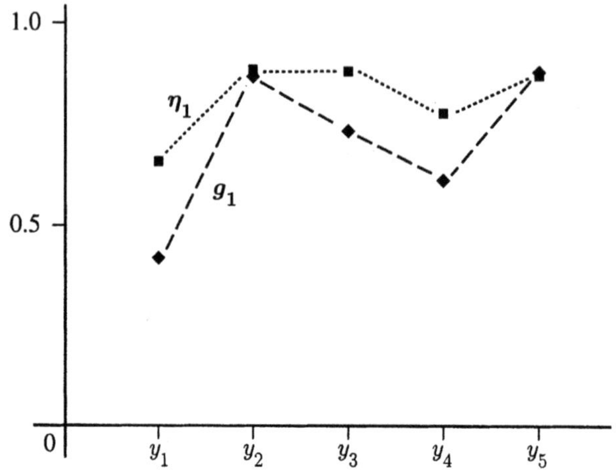


**FIGURE 7.9** Intraset correlation: cor($y_j$, $\eta_1$) and cor($y_j$, $g_1$)

cumulative redundancy between RDA($\boldsymbol{y}|\boldsymbol{x}$) and CRA($\boldsymbol{y}|\boldsymbol{f}$), and Fig. 7.6 the difference between RDA($\boldsymbol{x}|\boldsymbol{y}$) and CRA($\boldsymbol{x}|\boldsymbol{g}$). With these results, one can realize the theoretical statement by (7.112).

Let us focus on the indices in connection with the first composite variates in the analyses above. Figure 7.7 shows intraset correlation coefficients

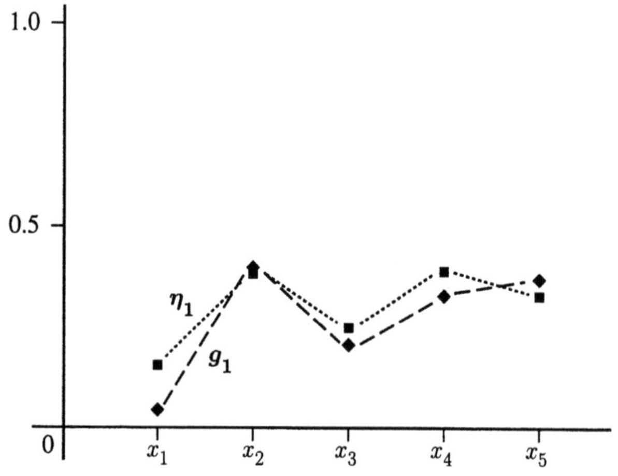


**FIGURE 7.10** Interset correlation: cor($x_j$, $\eta_1$) and cor($x_j$, $g_1$)

given by RDA($\boldsymbol{y}|\boldsymbol{x}$) in contrast to those given by CRA($\boldsymbol{y}|\boldsymbol{f}$). In the same way, Fig. 7.8 contrasts interset correlation coefficients given by the two analyses. Similarly, Fig. 7.9 shows intraset correlation coefficients given by RDA($\boldsymbol{x}|\boldsymbol{y}$) in contrast to those given by CRA($\boldsymbol{x}|\boldsymbol{g}$), and Fig. 7.10 interset correlations given by the two analyses. The four figures are provided for comparing the result of RDA with that of CRA. With attention to the asymmetrical relationship between sets, it would be interesting, for example, to compare cor($y_j$, $\xi_1$) in Fig. 7.8 with cor($x_j$, $\eta_1$) in Fig. 7.9.

# Bibliography

Akaike, H. (1974). A new look at the statistical model identification. *IEEE Transactions and Automatic Control* 19: 716–723.

Akaike, H. (1980). Likelihood and the Bayes procedure. In: Bernardo, J. M. et al., eds. *Bayesian Statistics*. Valencia: University Press, pp. 143–166.

Arabie, P., Schleutermann, S., Daws, J., Hubert, L. (1988). Marketing applications of sequencing and partitioning of nonsymmetric and/or two-mode matrices. In: Gaul, W., Schader, M., eds. *Data Analysis, Decision Support and Expert Knowledge Representation in Marketing*. Heiderberg: Springer Verlag, pp. 215–224.

Arbuckle, J., Nugent, J. H. (1973). A general procedure for parameter estimation for the law of comparative judgment. *Br. J. Mathemat. Statist. Psychol.* 26: 240–260.

Bass, F. M., Pessemier, E. A., Lehmann, D. R. (1972). An experimental study of relationships between attitudes, brand preference, and choice. *Behav. Sci.* 17: 532–541.

Bellman, R. (1970). *Introduction to Matrix Analysis*. New York: McGraw Hill.

Berlin, T. H., Kac, M. (1952). The spherical model of a ferromagnet. *Physical Rev.* 86: 821–835.

Bock, R. D., Jones, L. V. (1968). *The Measurement and Prediction of Judgment and Choice*. San Francisco: Holden-Day.

Borg, I., Groenen, P. (1997). *Modern Multidimensional Scaling: Theory and Applications*. New York: Springer.
Both, M., Gaul, W. (1986). Ein vergleich zweimodaler clusteranalyseverfahren. *Meth. Op. Res.* 57: 593–605.
Bove, G. (1992). Asymmetric multidimensional scaling and correspondence analysis for square tables. *Statistica Applicata* 4: 587–598.
Bove, G., Critchley, F. (1993). Metric multidimensional scaling for asymmetric proximities when the asymmetry is one-dimensional. In: Steyer, R., Wender, K. F., Widaman, K. F., eds. *Psychometric Methodology: Proceedings of the 7th (1991) European Meeting of the Psychometric Society*. Germany: Gustav Fischer Verlag.
Brossier, G. (1982). Classification hiérarchique à partir de matrices carrées non-symétriques. *Statistiques et Analyse des Donnees* 7: 22–40.
Caussinus, H., de Falguerolles, A. (1987). Tableaux carrés: modélisation et méthodes factorielles. *Revue de Statique Appliquée* 35: 35–52.
Chino, N. (1978). A graphical technique for representing the asymmetric relationships between N objects. *Behaviormetrika* 5: 23–40.
Chino, N. (1980). A unified geometrical interpretation of the MDS techniques. Paper presented at Meeting of the Psychometric Society, Iowa, 1980.
Chino, N. (1990). A generalized inner product model for the analysis of asymmetry. *Behaviormetrika* 27: 25–46.
Chino, N., Shiraiwa, K. (1993). A geometrical structure of some non-distance models for asymmetric MDS. *Behaviormetrika* 20: 35–47.
Coombs, G. H. (1964). *A Theory of Data*. New York: Wiley.
Constantine, A. G., Gower, J. C. (1978). Graphical representation of asymmetric matrices. *Appl. Statist.* 27: 297–304.
Cox, T. F., Cox, M. A. A. (1994). *Multidimensional Scaling*. London: Chapman & Hall.
Davies, P. T., Tso, M. K.-S. (1982). Procedures for reduced-rank regression. *Appl. Statist.* 31: 244–255.
Dawson, M. R. W., Harshman, R. A. (1986). The multidimensional analysis of asymmetries in alphabetic confusion matrices: Evidence for global-to-local and local-to-global processing. *Perception & Psychophysics* 40: 370–383.
DeSarbo, W. S. (1982). GENNCLUS: New model for general nonhierarchical clustering analysis. *Psychometrika* 47: 449–475.
DeSarbo, W. S., Johnson, M. D., Manrai, A. K., Manrai, L. A., Edwards, E. A. (1992). TSCALE: A new multidimensional scaling procedure based on Tversky's contrast model. *Psychometrika* 57: 43–69.
DeSarbo, W. S., De Soete, G. (1984). On the use of hierarchical clustering for the analysis of nonsymmetric proximities. *J. Consumer Res.* 11: 601–610.

DeSarbo, W. S., Manrai, A. K. (1992). A new multidimensional scaling methodology for the analysis of asymmetric proximity data in marketing research. *Marketing Sci.* 11: 1–20.

DeSarbo, W. S., Manrai, A. K., Burke, R. R. (1990). A nonspatial methodology for the analysis of two-way proximity data incorporating the distance–density hypothesis. *Psychometrika* 55: 229–253.

De Soete, G., Carroll, J. D. (1996). Tree and other network models for representing proximity data. In: Arabie, P., Hubert, L. J., De Soete, G., eds. *Clustering and Classification*. River Edge: World Scientific Publication, pp. 157–197.

De Soete, G., DeSarbo, W. S., Furnas, G. W., Carroll, J. D. (1984). The estimation of ultrametric and path length trees from rectangular proximity data. *Psychometrika* 49: 289–310.

Digby, R. G. N., Kempton, R. A. (1987). *Multivariate Analysis of Ecological Communities*. London: Chapman and Hall.

Eckes, T., Orlik, P. (1990). An agglomerative method for two-mode hierarchical clustering. In: Bock, H., Ilm, P., eds. *Classification, Data Analysis, and Knowledge Organization*. Berlin: Springer-Verlag, pp. 3–8.

Eckes, T., Orlik, P. (1993). An error variance approach to two-mode hierarchical clustering. *J. Classification* 10: 51–74.

Ekman, G. (1963). A direct method for multidimensional ratio scaling. *Psychometrika* 28: 33–41.

Escoufier, Y., Grorud, A. (1980). Analyse factorielle des matrices carrées non-symétriques. In: Diday, E., ed. *Data Analysis and Informatics*. New York: North Holland, pp. 263–276.

Espejo, E., Gaul, W. (1986). Two-mode hierarchical clustering as an instrument for marketing research. In: Gaul, W., Schader, M., eds. *Classification as a Tool of Research*. New York: North Holland, pp. 121–128.

Ferligoj, A., Batagelj, V. (1983). Some types of clustering with relational constraints. *Psychometrika* 48: 541–552.

Fershtman, M. (1997). Cohesive group detection in a social network by the segregation matrix index. *Social Networks* 19: 193–207.

Fujiwara, H. (1980). Methods for cluster analysis using asymmetric measures and homogeneity coefficient. *The Japanese Journal of Behaviormetrics* 7: 12–21 (in Japanese).

Furnas, G. W. (1980). Objects and their features: The metric representation of two class data. Unpublished Doctoral Dissertation, Stanford University.

Goldthorpe, J. H. (1980). *Social Mobility and Class Structure in Modern Britain*. Oxford: Clarendon Press.

Gower, J. C. (1977). The analysis of asymmetry and orthogonality. In: Barra, J. R. et al., eds. *Recent Developments in Statistics*. Amsterdam: North Holland, pp. 109–123.

Gower, J. C. (1984). Multivariate analysis: ordination, multidimensional scaling and allied topics. In: Lloyd, E., ed. *Handbook of Applicable Mathematics*. Vol. 6. Part B. New York: John Wiley, *Statistics*, Chapter 6, pp. 727–781.

Gower, J. C., Legendre, P. (1986). Metric and Euclidean properties of dissimilarity coefficients. *J. Classification* 3: 5–48.

Gregson, R. M. M. (1975). *Psychometrics of Similarity*. New York: Academic Press.

Guilford, J. P. (1954). *Psychometric Methods*. New York: McGraw Hill.

Harris, W. P. (1957). A revised law of comparative judgment. *Psychometrika* 22: 189–198.

Harshman, R. A. (1978). Models for analysis of asymmetrical relationships among *N* objects or stimuli. Paper presented at the *First Joint Meeting of the Psychometric Society and the Society of Mathematical Psychology*, Hamilton, Ontario.

Harshman, R. A., Green, P. E., Wind, Y., Lundy, M. E. (1982). A model for the analysis of asymmetric data in marketing research. *Marketing Sci.* 1: 204–242.

Harshman, R. A., Kiers, H. A. L. (1987). Algorithms for DEDICOM analysis of asymmetric data. Paper presented at the *European Meeting of the Psychometric Society*, Enschede.

Hellström, Å. (1978). Factors producing and factors not producing time errors. *Perception & Psychophysics* 23: 433–444.

Holman, E. W. (1979). Monotonic models for asymmetric proximities. *J. Mathemat. Psychol.* 20: 1–15.

Hubert, L. (1973). Min and max hierarchical clustering using asymmetric similarity measures. *Psychometrika* 38: 63–72.

Hutchinson, J. W. (1989). NETSCAL: A network scaling algorithm for nonsymmetric proximity data. *Psychometrika* 54: 25–51.

Imaizumi, T. (1987). Fitting a restricted asymmetric dissimilarity model to a symmetric dissimilarity matrix. *Behaviormetrika* 22: 1–10.

Israëls, A. Z. (1984). Redundancy analysis for qualitative variables. *Psychometrika* 49: 331–346.

Israëls, A. Z. (1986). Interpretation of redundancy analysis: rotated vs. unrotated solutions. *Appl. Stochastic Models and Data Anal.* 2: 121–130.

Izenman, A. J. (1975). Reduced-rank regression for the multivariate linear model. *J. Multivar. Anal.* 5: 248–268.

Johansson, J. K. (1981). An extension of Woolenberg's redundancy analysis. *Psychometrika* 46: 93–103.

Kendall, M. G. (1962). *Rank Correlation Methods*. London: Charles Griffin.

Kendall, M. G., Babinton Smith, B. (1940). On the method of paired comparisons. *Biometrika* 31: 324–345.

Kiers, H. A. L. (1989). An alternating least squares algorithm for fitting the two- and three-way DEDICOM model and the IDIOSCAL model. *Psychometrika* 54: 515–521.

Kiers, H. A. L., ten Berge, J. M. F., Takane, Y., de Leeuw, J. (1990). A generalization of Takane's algorithm for DEDICOM. *Psychometrika* 55: 151–158.

Kiers, H. A. L., Takane, Y. (1994). A generalization of GIPSCAL for the analysis of asymmetric data. *J. Classification* 11: 79–99.

Krumhansl, C. L. (1978). Concerning the applicability of geometric models to similarity data: The interrelationship between similarity and spatial density. *Psychol. Rev.* 85: 445–463.

Kruskal, J. B. (1964). Multidimensional scaling by optimizing goodness of fit to a nonmetric hypothesis. *Psychometrika* 29: 1–27.

Lambert, Z. V., Wildt, A. R., Durand, R. M. (1988). Redundancy analysis: An alternative to canonical correlation and multivariate multiple regression in exploring interset association. *Psychol. Bull.* 104: 282–289.

Lance, G. N., Williams, W. T. (1967). A general theory of classificatory sorting strategies, I. Hierarchical systems. *Comput. J.* 9: 373–380.

Landau, H. G. (1951). On dominance relations and the structure of animal societies: effect of inherent characteristics. *Bull. Mathemat. Biophys.* 13: 1–19.

Lauro, C., Esposito, V. (2000). Non-symmetrical data analysis approaches: Recent developments and perspectives. In: Gaul, W., Opitz, O., Schader, M. S., eds. *Data Analysis*. New York: Springer, pp. 219–232.

Levin, J., Brown, M. (1979). Scaling a conditional proximity matrix to symmetry. *Psychometrika* 44: 239–243.

Masure, R. H., Allee, W. C. (1934). The social order in flocks of the common chicken and pigeon. *Auk* 51: 306–325.

McCormick, W. T., Schweitzer, P. J., White, T. W. (1972). Problem decomposition and data reorganization by a clustering technique. *Opns Res.* 20: 993–1009.

Mehler, J. (1963). Some effects of gramatical transformations on the recall of English sentences. *Verbal Learning and Verbal Behavior* 2: 346–351.

Micko, H. C. (1970). A halo-model for multidimensional ratio scaling. *Psychometrika* 35: 199–277.

Muller, K. E. (1981). Relationships between redundancy analysis, canonical correlation and multivariate regression. *Psychometrika* 46: 139–142.
Nakajima, A., Saito, T. (1996). Detection of ordinal structure inherent in asymmetric data. *Bulletin of The Computational Statistics Society of Japan* 9(2): 35–45 (in Japanese with English summary).
Nosofsky, R. M. (1991). Stimulus bias, asymmetric similarity, and classification. *Cognitive Psychol.* 23: 94–140.
Okada, A., Imaizumi, T. (1987). Nonmetric multidimensional scaling of asymmetric proximities. *Behaviormetrika* 21: 81–96.
Okada, A., Iwamoto, T. (1995). An asymmetric cluster analysis study on university enrollment flow among Japanese prefectures. *Sociological Theory and Methods* 10: 1–13 (in Japanese).
Okada, A., Iwamoto, T. (1996). University enrollment flow among the Japanese prefectures: A comparison before and after the joint first stage achievement test by asymmetric cluster analysis. *Behaviormetrika* 23: 169–185.
Ozawa, K. (1983). Classic: A hierarchical clustering algorithm based on asymmetric similarities. *Pattern Recognition* 16: 201–211.
Rao, C. R. (1964). The use and interpretation of principal components analysis in applied research. *Sankhya A* 26: 329–358.
Reitz, K. P. (1988). Social groups in a monastery. *Social Networks* 10: 343–357.
Rosch, E. (1975). Cognitiver reference points. *Cognitive Psychol.* 74: 532–547.
Rothkopf, E. Z. (1957). A measure of stimulus similarity and errors in some paired associate learning. *J. Exper. Psychol.* 53: 94–101.
Saaty, T. L. (1980). *The Analytic Hierarchy Process*. New York: McGraw-Hill.
Saaty, T. L., Vargas, L. G. (1984a). Comparison of eigenvalue, logarithmic least squares and least squares methods in estimating ratios. *Mathemat. Modelling* 5: 309–324.
Saaty, T. L., Vargas, L. G. (1984b). Inconsistency and rank preservation. *J. Mathemat. Psychol.* 28: 205–214.
Saito, T. (1983). A method of multidimensional scaling to obtain a sphere configuration. *Hokkaido Behavioral Science Report* M4: 1–42.
Saito, T. (1986). Multidimensional scaling to explore complex aspects in dissimilarity judgment. *Behaviormetrika* 20: 35–62.
Saito, T. (1991). Analysis of asymmetric proximity matrix by a model of distance and additive terms. *Behaviormetrika* 29: 45–60.
Saito, T. (1992). Measurement of the asymmetry observed in comparative judgment. *Hokkaido Behavioral Science Report* M-19: 1–29.
Saito, T. (1993a). Multidimensional scaling for asymmetric proximity data. In: Steyer, R., Wender, K. F., Widaman, K. F., eds. *Psychometric Method-*

*ology: Proceedings of the 7th European Meeting of the Psychometric Society*, Germany: Gustav Fischer Verlag, pp. 452–456.

Saito, T. (1993b). Multivariate analysis of environmental data with asymmetrical relationship among locations. *Hokkaido Behavioral Science Report* M-22: 1–12.

Saito, T. (1994). Psychological scaling of the asymmetry observed in comparative judgement. *Br. J. Mathemat. Statis. Psychol.* 47: 41–62.

Saito, T. (1995). Relationships between redundancy analysis and canonical regression analysis. *J. Japan Statistical Society* 25: 45–53.

Saito, T. (1997). Line structure derived from decomposition of asymmetric matrix. *Journal of the Japanese Society of Computational Statistics* 10: 47–57.

Saito, T. (2000). Unidimensional structure detected by analysis of an asymmetric data matrix. In: Gaul, W., Opitz, O., Schader, M. S., eds. *Data Analysis: Scientific Modeling and Practical Application.* New York: Springer, pp. 347–357.

Saito, T. (2002). Circle structure derived from decomposition of asymmetric matrix. *Journal of the Japanese Society of Computational Statistics* 15: 1–23.

Saito, T. (2003a). Utilization of two procedures for analysis of asymmetry with distance properties. *Proceedings of the 31st Annual Meeting of the Behaviormetric Society of Japan*, 278–281.

Saito, T. (2003b). Analysis of asymmetry by the slide vector model. *VALDES Research Paper E-03-07*, 1–12.

Saito, T., Takeda, S. (1990). Multidimensional scaling for asymmetric proximities: Model and method. *Behaviormetrika* 28: 49–80.

Sato, Y. (1988). An analysis of sociometric data by MDS in Minkowski space. In: Matsusita, K., ed. *Statistical Theory and Data Analysis II.* North Holland, pp. 385–396.

Scheffe, H. (1952). An analysis of variance for paired comparisons. *J. Am. Statis. Assoc.* 1: 381–400.

Scholten, H. J. (1984). Residential mobility and log-linear Modeling. In: Bahrenbert, G., Fischer, M. M., Nijkamp, P., eds. *Recent Developments in Spatial Data Analysis: Methodology, Measurement, Models.* Hantsm: Gower, pp. 271–287.

Schönemann, P. H. (1972). An algebraic solution for a class of subjective metrics models. *Psychometrika* 37: 441–451.

Shepard, R. N. (1963). Analysis of proximities as a technique for the study of information processing in man. *Human Factors* 5: 33–48.

Shepard, R. N. (1964). Attention and the metric structure of the stimulus space. *Journal of Mathematical Psychology* 1: 54–87.

Shepard, R. N., Arabie, P. (1979). Additive clustering: representation of similarities as combinations of discrete overlapping properties. *Psychol. Rev.* 86: 87–123.

Sjöberg, L. A. (1972). A cognitive theory of similarity. *Göteberg Psychological Reports*, No. 10.

Stewart, D., Love, W. (1968). A general canonical correlation index. *Psychol. Bull.* 70: 160–163.

Stigler, S. M. (1994). Citation pattern in the journals on statistics and probability. *Statist. Sci.* 9: 94–108.

Strauss, D., Ikeda, M. (1990). Pseudolikelihood estimation for social networks. *J. Am. Statist. Assoc.* 85: 204–212.

ten Berge, J. M. F. (1985). On the relationship between Fortier's simultaneous linear prediction and van den Wollenberg's redudancy analysis. *Psychometrika* 50: 121–122.

ten Berge, J. M. F. (1997). Reduction of asymmetry by rank-one matrices. *Comput. Statis. Data Anal.* 24: 357–366.

Thurstone, L. L. (1927). A law of comparative judgments. *Psychol. Rev.* 34: 273–286.

Tobler, W. (1976–77). Spatial interaction patterns. *J. Environ. Syst.* 6: 271–301.

Torgerson, W. S. (1952). Multidimensional scaling: I. Theory and method. *Psychometrika*, 17: 401–419.

Torgeerson, W. S. (1965). Multidimensional scaling of similarity. *Psychometrika* 30: 379–393.

Tso, M. K.-S. (1981). Reduced rank regression and canonical analysis. *J. Roy. Statist. Soc. Series B* 43: 183–189.

Tversky, A. (1977). Features of similarity. *Psychol. Rev.* 84: 327–352.

Tyler, D. E. (1982). On the optimality of the simultaneous redundancy transformations. *Psychometrika* 47: 77–86.

United Nations (1992). International Trade Statistics Yearbook. New York: United Nations.

van den Wollenberg, A. L. (1977). Redundancy analysis: an alternative for canonical correlation analysis. *Psychometrika* 42: 207–219.

van der Burg, E., de Leeuw, J. (1990). Nonlinear redundancy analysis. *Br. J. Mathemat. Statis. Psychol.* 43: 217–230.

van der Heiden, P. G. M., Mooijaart, A. (1995). Some new log-bilinear models for the analysis of asymmetry in a square contingency table. *Sociological Methods and Research* 24: 7–29.

Vickers, M., Chan, S. (1981). *Representing Classroom Social Structure*. Melbourne: Victoria Institute of Secondary Education.

Wasserman, S., Faust, K. (1994). *Social Network Analysis: Methods and Applications*. Cambridge: Cambridge University Press.

Wasserman, S., Pattison, P. (1996). Logit models and logistic regressions for social networks: I. An introduction to Markov graphs and $p^*$. *Psychometrika*, 61: 401–425.

Weeks, D. G., Bentler, P. M. (1982). Restricted multidimensional scaling models for asymmetric proximities. *Psychometrika* 47: 201–208.

Willekens, F. J. (1983). Log-linear modeling of spatial interaction. *Papers of the Regional Science Association* 52: 187–205.

Winsberg, S., Carroll, J. D. (1989). A quasi-nonmetric method for multidimensional scaling via an extended Euclidean model. *Psychometrika* 54: 217–229.

Yadohisa, H. (1998). Visualization of proximity data by multidimensional scaling with residual from space reduction. *Machine GRAPHICS and VISION* 7: 681–686.

Yadohisa, H. (2002). Formulation of asymmetric agglomerative hierarchical clustering and graphical representation of its result. *Bulletin of The Computational Statistics Society of Japan* 15: 309–316 (in Japanese with English summary).

Yadohisa, H., Niki, N. (1999). Vector field representation of asymmetric proximity data. *Commun. Statist: Theory and Method* 28: 35–48.

Yadohisa, H., Niki, N. (2000). Vector representation of asymmetry in multidimensional scaling. *J. Japanese Society of Computational Statistics* 13: 1–14.

Young, F. W. (1975). An asymmetric Euclidean model for multi-process asymmetric data. Paper presented at U.S.–Japan seminar om MDS, San Diego.

Young, G., Householder, A. S. (1938). Discussion of a set of points in terms of their mutual distances. *Psychometrika* 3: 19–22.

Zielman, B., Heiser, W. J. (1993). Analysis of asymmetry by a slide-vector. *Psychometrika* 58: 101–114.

Zielman, B., Heiser, W. J. (1996). Models for asymmetric proximities. *Br. J. Mathemat. Statist. Psychol.* 49: 127–146.

# Index


ABIC, 94
Actor, 200
ADCLUS, 165, 194
Additive clustering, 194
Additive constant, 144
Additive structure, 131
Additive tree, 165, 185
Additivity, 47, 60, 127, 130, 133, 137, 141, 143, 144, 152
Affinity, 2
Agglomerative clustering, 5
Aggregated level, 104
AHP, 41
AIC, 35
Akaike's Bayesian Information Criterion, 94
Alternating least squares approach, 194, 196
Analysis of asymmetry, 222
Analysis of variance, 15, 16
Analytic hierarchy process, 41
Angular response function, 40
ANOVA, 20
  approach, 49
Arc, 200, 207
Arc length, 207
Association, 2, 104
Asymmetric data, 1, 51, 103, 164, 199
Asymmetric hierarchical clustering, 164
  algorithm, 165, 169
Asymmetric MDS, 82
Asymmetric relationship, 223
Asymmetric structure, 81, 82
Asymmetry, 1, 8, 25, 103, 199, 211, 215, 234
ASYMSCAL, 105, 106, 113
Attention shift, 105

Bayesian point of view, 92
Bias component, 126
Bi-cluster, 191
Binary relation, 175
Binary response, 9
Biplot, 115
Bond energy algorithm, 5, 165, 188
Brand loyalty, 154

Brand switching, 2, 104
Brossier algorithm, 180

Canonical analysis, 71
Canonical correlation, 221
  analysis, 216, 218
  coefficient, 218
Canonical equation, 218
Canonical regression analysis, 5, 216, 225
Canonical variate, 218, 226
Car switching data, 120, 123, 158
Cartesian product, 191
Case 2 assumption, 32
Case 3 assumption, 34
Case 5 assumption, 28, 31, 34
CCA, 218, 225, 238
Centered data, 217
Centered symmetry condition, 130, 144
Centering matrix, 139, 217
Centroid effect, 192
  algorithm, 5, 165, 190
Chi-square distribution, 19, 31
Choice, 202
Circle pattern, 61
Circle structure, 62, 63
Circular triad, 9, 12
CLASSIC, 173, 179
Classification, 164
Clique, 201
Cluster analysis, 5
Cluster structure, 2
Clustering, 164
Clustering algorithm, 164
Coefficient of consistence, 11
Coefficient of consistence, 48
Cohesive group, 201
Column bias model, 126
Combination effect, 16, 20
Combined distance, 171
Communication flow, 104
Comparatal dispersion, 26, 33
Comparative judgment, 9, 15, 25, 40
Complete graph, 206
Complete linkage algorithm, 164
Complex data matrix, 118
Complex vector space, 75
Composite variate, 216, 222, 225
Composition of relations, 176
Confidence interval, 34, 38
Confidence region, 20
Configuration of objects, 54
Configuration of stimuli, 103
Confusion, 2, 104, 208
Confusion data, 100, 142, 154
Consistency, 42
Consistent matrix, 47
Constant ratio condition, 130, 136, 137, 141
Construction of a network, 206
Content model, 105
Context effect, 105
Contingency table, 4, 107, 127, 134
Contour map, 100
Contrast model, 159
Correlation matrix, 217
Correspondence analysis, 128
Covariance matrix, 217
CRA, 225, 229, 231, 234
Cross-citation data, 171
Cumulative redundancy, 221, 223, 227, 230
Cyclic matrix, 61, 62, 113

De Soete et al. algorithm, 185
Decomposition, 51, 80, 108, 111, 112, 149
DEDICOM, 106, 109, 115, 117, 121
Deficit, 181
Degree of asymmetry, 55, 134
Degree of inconsistence, 12, 45, 48
Dendrogram, 164
Density, 27, 140, 145
Density of choices, 202
Dependence, 2, 215, 220
Dependency, 5

Diagonal element, 51
Diagonal entries, 118, 128, 134
Dichotomous directed network, 208
Dichotomous network, 201, 202
Digraph, 200, 207
Dimensional structure, 161
Direct clustering, 165, 188, 191
Directed arc, 200
Directed graph, 200
Directional analysis, 215
Directional relation, 109
Directional tie, 202, 209
Discriminal process, 23, 26
Dissimilarity, 2, 51, 104, 160, 168
  data, 147
  matrix, 81, 128
  measure, 108
Distance density model, 106
Distance matrix, 185, 200
Distance model, 4, 103, 105, 128
Distance-density hypothesis, 142, 154
Distance-density model, 105, 140
Distance-like, 108
  measure, 206
Divergence, 84
Dominance, 125
  matrix, 8–10
Dominance relation, 9, 48
Dominant-subordinate relationship, 11, 69
Dual domain DEDICOM, 112
Dynamical interpretation, 82

E-G, 76, 77, 114, 117
Equal arcs, 62
Equivalence relation, 176, 177
Escoufier and Grorud's procedure, 69
Euclidean distance, 152
Exchange, 2
Extended dendrogram, 170
Extended updating formula, 167, 169

F distribution, 19
Feature matching model, 4, 105, 159
Finite Hilbert space, 120
Fisher's scoring method, 34
Flow, 82
Frame of reference, 105
Frequency of citation, 156

GDM-I, 132, 152
GDM-II, 136, 154
Generalized distance model, 131, 136
Generalized GIPSCAL, 106, 111, 114, 123
Generalized triangular inequality, 129, 133, 137, 141, 143, 144
GENNCLUS, 5, 165, 194, 195
Geodesic, 200, 211
GIPSCAL, 106, 111, 113
Gower diagram, 54
Gower's procedure, 52
Graph statistics, 211
Graph theoretic distance, 202
Graphical representation, 4, 52

Hermitian form, 74
Hermitian matrix, 70, 71, 120
Heterogeneity measure, 192
Hierarchical tree representation, 180
Hierarchy index, 14
Homogeneity criterion, 188
Hyperparameter, 92

Inclusion of relations, 176
Inconsistent matrix, 47
Independence, 107
Indeterminacy, 55, 110, 114, 133, 142, 143, 147, 149, 163
Index of consistency, 45
Index of inconsistency, 48
Inner product, 75, 118
Instrumental variable, 227
Interpoint distance, 56, 73
Interset correlation, 221
  coefficient, 242

Interset structural correlation, 220, 224
Interval scale, 58, 108
Intraset correlation coefficient, 241
Intraset structural correlation, 220, 224
IPCA, 227, 229, 230
Irreducible representation, 207

Jet-stream model, 100
Journal citation, 104
Judgment of proximity, 104

Lance and Williams updating formula, 168
Latent axis, 22
Latent space, 104
Law of comparative judgment, 24
Likelihood function, 32, 212
Line pattern, 59, 152
Line structure, 151
Linear flow, 83
Linear hierarchy, 10, 80
Linear regression, 215
Logit, 211
Logit model, 210
Logit p* model, 211
Log-likelihood function, 32, 33
Log-linear model, 127

Main effect, 20, 107
MAPCLUS, 196
Marketing research, 64
Matrix decomposition, 180
MDS, 103, 201
Mean-squared deviation, 192
Measure of relationship, 2
Measurement level, 108
Measurement of inconsistence, 12
Metric, 151
Metric axioms, 107, 130, 144, 152
Metric MDS, 108, 140
Metric space, 118
Migration, 2
Migration rate, 104
Min and max clustering, 164
Minimality, 107
Minimum pathlength metric, 200, 207
Minkowsky distance, 136
Mixed model, 145
Morse code confusion, 14
MRA, 217, 225
Multidimensional scaling, 4, 81, 103
Multiple regression model, 218
Multiple relation, 200
Multirelational network, 214
Multivariate analysis, 215
Multivariate data, 2, 216
Multivariate regression, 225
  analysis, 217
Mutual interaction, 2

Nearest neighbors relation, 173, 177
NETSCAL, 206, 207
Network analysis, 5, 199
Network data, 199
Network scaling, 5, 206
Network statistics, 208
Network structure, 2
NNR, 173, 177, 178
Nonmetric approach, 108
Nonmetric MDS, 105
Nonmetric model, 140
Nonmetric procedure, 107, 147
Nonmetric treatment, 107
Non-negativity, 186
Nonoverlapping, 188, 194
Nonsymmetrical data analysis, 216, 229
Normal deviate, 27
Normal distribution, 15, 23, 27, 31, 231
Normal matrix, 52
Normal response function, 40
Normalized data, 217
Null diagonal entries, 77

Observed deviate, 28
Occupation mobility, 14
Odds ratio, 211

One-class ultrametric inequality, 187
One-dimensional structure, 180
One-mode approach, 165, 173, 180
One-mode two-way data, 2, 106, 164, 185
Operational approach, 49
Operational scaling, 41
Order effect, 16, 20, 25, 39
Ordered categories, 25
Ordinal scale, 9
Ordinal structure, 8, 9, 13
Oriented paralleotope, 73
Oriented triangle, 56, 68
Orthogonal projection operator, 218
Overlapping, 188, 194

P* model, 210
Paired comparison, 3, 8, 9, 15
Paired comparison matrix, 42
PC judgment, 25, 42, 104
PCA, 216, 229
Pecking data, 68, 79
Pecking frequency, 11, 14, 104, 128
PENCLUS, 165
Perron root, 44
Perron-Frobenius root, 44
Positive matrix, 42, 43
Positive reciprocal consistent matrix, 44
Positive reciprocal matrix, 42, 44, 45, 47
Posterior likelihood, 92
Preference, 9
Principal component, 227
  analysis, 5, 216
Principal eigenvector, 44, 48
Prior distribution, 92
Probabilistic model, 23
Proximity, 2, 201
  data, 164
  measure, 103
Pseudo-likelihood function, 213
Psychological continuum, 22, 25, 50
Psychological difference, 23
Psychological distance, 2
Psychological effect, 145
Psychological scaling, 8, 21

Quasi-asymmetry, 107
Quasi-independence, 107, 128
Quasi-nonmetric, 146
Quasi-symmetry, 128

Random variable, 15, 23, 210
Rank order, 9, 13, 47, 69
RANKOR, 178
Ratio scale, 41, 108
RDA, 222, 229, 231, 234
Rectangular distance matrix, 186
Reduced rank regression, 216, 229, 231
  analysis, 5
Redundancy, 220, 221, 238
  analysis, 5, 216, 220, 222
  index, 222
Reference point, 149
Regression model, 227
Relation, 200
Relational matrix, 200
Relational tie, 200
Relationship, 2
Residential mobility, 128
Revised law of comparative judgment, 26, 39
Rotation, 83
Rotational indeterminacy, 55, 110
Row bias model, 126
RRR, 229, 231

Scalar potential, 91
  of the vector field, 100
Scalar product model, 4, 103, 106, 109, 118
Scale, 41
Scaling approach, 49
Scheffe's method, 15
S-clique, 203, 204
Segregation matrix index, 202, 203
Self-dissimilarity, 108, 141, 147
Self-similarity, 108

Set-theoretic approach, 105, 159
Similarity, 2, 51, 104, 118, 159, 174, 195
  bias model and, 4, 107, 124
  choice model, 126
  matrix, 124, 166
  measure, 109
  order matrix, 175, 178
Single domain DEDICOM, 112
Single linkage algorithm, 164
Single relation, 201
Singular value, 52, 111, 114
  decomposition, 52, 53, 111
Singular vector, 52, 114
Skew-symmetric component, 120, 126
Skew-symmetric matrix, 47, 51, 52, 180, 181
Skew-symmetric part, 150
Skew-symmetric relation, 56
Skew-symmetry, 16, 52, 125, 161
  analysis, 54
Slide-vector model, 106, 128, 147, 158
SMI, 203
Smoothness, 91
Social actor, 200
Social exchange, 181
Social hierarchy, 10
Social mobility, 2, 104, 128
Social network, 2, 5
Social network, 200, 204, 208
Sociomatrix, 200, 208
Soft drinks brand switching data, 64, 79, 152, 188, 190, 194, 198
Spatial structure, 2
Spectral decomposition, 54, 70, 71, 117
SSA, 54, 59, 64, 111
Stimuli, 8
Stimulus identification, 104
Stochastic model, 17
Structural correlation, 224, 227
  coefficient, 220
Studentized range, 20
Substitutability, 2
Successive presentation, 25
Supersphere, 145
Surplus, 181
SVD, 53
Symmetric component, 120
Symmetric matrix, 51
Symmetric matrix, 180
Symmetric MDS, 81, 104
Symmetric part, 151
Symmetric similarity matrix, 181
Symmetrization, 166, 167, 201
Symmetry, 107

Threshold parameter, 27
Thurstone's scaling, 8, 22
Tie, 200
Time order error, 35, 38
TOE, 35
Torgerson's procedure, 101
Total redundancy, 222
Trade exchange, 182
Transformation of data, 57, 58, 66, 73
Transformed data, 74
Transitive closure, 176
Transitive relation, 124
Transitivity, 47, 125, 127
Tree structure, 2, 165
Tree-fitting algorithm, 142
Tree-fitting method, 191
Triad, 9, 60
Triangle, 56
Triangular inequality, 107, 144, 186
Trinomial distribution, 30
TSCALE, 106, 159, 161
Two-class ultrametric condition, 186
Two-class ultrametric inequality, 186
Two-mode approach, 185, 188, 190
Two-mode array, 191
Two-mode cluster, 191, 193
Two-mode data matrix, 191
Two-mode error variance, 192
Two-mode hierarchical clustering, 191
Two-mode submatrix, 191
Two-mode three-way data, 3

Two-mode three-way dissimilarity, 162
Two-mode two-way data, 165, 185

Ultrametric distance, 185
Ultrametric inequality, 186
Ultrametric matrix, 180, 181
Ultrametric tree, 165, 185
Unfolding model, 106
Unfolding procedure, 106
Unidimensional judgment, 9
Unidimensional latent scale, 9
Unidimensional scale, 13, 48
Unidimensional scaling, 24
Unidimensional structure, 127
Unidimensionality, 60
Union of relations, 176
Unitary matrix, 70
Updating formula, 164

Valued network, 201, 203, 204
Variance extracted, 221, 239
Vector field model, 89
Vector model, 80
Vector space, 118
Vertex, 200

Ward's method, 194
Weeks and Bentler's model, 143
Weighted average algorithm, 164
Weighted Euclidean distance, 105
Weighted sum of squared distances, 74
World trade data, 87

Young and Householder's theorem, 118, 119

Milton Keynes UK
Ingram Content Group UK Ltd.
UKHW020025071024
449327UK00032B/2931

9 780367 393373